Quantitative Molecular Pharmacology and Informatics in Drug Discovery

Quantitative Molecular Pharmacology and Informatics in Drug Discovery

MICHAEL LUTZ
Section Head, Cheminformatics Group, Glaxo Wellcome Research and Development, Research Triangle Park, NC, USA

and

TERRY KENAKIN
Principal Research Scientist, Glaxo Wellcome Research and Development, Research Triangle Park, NC, USA

JOHN WILEY & SONS, LTD
Chichester • New York • Weinheim • Brisbane • Singapore • Toronto

National 01243 779777
International (+44) 1243 779777
e-mail (for orders and customer service enquiries): cs-books@wiley.co.uk
Visit our Home Page on http://www.wiley.co.uk
or http://www.wiley.com

Other Wiley Editorial Offices

John Wiley & Sons, Inc., 605 Third Avenue,
New York, NY 10158-0012, USA

VCH Verlagsgesellschaft mbH, Pappelallee 3,
D-69469 Weinheim, Germany

Jacaranda Wiley Ltd, 33 Park Road, Milton,
Queensland 4064, Australia

John Wiley & Sons (Asia) Pte Ltd, Clementi Loop #02-01,
Jin Xing Distripark, Singapore 129809

John Wiley & Sons (Canada) Ltd, 22 Worcester Road,
Rexdale, Ontario M9W 1L1, Canada

Library of Congress Cataloging-in-Publication Data

Lutz, Michael William, 1960–
Quantitative molecular pharmacology and informatics in drug discovery/Michael Lutz and Terry Kenakin.
p. cm.
Includes bibliographical references and index.
ISBN 0-471-98861-8 (alk. paper)
1. Molecular pharmacology — Computer simulation. 2. Pharmacology, Experimental–Computer simulation. I. Kenakin, Terrance P. II. Title.
[DNLM: 1. Pharmaceutical Preparations — standards. 2. Drug Design. 3. Drug Evaluation — methods. QV 744 L975q 1999]
RM301.65.L88 1999
615'.19 — dc21
DNLM/DLC 99-29628
for Library of Congress CIP

British Library Cataloguing in Publication Data

A catalogue record for this book is available from the British Library

ISBN 0 471 98861 8

Typeset in 10/12pt Times from the author's disks by Laser Words, Madras, India
Printed and bound in Great Britain by Biddles, Guildford, Surrey
This book is printed on acid-free paper responsibly manufactured from sustainable forestation, for which at least two trees are planted for each one used for paper production.

To Michelle, my parents and in memory of Sharon and, to Debbie, for the idea.

Contents

Preface

Screening is the process by which the biological effects of chemical compounds are initially assessed on stable biological systems to determine possible utility as medicines for humans. Before the advent of high-volume robotic screening procedures, a limited number of tests yielded a certain 'hit' rate, i.e. if 1200 targeted molecules were screened, perhaps two were found to have activity (0.17%). The notion that the success of drug screening programs would necessarily increase as the net cast out into chemical space gets larger came with the promise of the larger net, namely chemical libraries and robotic screening. Thus, by implication, if 1 200 000 molecules were screened, then 2000 'hits' would result. Coincidentally came the ability, through molecular biology, to engineer drug receptor systems and utilize human biological material. These advances together reshaped the drug discovery process from that seen just 10 years ago.

Perhaps not anticipated with this increased technological capability were new challenges appearing directly as a result of this new technology, namely the potential biological dissimulations observed in some recombinant systems (see Chapters 4 and 5) and the added requirements for data collection and statistical analysis resulting from the huge volumes of data that need to be assimilated and understood (see Chapters 6 and 9). Technological advances in computing have kept pace with advances in biochemistry, chemistry and molecular biology. The processing speed and storage capacities of computers are well suited for the large amounts of data. The Internet provides a facile link to databases of biochemical results from laboratories around the world. The convergence of biochemical and computer technologies has led to the development of methods to extract information from large-scale experiments, including screening, and from the vast amount of information from publicly available electronic databases and literature.

The term informatics, in its most general sense, indicates the use of information about a defined problem or area of science to deduce patterns, structure or function from data. In cheminformatics, the sources of information include chemical structure (two-dimensional and/or three-dimensional), calculations made on the structure (topological, electronic, quantum mechanical) and observations associated with the structure (biochemical testing results or chemical properties). For bioinformatics, the sources start with the genotypes and phenotypes of individuals, the nucleotide sequences of proteins and the motifs and patterns deduced using the tools of genetics and genomics. In many cases, computer-based algorithms are needed to analyze the data, both to automate extremely laborious calculations and to summarize the results of the analysis in an understandable form. Effective use of informatics in drug discovery requires contributions from several disciplines: mathematics to develop algorithms and methods to identify patterns in data; statistics to address issues of variability and to extract signal from noise; and computer science to handle the technical problems of storing and retrieving large quantities of data. Parts of this book (Chapters 6–9) address the mathematical and statistical issues associated with the use of informatics in drug discovery. These parts of the book build on classical statistical methods (understanding variability, experimental design, regression models), illustrating their application with specific examples and extending the methods to the larger datasets associated with screening. Components of the theory and different applications form a statistical framework for effective design of experiments and interpretation of the vast amount of data obtained in the drug discovery process.

The types of information available to scientists working in drug discovery have expanded to include genetic and genomic information about proteins and receptors and electronic databases of biological information about the genome and signal transduction systems of many species, including humans. Clinical trials are now both the start and culmination of the drug discovery process. Information from clinical trials is being used increasingly to identify new biochemical targets and to define the genetic basis for increased susceptibility to disease. Publication of the human genome and high-performance bioinformatics tools adds new considerations for target selection and validation. Chapter 7 of this book addresses the use of the genetic and genomic information that is becoming available to research project teams and the methods to analyze these data at the level of being able to utilize reports in the literature and the tools of genetics and genomics in problems of target validation.

Although many thousands of molecules now can be synthesized and tested for biological activity, diversity in chemical space is still the key, i.e. many thousands of similar structures fall short of presenting the biological target, i.e. the required range of conformations. Experimental designs that are useful for design and optimization of compounds, and libraries and statistical methods to extract information from screening results in order to maximize the diversity of molecules, are the subjects of part of Chapters 8 and 9.

The connection between quantitative methods in molecular pharmacology and the mathematical and statistical framework must be emphasized. Drug discovery involves problems in optimization (designing molecules, combinatorial libraries and biochemical assays), problems in pattern recognition (structure–activity relationships and quantitation of biological response) and problems in modeling (understanding signal transduction systems and the limitations of specific assay systems). Optimization of a test system involves both considerations about the pharmacological basis for the screen (Chapter 4) and experimental design (Chapter 8). Extension of classical methods in statistics and pharmacology to problems in compound screening and drug discovery is grounded in an understanding of the underlying theory of these areas: affinity, efficacy and analysis of variance (Chapters 2, 3 and 6). The biochemistry, pharmacology, mathematics and statistics are linked by the experimental designs, information obtained from the experiments and specific decisions that are made to identify biochemical targets for disease and to identify and optimize novel drugs.

Acknowledgements

We wish to thank Glaxo Wellcome for support and encouragement during the preparation of this book. Also, we are grateful to our colleagues who have contributed to this book through discussions and project work. It is not possible to mention all of the people who have contributed to this work, but it is important to acknowledge the contributions of several individuals to specific chapters. Margaret Gelder-Ehm, Richard Buckholtz and Don Wallace contributed information about genetics, the yeast two-hybrid system and differential gene expression. Laura Giovanelli assisted with the literature review for Chapter 7. Peter Leightner, Paul Domanico, Becky Laskody and David Saussy contributed to examples on the use of experimental design for high-throughput screening assays. Alan Menius, Stan Young and Steve Blanchard provided insight about statistical analysis of large datasets.

1 Drug Discovery

1.1 INTRODUCTION

The modern drug discovery process utilizes science beyond the realms of just pharmacology and medicine. As the power of technology increases to allow the use of formerly unrealistic sample sizes for experiments, there comes a need for statistics to assist in the determination of what is real and what is apparent difference. The types of information available to scientists working in drug discovery have expanded to include genetic and genomic information about different species, electronic literature, databases of millions of chemical structures and results from high-throughput screening. Mathematical and statistical methods for analysis of data have extended the classical regression and confirmatory statistics to database mining. New questions can be asked of data. In addition to confirming patterns in data, algorithms are written to *find* interesting patterns in data. Although new algorithms, faster computers and improved software are released at an increasingly fast rate, there are fundamental principles for the collection and interpretation of data that transcend methods. This book addresses statistical and informatics issues that a scientist with the task of finding or optimizing a lead molecule for a biological target should consider before setting up a screen and choosing sets of compounds or chemical libraries to test. Schematic maps of the drug discovery process, from target selection to optimization of lead compounds, are used to indicate decision points and the data necessary to make the decisions.

1.2 THE ORGANIZATIONAL STRUCTURE OF THIS BOOK

This first chapter is designed to orient the reader to the drug discovery process. Chapter 2 describes the tools used to measure the affinity of ligands for biological targets. Chapter 3 discusses a special subclass of ligands, namely those that cause the biological target to change its behavior within the cellular host. These latter ligands are described as having efficacy, i.e. the ability to produce physiological response. Chapter 4 then goes on to outline the various measuring systems that are used to detect ligand affinity and efficacy and the construction of human recombinant cellular assays. Chapter 5 discusses advanced concepts required to optimize drug screening assays.

Chapter 6 builds on this introduction by describing a statistical framework for analysis and interpretation of data in terms of the sample data and statistical methods available for problems in drug discovery, specifically identification and optimization of a drug molecule. Chapter 7 describes methods that are available to solve a different problem: identification and validation of biochemical targets. This chapter addresses the use of the genetic and genomic information that is becoming available to research project teams and the methods to analyze these data. Chapter 8 introduces statistical experimental design for chemical and biological application. Design is used to gain the maximal amount of statistical information from a set of experiments. Chapter 9 focuses on high-throughput screening data, describing the advantages and limitations of these data and the current algorithms for finding structure–activity patterns in large datasets. This chapter also describes evaluation of the quality of data from different phases in lead identification and optimization, and describes assays and informatics methods for the optimization of compounds as drugs that will be effective in humans.

1.3 THE DRUG DISCOVERY PROCESS

There are two apparently divergent views of the drug discovery process. The first is to assume that biological and chemical diversity can be sampled adequately in a random manner and that random high-volume screening will furnish new drug entities. The second is that drugs can be designed in a rational manner and that focused libraries of compounds offer a better approach to drug discovery. There are examples of success in both modes of screening and it may be that the nature of the biological target determines which approach works best. This book is designed to offer the tools required for the optimal use of pharmacological and statistical tools in the drug discovery process. Modern automated methods of drug screening produce enormous datasets from which biologists and chemists must glean weak biological signals.

A high-level map of the drug discovery process is shown in Fig 1.1. The map is a good starting point for a discussion about informatics, because it shows the sequence of events that transform an idea gained from clinical research to a marketed drug. At each step, there are different types of information and statistical analysis. More recently, many companies have recognized that the sequential nature of this process is not efficient. For example, it is more efficient to design compounds with good pharmacokinetic profiles earlier in the process than after a compound is optimized for potency *in vitro*.

The idea for a biochemical target initiates in therapeutic research. Traditionally, ideas have come from medical practice. The need for compounds to lower blood pressure in hypertensive patients, to develop antibiotics for infectious diseases and to discover antineoplastic agents has provided the clinical basis for common drugs. More recently, ideas are being generated from clinical genetics.

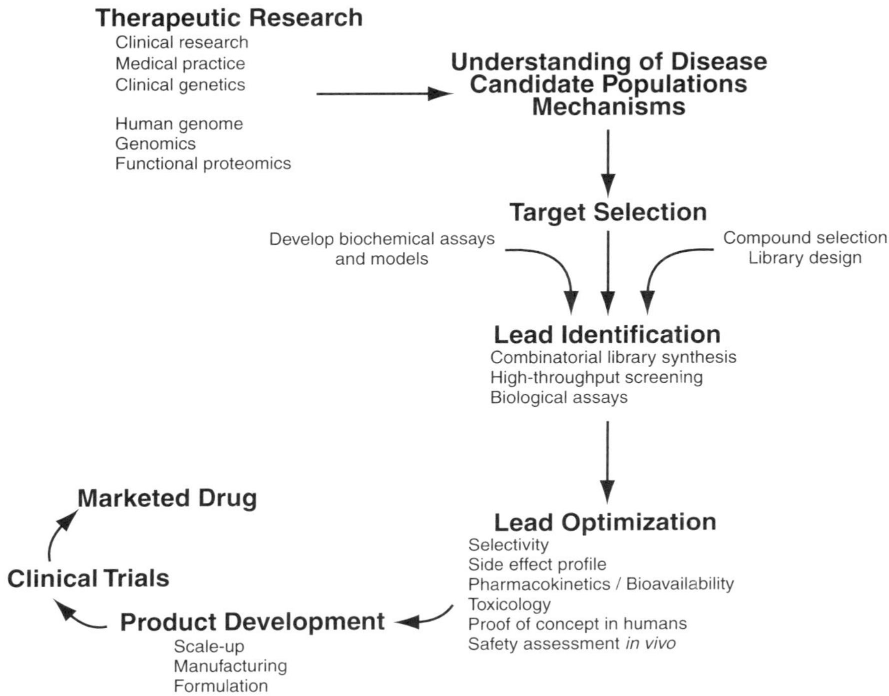

Fig. 1.1. High-level map of drug discovery. Stages of drug discovery involve the interaction of numerous scientific and clinical disciplines. Although drug discovery is often portrayed as a linear process, optimization of the process is achieved by recognizing that design, assay and clinical research fit together in an iterative web, e.g. pharmacokinetic profiles are considered during design of combinatorial libraries for screening

1.3.1 CLINICAL GENETICS AND TARGET SELECTION

In Chapter 7, the informatics issues associated with target validation and selection are considered in detail. This section describes the basis for using clinical genetics for target selection. A schematic of the process is shown in Fig. 1.2. It is known that people are predisposed to certain diseases based on their family history. With the discovery that the ApoE gene is associated with susceptibility to Alzheimer's disease, there is the recognition that susceptibility genes underlie many common, chronic diseases such as cardiovascular disease and cancer. In the case of Alzheimer's disease, a clear linkage is shown between five genotypes of the ApoE gene and a decreased age for onset of the disease. A specific allele, ApoE4, is associated with the elevated risk of Alzheimer's disease. Three factors associated with the transition to genetic-based drug discovery are: more of the human genome is known, although the linkage to disease remains largely

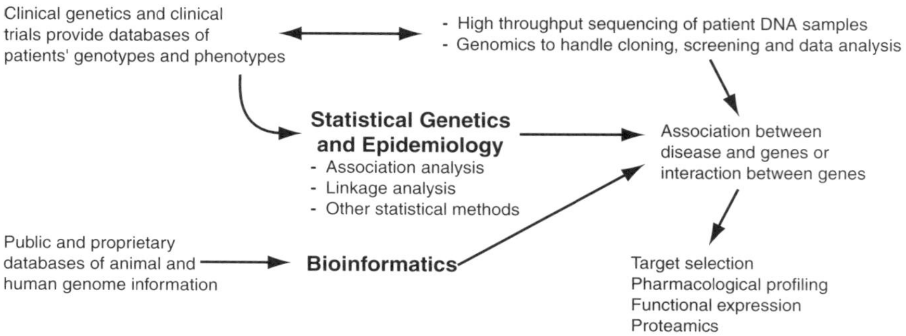

Fig. 1.2. Map of clinical genetics in the drug discovery process. The goal of clinical genetics is to identify susceptibility genes that predispose humans to disease. The tools involved include statistical tools to perform association and linkage analysis on patient samples and genomic tools to process the DNA samples. Public and proprietary databases contain information that can be mined using statistical methods. Results are used for target selection

unknown; methods for genomics, processing data from clinical samples and DNA clones continue to improve, giving more data about the basis for the disease in the general population; and statistical methods for analysis of genetic data continue to improve, allowing investigators to understand the genetic basis for diseases and to identify proteins as biochemical targets for drugs. It is worth making the distinction between genetics, genomics and proteomics. Genetics is the study of heredity and variation in inherited traits, including disease. Some basic terms include gene, allele, genotype and phenotype. A gene is a specific coding sequence of DNA; in turn, it will be associated with some function or characteristic. Genes can exist in different forms known as alleles. Genotype is determined by using blood samples to identify the specific alleles. A phenotype is an observable expression of a specific genotype, e.g. the presence or absence of a disease or a physical characteristic. However, observed characteristics and diseases are often associated with the interaction of multiple genes, different allelic forms of the same gene and environmental factors. Sorting out the complex interactions and temporal order of events is a difficult task and is the study of research in genetics. In genetic studies, the genotypes of individuals, obtained from sequencing DNA from blood samples, are compared to their phenotypes, i.e. whether or not they have a disease or trait. The genome is the entire collection of genetic information for an organism. In humans, the genome is estimated to encode approximately 100 000 different genes. Genomics is a set of research processes to handle the sequencing and interpretation of data from DNA samples. Finally, proteomics is concerned with the study of the proteins encoded by the genome. Ultimately, the proteins are the targets for drugs. Just as the genome is the collection of all the genes for an organism, the proteome is the collection of all of the proteins. In proteomics, high-throughput two-dimensional gel electrophoresis and mass spectroscopy are used to identify the target proteins.

As a consequence of high-throughput genomics, the number of potential targets for new drugs has increased significantly. Figure 1.3 shows the growth in the number of novel genes identified by the genome project, available in the public literature and derived from expressed sequence tags (ESTs) projected to the year 2005 when the entire human genome is expected to be sequenced completely. The size of the human genome is estimated to be 3×10^9 base pairs, of which a small 5% fraction code for genes. It is estimated that there are 100 000–150 000 genes in the human genome. From this set of genes, it is anticipated that there will be between 3000 and 10 000 potential new drug targets, of which 20–30 will be most relevant to human diseases. Selecting the relevant targets is a cross-disciplinary process that involves understanding both the linkage between genes and disease and the pathophysiology of the disease.

The oldest approach for assessing a linkage between genes and disease is linkage analysis (Fig. 1.4). Linkage analysis requires families that exhibit an inherited susceptibility to a disease. For each individual in the family, the observed

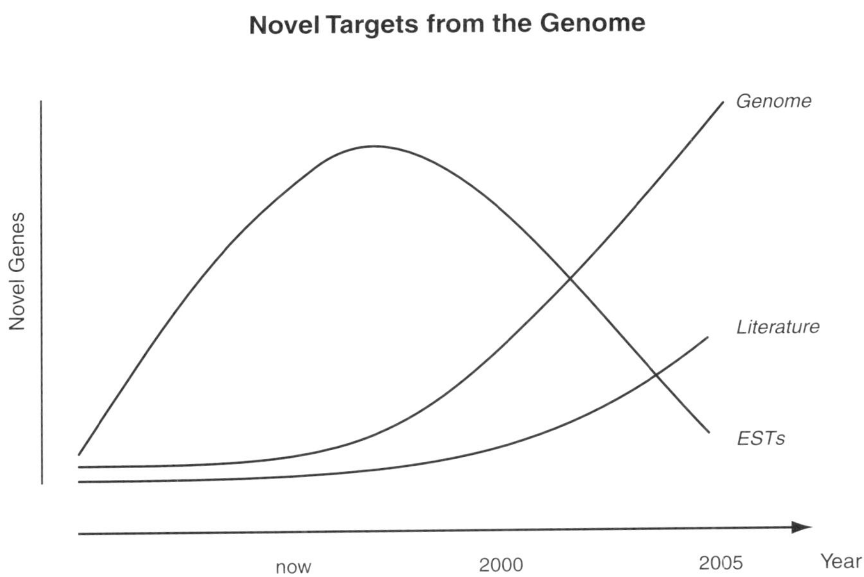

Fig. 1.3. Number of new targets identified from the human genome, published literature and expressed sequence tags (ESTs). As more of the human genome is sequenced, more targets will be identified from the genome itself using the methods outlined in Fig. 1.2. More targets will also be identified from the literature as scientists publish novel work in genetics. The number of targets identified by ESTs will decrease as these other methods are used for target selection. (From Steve Foord, Glaxo Wellcome.)

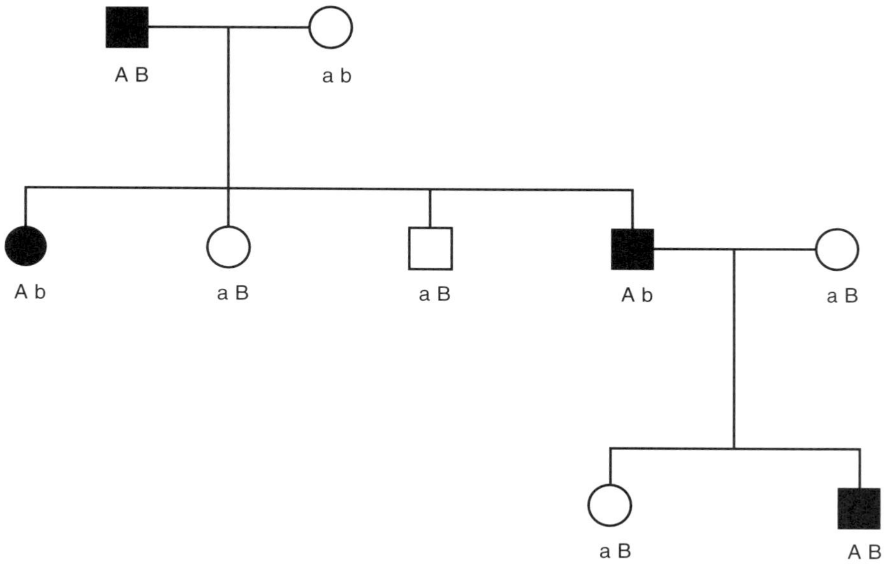

Fig. 1.4. Linkage analysis. Linkage analysis is based on the availability of familial data. In a pedigree tree, squares represent males and circles represent females. The tree represents family relationships between individuals. Shaded squares and circles indicate that the individual has the disease phenotype. It is clear from the pedigree that there is a linkage between the disease and the dominant A allele. Linkage analysis is based on observing recombinations: the exchange of genetic information between pairs of chromosomes

phenotype or presence and severity of the disease is recorded. The pedigree or relationship between the family members is depicted as in Fig. 1.4. Finally, DNA samples are obtained from the family members and genetic markers are established on the entire genome. Markers are a common term using in genetic mapping. For the portion of the genome being considered, the markers are locations that show a high degree of polymorphism, e.g. when the genes for many subjects are considered there is considerable variability at the site. Blood type (ABO) is an example of a common marker site. Markers are identified in clinical genetics by sequencing segments of DNA that show polymorphism for the subjects in the trial. A genetic map is constructed by placing the markers on the chromosomes in order to indicate distance between the markers. The distance between the markers is important, because markers located in close proximity are more likely to be linked. The unit of distance is a centimorgan, defined as a separation of markers that have a 1% frequency of recombination during meiosis. In humans, 1 centimorgan is approximately 1 million base pairs. The object of the genetic map is to place the disease loci on specific chromosomal regions using genetic markers or sites that show a high degree of polymorphism, i.e. locations that show a moderate to high (1–50%)

degree of DNA sequence variability. Once the map is available, the statistical analysis involves correlating mutations at the polymorphic sites with the presence of disease. There are several approaches to linkage analysis, many involving pairs of siblings. For example, if pairs of siblings both have the disease, alleles (different forms of a gene) at specific markers are examined to determine if the same genotype occurs frequently. Different alleles may be associated with different physical characteristics or the presence of a disease. If marker loci are close to the disease gene, they will group individuals in the same family with the disease together. Markers that are far from the disease gene do not group individuals by disease status. Once markers that show strong linkage to the disease are located, positional cloning (Fig. 1.5) is done to identify the mutated gene or genes associated with the disease. In linkage analysis, no a *priori* assumptions are made about the linkage between a gene and the disease. Both parametric and non-parametric statistical methods are used in linkage analysis, each having different sets of assumptions about the genetic model of inheritance.

In genetic terms, loci are linked if they are on the same chromosome and are physically located close enough so that their alleles do not separate independently at the time when gametes are formed. To search for disease linkage, investigators determine if genetic markers associated with a phenotype are linked to a gene located within 10–20 centimorgans that has allelic variation. The success of the method is due, in part, to a finely spaced set of markers so that the disease gene will be located near a marker.

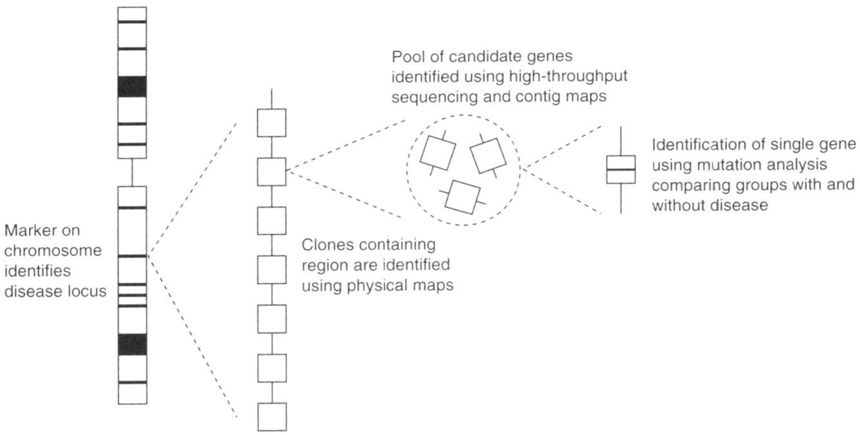

Fig. 1.5. Positional cloning. Positional cloning is a sequence of steps to identify single genes associated with disease. A marker (polymorphic locus) identifies the locus or region that contains the disease susceptibility gene. Clones are made that contain this region using physical maps of the genome. From the clones, a pool of candidate genes is constructed using high-throughput sequencing and contig maps. Finally, the disease gene is identified using mutation analysis and comparing groups with and without the disease

Results are often reported using a lod score: the common logarithm of the ratio of the likelihood of linkage with the likelihood of free recombination (no linkage).

An alternative method for detecting and assessing the strength of linkage between a gene and disease is association analysis (Fig. 1.6). Association analysis does not require familial data. Instead, DNA samples and phenotype information are collected from large (thousands) samples of the population at random. Polymorphism in specific genes is correlated with presence of the disease. Association genetics builds on the capability of rapid genotyping of patients. If a subset of the population is identified as having susceptibility to the disease, other members of the subset population can be tested to give further support to the hypothesis of an association between the gene or genes and the disease. Unlike linkage analysis, association analysis does assume a linkage between the disease and a gene or genes at the start of the analysis.

A more recent approach for identifying linkages between genes and disease is differential gene expression (Fig. 1.7). This method is particularly well suited to the problem of screening against thousands of potential genes at the same time. For differential gene expression, messenger RNA (mRNA) from normal and diseased tissue is screened against thousands of potential genes using grids made from complementary DNA (cDNA) from the species of interest. Sophisticated imaging systems are used to detect levels of gene expression in the same manner that electrophoretic gels are read. An increase in gene expression level is associated with an increase in brightness. Because the grid is made from many different genes, one advantage of this method is that it can identify previously unknown genes associated with a disease process.

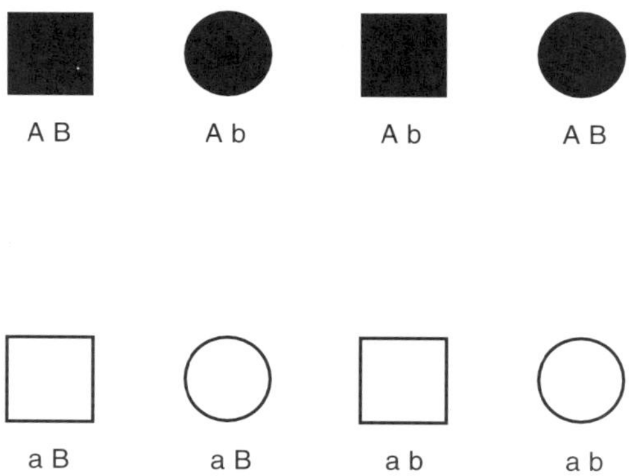

Fig. 1.6. Association analysis. Association analysis is an alternative to linkage analysis. Familial data are not required. In association analysis, alleles at a marker locus close to the disease gene are tabulated to determine if they are associated with the presence of the disease. Both phenotype and genotype data are required

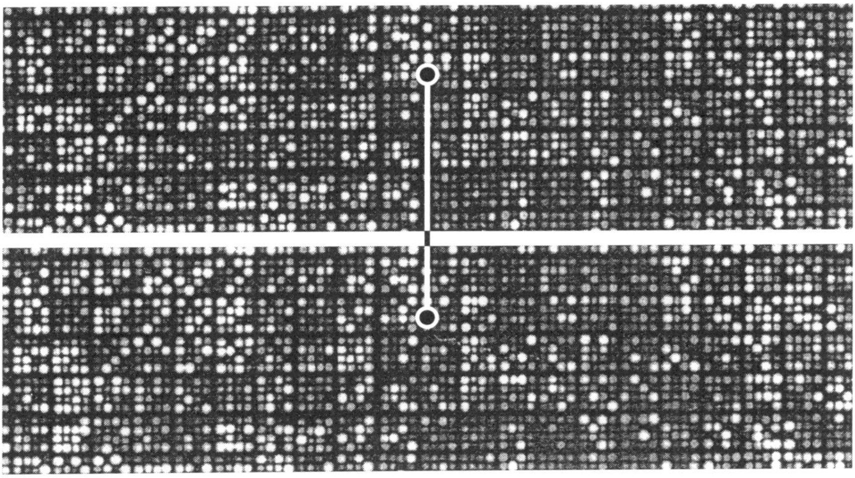

Fig. 1.7. Differential gene expression. The picture shows two gene grids. Each grid contains cDNA from over 10 000 different genes. Each spot on the grid represents a different gene. The intensity of the spot is a measure of the gene's activity. A different probe is applied to each grid: one probe is from tissue treated with a test compound; the other probe is from untreated tissue. Two corresponding spots are highlighted where the intensity is different between the two grids. This difference in intensity is indicative of differential gene expression in the treated and untreated tissue. Advanced imaging techniques are used to construct and compare the grids. (From Don Wallace, Glaxo Wellcome.)

1.3.2 ASSAY AND SCREEN DEVELOPMENT

Following selection of the biochemical target, a plan for testing potential drugs is developed. There are a variety of methods to measure drug effects on a biological system; these are covered in detail in Chapter 4. Screens are developed to measure binding to the target receptor or a functional response. A functional response includes some part of the signal transduction process after the binding of the drug to the receptor. Choice of a screen is based on aspects of the biological system, e.g. can a functional assay be developed for the protein of interest and on the characteristics of the desired drug.

Regardless of the assay system, there are steps that must be taken in order to interpret data from the assay in the statistical framework described in Chapter 6 and to develop the assay to provide the highest quality data. An outline of these steps is shown in Fig. 1.8. Estimates of precision, accuracy and variability must be obtained for the assay. Initially, these estimates are obtained from the assay development process and are derived from a limited amount of data. Often, only a curve for the radioligand and possibly some standard compounds can be obtained. After the assay is optimized and compounds are being screened, estimates of variability are rechecked to determine if variability is a function of

Develop Screening Assays

- Determine accuracy, precision and estimates of variability
- Develop quality control plan
  - Identify standards for QC tests: controls, standard compounds
  - Identify data to be collected for each plate, run and what data is to be tracked over time
  - Decide QC limits
- Test for systematic errors: robotic liquid handling, edge effects on plates, degradation of reagents over time
- Choose testing strategy: number of replicates, test in pools or singletons
- Experimental design to optimize screen: better signal to noise, lower variability

Develop Secondary Assays

- Determine accuracy, precision and estimates of variability
- Develop quality control plan
  - Identify standards for QC tests: controls, standard compounds
  - Identify data to be collected for each plate, run and tracked over time
  - Decide QC limits
- Test for systematic errors: dosing, carry over, plate effects
- Decide treatment and experimental design
- Sample size calculations
- Experimental design to lower variability and optimize assay

signal or affinity itself. The precision and accuracy are tracked with respect to time. These estimates may change as instruments or reagents are changed and it is important for the investigator to be able to note the magnitude of changes over time. For quality control, it is important to note what data are collected on each plate versus summary-level data for the run. Based on experience with similar systems, limits are placed on the quality control data. When the screen is being run, if the quality control data are not within the limits, the source of the variation is determined before screening is restarted.

A comprehensive report on a screening assay contains summary quality control information (Fig. 1.9). Variation within a run (set of plates run on the same day) and between runs (plates run on different days) is recorded. These values are used to set quality control limits and to aid in the interpretation of significant differences in test compounds. The trend of assayed values for standards is plotted with respect to time in order to determine if there are changes in assay performance due to reagent batches, biological factors, instrumentation changes or other time-dependent factors.

For high-throughput screening assays where compounds are tested at a single concentration, the signal window (Fig. 1.10) is used to define the range of activities that constitute an active compound (Sittampalam, 1997). For the signal window, the variability of the maximal response, minimal response and response at the IC_{50} or EC_{50} are combined mathematically to determine the range that defines an active compound in an inhibition or activation assay. For an inhibition assay, the size of the signal window to detect a hit is:

$$Window = \frac{(x_{max} - k\sigma_{max}) - (x_{IC50} + k\sigma_{IC50})}{\sigma_{max}}$$

where *Window* is the size of the signal window to detect an active inhibitor, x_{max} is the mean maximal response, x_{IC50} is the mean value of percentage inhibition at the IC_{50} (50%), σ_{max} is the standard deviation of the maximal response, σ_{IC50} is the standard deviation of the response at the level of the IC_{50} and k is an arbitrary multiple on the standard deviation for determining the width of the window. If $k = 1$, the windows form narrow bounds around the mean response. If $k = 2$ or 3, the bounds are wider. After the size of the window is calculated, the probability of detecting an active compound is calculated:

$$Pr(active) = Pr(x < thresh) = \Phi\left[\sqrt{n} \bullet \frac{(thresh - \mu)}{\sigma}\right]$$

Fig. 1.8. Maps for developing screening assays and secondary assays. These maps include the design and statistical issues to consider when developing biochemical assays: measures of accuracy and precision, quality control (QC) and tests for systematic errors

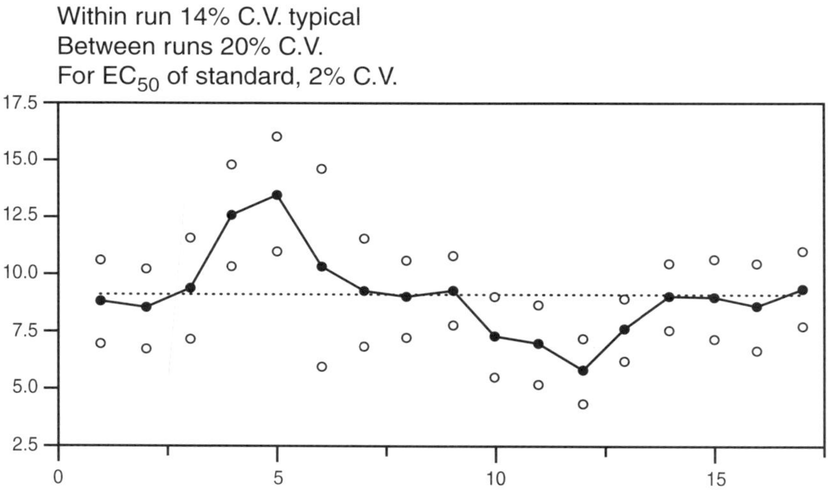

Fig. 1.9. Assay variability of a typical screen. The four key types of assay variability to record are: variability within a run; variability between runs; trend with time; and variability of a control. C.V. = coefficient of variation. (Data from Diane Ignar, Glaxo Wellcome.)

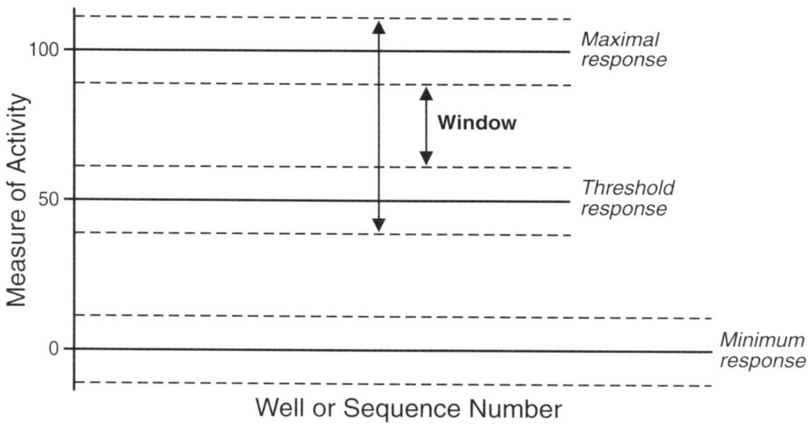

Fig. 1.10. Signal window for a high-throughput screening assay. To evaluate an individual single-point screening result requires estimates of the variability for different levels of response (percentage inhibition at the minimum, maximum and IC_{50}). For an inhibition assay, the window for a signal showing a response greater than 50% is defined by the upper confidence limit for response at the IC_{50} and the lower confidence limit for the maximal response. (Reprinted with permission from Sittampalam G *et al* (1997) Design of signal windows in high-throughput screening assays for drug discovery. *J. Biomolec. Screening* **2**: 159–169. Copyright 1997, *Journal of Biomolecular Screening*

where $Pr(active)$ is the probability of declaring a test compound active if the compound's activity is μ, the standard deviation is σ and the number of measurements is n. *Thresh* is the threshold that the investigator sets as being active (often 50%) and Φ is the cumulative normal distribution.

All biological assay systems have the potential to introduce systematic errors. Robotic systems have a propensity for systematic errors in that the same operations happen repetitively without a human to introduce random effects. It is important, particularly during assay development, to test for systematic effects due to a variety of sources, including carry-over from the liquid-handling system's tips, degradation of reagents over time and edge effects on microtiter plates. Systematic effects are identified on an ad hoc basis by assaying homogenous plates that contain a standard in all wells and plotting the results against row, column and other coordinate identifiers (Fig. 1.11). For automated data processing, analysis of variance (ANOVA) is used to test for systematic effects.

The testing strategy is developed using some of the same criteria as choice of the assay format. Based on the experimental objectives, variability of the assay, technical considerations about the assay format (e.g. toxicity in cell-based assay systems) and the collection of compounds to be screened, decisions are made about screening concentrations, pooling and replication. Compounds can be tested at a single concentration or in concentration–response format, and multiple compounds can be tested in a pool. As described in Chapter 4, pools of biological receptors can be constructed in the melanophore assay.

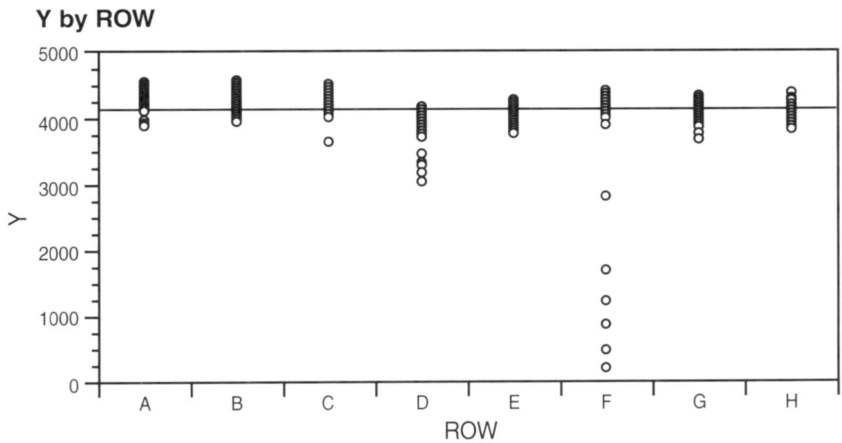

Fig. 1.11. Example of systematic error. This is a profile of maximal binding (B_0, counts per minute) for 60 plates of data grouped by plate row. It is clear that for row F some systematic effect is responsible for decreased counts. It is possible that a clogged pipette tip is responsible for the systematic effect, because the counts are not only different from the remainder of the plates but they also decrease systematically

1.3.3 COMPOUND SELECTION AND LIBRARY DESIGN

The choices of compounds to test in high-throughput screening or design of combinatorial libraries are critical issues in drug discovery as the size of compound collections grows and the capability to synthesize different libraries increases. Figure 1.12 shows some of the steps involved in identifying compounds and libraries to screen when a new biochemical target is available. There are numerous strategies for compound selection; some have a significant computational chemistry basis; others are based on an iterative screening approach; and others are based on selection of small subsets of a compound collection. Knowledge about the target molecule largely determines the compound selection strategy and the design of potential libraries.

There are three general types of libraries used to identify lead molecules: focused, targeted and primary screening libraries. Focused libraries are used when there is considerable knowledge about the target, a pharmacophore model or some structural information about the target protein. Focused libraries tend to be small (hundreds to thousands of compounds) and structurally homogenous. Focused libraries are often used during lead optimization in order to develop structure–activity relationships for binding affinity, efficacy and metabolic profiles. A good example of using structural information to design a focused library is found

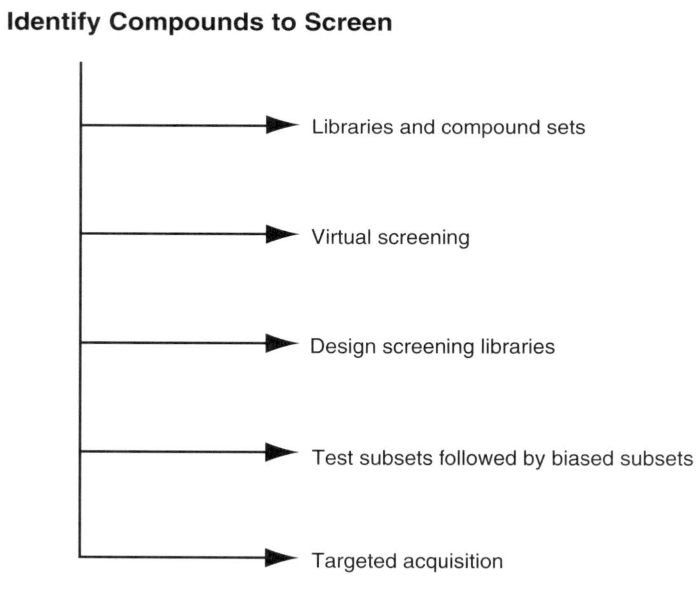

Fig. 1.12. Map for identifying compounds to screen. There are numerous computational strategies to assist in the choice of compounds to screen. Profiling compound sets and virtual libraries with respect to calculated descriptors is used to bias test sets toward specific pharmacophores or properties

in Kick *et al* (1997). For this work, the scaffold of a library for cathespin D was designed based on the crystal structure of several inhibitor–protein complexes.

Targeted libraries represent the next level of generality. The targeted libraries are developed for subsets of receptors and proteins. Examples include seven-transmembrane receptors and kinases. Some information is available about the target: pharmacological data about ligands that bind to similar proteins. The ligand data can give insight about pharmacophoric features, particularly if data are available on several ligands that can be modeled either in a two-dimensional or three-dimensional overlay. An example of a targeted library for seven-transmembrane receptors is the peptoid library reported by Zuckermann *et al* (1994). For this library, the peptide side chains were chosen based on similarity to seven-transmembrane ligands. From a computational standpoint, it is possible to select compounds from lists of drugs known to be active against similar receptors and use common features to design the scaffold and/or select monomers.

Primary screening libraries are an elusive target for the pharmaceutical industry. Theoretically, these libraries are tested against any new target when there are insufficient data to design either a focused or targeted library. By definition, a primary library must cover chemical space, an entity that defies description in terms of dimensions or size. Although claims of diversity of primary screening libraries are suspect, because they depend on an arbitrary definition of chemical space, it is possible to bias libraries towards molecules that are known drugs. Computational methods select compounds from the population of all known drugs. Selection is based on globally common features or features related to bioavailability and metabolism, instead of on common pharmacophoric features as in the design of targeted libraries. For example, there is a limited subset of ring systems that are present in common drugs (Fig. 1.13). These are starting points for the scaffolds of primary screening libraries. Any design work based on existing drugs and ligands must be tempered against identifying novel scaffolds and compounds that are possible to synthesize due to the availability of new chemistry or reagents. In the past, library chemistry was limited to a small number of scaffolds. New combinatorial techniques have greatly increased the number of frameworks available to construct libraries.

The role of computational methods in selecting a subset of a compound collection for testing, or in choosing monomers to reduce a combinatorial library to a size that can be synthesized, is to filter candidate structures. This process is often called virtual screening and specific examples and methods are discussed in Chapter 9. A map of where virtual screening fits in the discovery process is shown in Fig. 1.14. Objectives of virtual screening range from simply removing reactive and large molecules from the candidate pool to making a prediction on activity in a biochemical assay. The filters are based either on chemical properties (molecular weight, lipophilicity, solubility) or matching patterns against the chemical structure, e.g. removing all compounds that are alkylating agents. Prediction of biochemical activity is based on a statistical correlation between

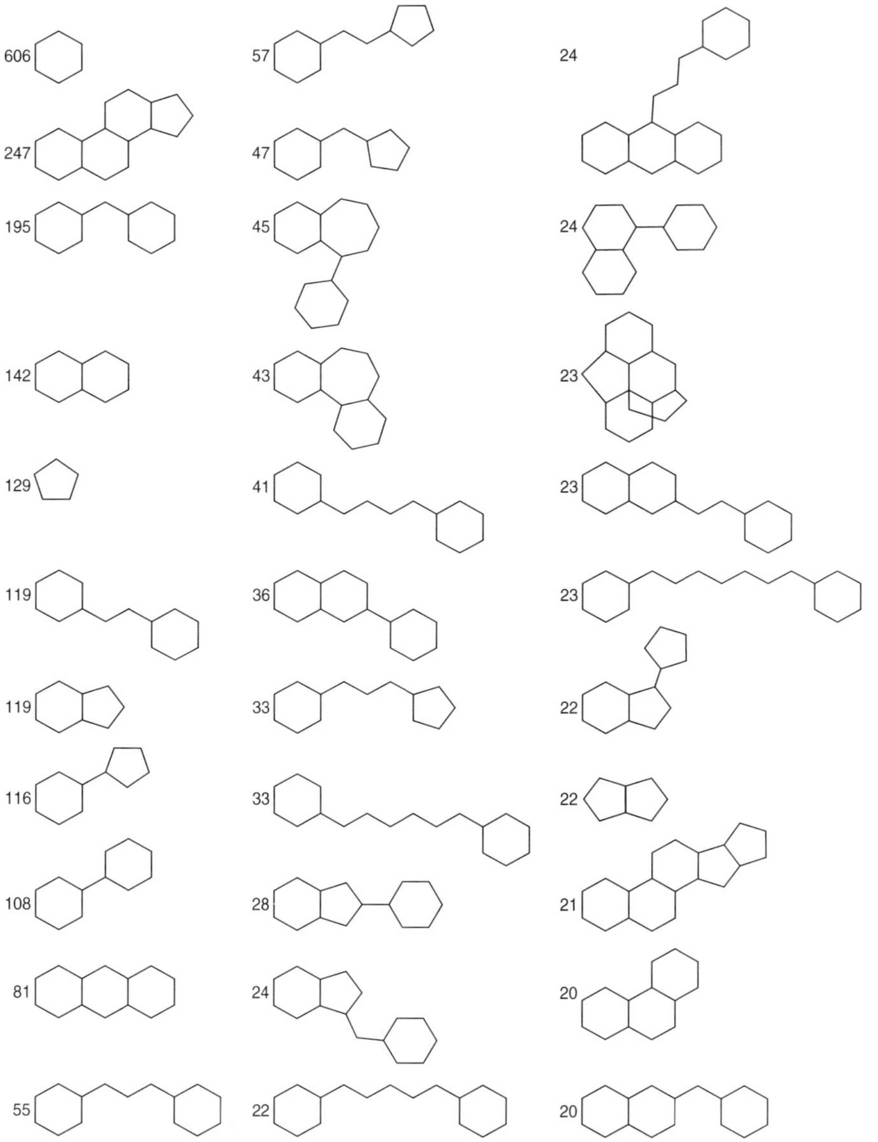

Fig. 1.13. Common scaffolds for drugs. Common drugs have a limited number of scaffolds. This figure shows 32 of the most common scaffolds found in a set of 2548 drugs. The numbers show how many drugs have been based on each framework in this set. (Reprinted from *Drug Discovery Today*, Vol. 3, Walters WP *et al*, Virtual screening — an overview, pp. 160–178. Copyright 1998, with permission from Elsevier Science.)

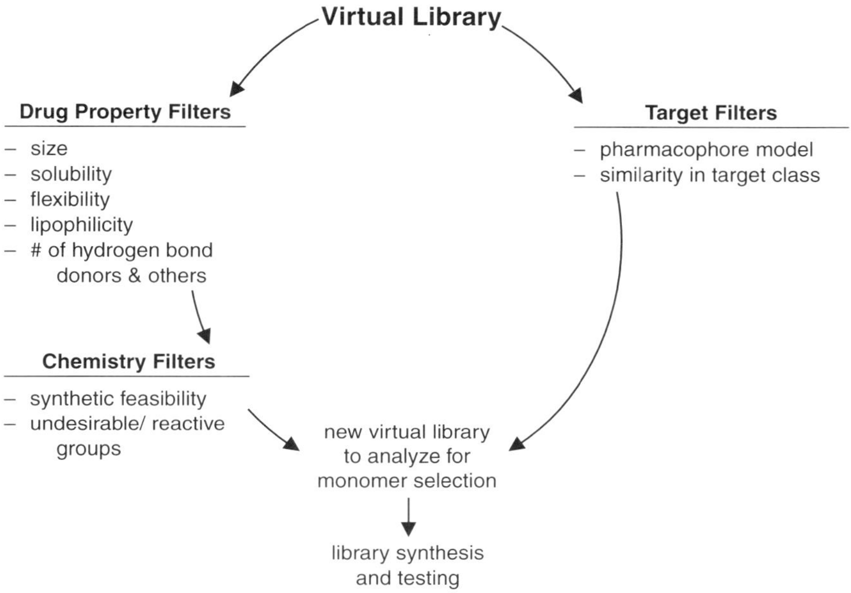

Fig. 1.14. Virtual libraries and computational filters for library design. Computational filters based on calculated properties and pharmacophore models can be used to design libraries to fit a specific profile or to eliminate undesirable compounds. The same tools are useful for selecting compounds from screening sets. (Reprinted from *Drug Discovery Today*, Vol. 3, Walters WP *et al*, Virtual screening — an overview, pp. 160–178. Copyright 1998, with permission from Elsevier Science.)

calculated properties of the compound and biological activity for a test set of compounds. Properties include molecular descriptors based on the physical nature of the compound (molecular weight, lipophilicity, topology) and fragment-based descriptors. The fragment-based descriptors indicate the presence or absence of combinations of atoms. The fragments can be large (ring systems, functional groups), small combinations of atoms (carbonyl or hydroxyl groups) or simply arbitrary pairs and triples of atoms. Various statistical methods correlate the structural features to activity: classification and regression trees, linear and non-linear models, neural networks and various heuristic models. The success of these methods depends on whether the training set and descriptors cover the space necessary to make predictions about the test compounds.

Design and compound or monomer selections are iterative processes. This discussion has centered on making an initial selection of a subset from a large virtual library or a physical compound store. After biochemical results are obtained, more information is available to select the next subset. The screening results and the computational methods are complementary and run in parallel. Choosing the number of compounds to test in a screening cycle depends on compound-handling logistics, screening throughput and a balance between providing more information to select the next set of compounds and the probability of finding leads of sufficient quality and quantity to begin optimization.

1.3.4 OPTIMIZATION OF LEAD COMPOUNDS FOR USE IN HUMANS

The challenge of drug discovery is not only to find a ligand that has a favorable binding affinity at a receptor or target of interest, but to optimize the ligand to have a profile that is suitable for an orally active compound for use in humans. The profile includes measures of selectivity at other receptors, toxicity and metabolic characteristics. After high-throughput screening is complete, possible outcomes are: there are a set of active compounds in diverse chemical classes to optimize, there are a limited number of active compounds in a restricted chemical class to optimize; or there are no compounds to follow up. Because optimization can be a longer process than screening, many pharmaceutical companies are adopting a strategy to perform testing on selectivity, toxicity and ADME (absorption, distribution, metabolism and excretion) properties earlier.

Calculated physical properties are used in conjunction with measured properties to optimize compounds. Because of the ease of obtaining the calculated properties, they can be used early in library design to bias compounds toward improved bioavailability. Figure 1.15 shows histograms of calculated $\log P$, molecular weight, number of hydrogen bond donors and number of hydrogen bond acceptors for the world drug index. For common drugs, it is seen that there is a narrow range of values for these parameters. Generally, the desirable profile for a drug is small molecular weight (less than 500), high aqueous solubility and hydrophilicity. Usually hydrophilic compounds are soluble in water. In addition to aqueous solubility, other physical parameters for design include the number of hydrogen-bonding functional groups. If a compound has a large number (greater than 5) of hydrogen bonds, the bonds are likely to produce a stable, high-energy lattice network. The breaking of these bonds can be rate limiting for transport across biological membranes.

Lipinski *et al* (1997) published a paper listing guidelines for five easily calculated physical properties in order to introduce design recommendations for libraries. The 'rules of 5', so named for the mnemonic nature as multiples of five, state that poor absorption or membrane permeation is more likely if: molecular weight is greater than 500; there are more than five hydrogen bond donors; there are more than 10 hydrogen bond acceptors; and calculated $\log P$ is less than 5. Note that although these rules are designed specifically for absorption and membrane permeation, by constraining $\log P$, it is also likely to lead to a better metabolism profile.

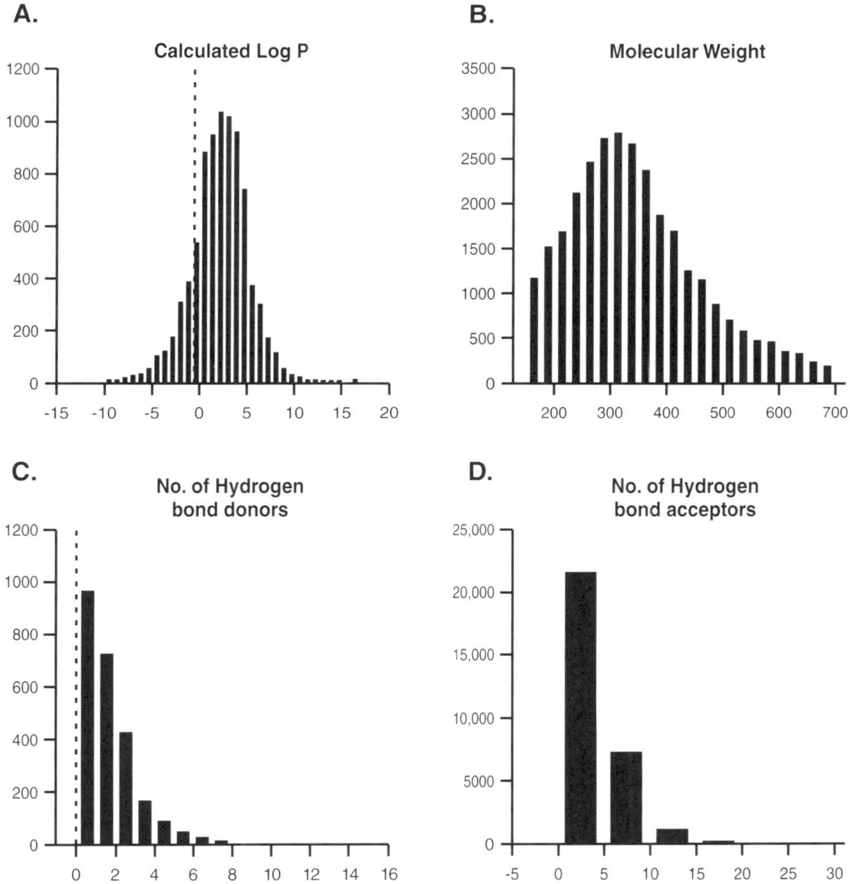

Fig. 1.15. Histograms of calculated properties for the world drug index. These histograms show the distributions for these properties for a series of common drugs. In order to bias chemical libraries to being more drug-like, monomers are chosen so that the combinatorial products have similar property distributions: (A) calculated log *P*; (B) molecular weight; (C) number of Hydrogen bond donors; (D) number of Hydrogen bond acceptors

Although the calculated parameters assist in the design of biologically active compounds, the measurement of absorption, metabolism and toxicity in surrogate systems is essential to approximate how the molecule will be absorbed and metabolized *in vivo*. Figure 1.16 is a list of biochemical assays and properties that are useful for the design of combinatorial libraries. Assays are used to measure lipophilicity, solubility and membrane permeability. Surrogate assay systems are used to model blood–brain absorption, metabolism (e.g. the cytochrome

Assays	**Descriptors**
Solubility	Molecular weight
Lipophilicity	Predicted solubility
Membrane permeability	Predicted lipophilicity
Genotoxicology	No. rotatable bonds
Metabolism	No. hydrogen bond donors
Blood brain barrier	No. hydrogen bond acceptors
Gastrointestinal absorption	

Fig. 1.16. Lists of assays and descriptors for lead optimization. Optimization of a compound for use as a drug is a complex process. In addition to the affinity and efficacy profile, side effects must be minimized and delivery and metabolism of the drug must be optimized. The list of assays shows some of the tests that are useful in the optimization process. The list of descriptors are variables that are often used to model and predict compound profiles from the chemical structure

P-450 system) and stability in the gastrointestinal tract. The CACO-2 model is an example of a commonly used system to measure gastrointestinal absorption that has become a higher throughput assay with a capacity of several hundred compounds per week due to advances in automation. Models for genetic toxicity continue to be low throughput and take several weeks to obtain data for a few compounds.

1.4 PHARMACOLOGICAL DEFINITIONS

It is useful, at this point, to describe some biological nomenclature unique to the pharmacology of drug discovery. Chemicals that interact with biological targets such as receptors, enzymes and transport processes are generally referred to as drugs. A drug can be an incredibly potent agent for the initiation of change of state of a complete biological system. The root of this behavior is the interaction of the molecule with its initial target. Much of this book is written from the point of view of this initial target being specialized proteins on the cell surface termed receptors. These are the portals through which cells communicate with their extracellular world and adjust their state to the environment. However, the chemical, biological and statistical concepts described for receptors very often apply to other biological targets as well, namely enzymes, ion channels and transport processes; all involve the initial interaction of small molecules with proteins. The first step in this characterization is the definition, in generic terms, of the process of drug interaction with proteins.

The study of drug effect on drug receptors is complicated by the indirect nature of the measurements made. Thus, the affinity and/or efficacy of drugs is observed indirectly on varying scales. These scales vary because physiological systems alter the initial biological signal and translate the information in different ways, often unique to the specific organ system. This can cause dissimulation between the effects seen in the pharmacological test system and the therapeutic arena in humans. The translation of chemical information into various forms by biological systems leads to the concept of system-dependent and system-independent scales of drug activity. Clearly system-dependent indices of activity are limited unless the testing for therapeutic activity is done under the exact circumstances as the proposed therapy (i.e. in humans under the same pathological conditions), and then the system-dependent nature of the test data can be altered when the drug is used under different circumstances. *The major thesis of this book is the need to obtain chemical estimates of drug activity that transcend the test system in which they are estimated and thus can be used to predict activity in all systems.*

There are basically two system-independent indices of drug activity. These are ***affinity*** and ***intrinsic efficacy***. The first property is the one that brings the drug into the realm of influence of the receptor and allows the drug to bind to it. Affinity is a chemical term and usually is described as the equilibrium dissociation constant of the drug–receptor complex (the concentration of drug that binds to 50% of the receptor population). By definition, a molecule with zero affinity is not a drug because it does not associate with the receptor. The scale for affinity is values greater than zero. The second, efficacy, is the property of the drug that causes the receptor to change its behavior once the drug is bound. It should be noted that efficacy can be negative (inducing blockade of receptor constitutive activity), zero (the ligand simply interferes with the action of other drugs but itself does not alter the behavior of receptors) or positive (the drug induces a physiological excitation of the receptor). This book will primarily focus on the operational methods available to measure drug affinity and efficacy and the application of these estimates to the prediction of drug activity in all receptor systems. As a preface to the description of the measurement of system-independent drug activity, it is useful to define some critical pharmacological terms.

Chemicals that bind to biological targets (generally termed receptors) can be referred to generically as ***ligands***. Those that bind but do not alter receptor function and/or behavior but do interfere with the action of other ligands are termed ***antagonists***. Ligands that change the function and/or behavior of receptors are termed ***agonists***. Usually, the interaction of agonists with receptors results in physiological response. This can take the form of an excitation or augmentation of cellular function or inhibition and/or modulation of cellular function. The excitation of pharmacological preparations is referred to as agonism; this can take on a spectrum of magnitudes. Zero agonism refers to antagonism (as described above) and full agonism is the production of the maximal response capable of the preparation. Agonists that produce a maximal response, the magnitude of which is less than the tissue maximum (and the maximum produced by a ***full agonist***), are referred to as ***partial agonists***. The dose–response curves to these types of ligands are shown in Fig. 1.17A.

If a drug possesses efficacy, then it will impart a signal to the receptor to change its subsequent behavior; this signal is referred to as the receptor ***stimulus***.

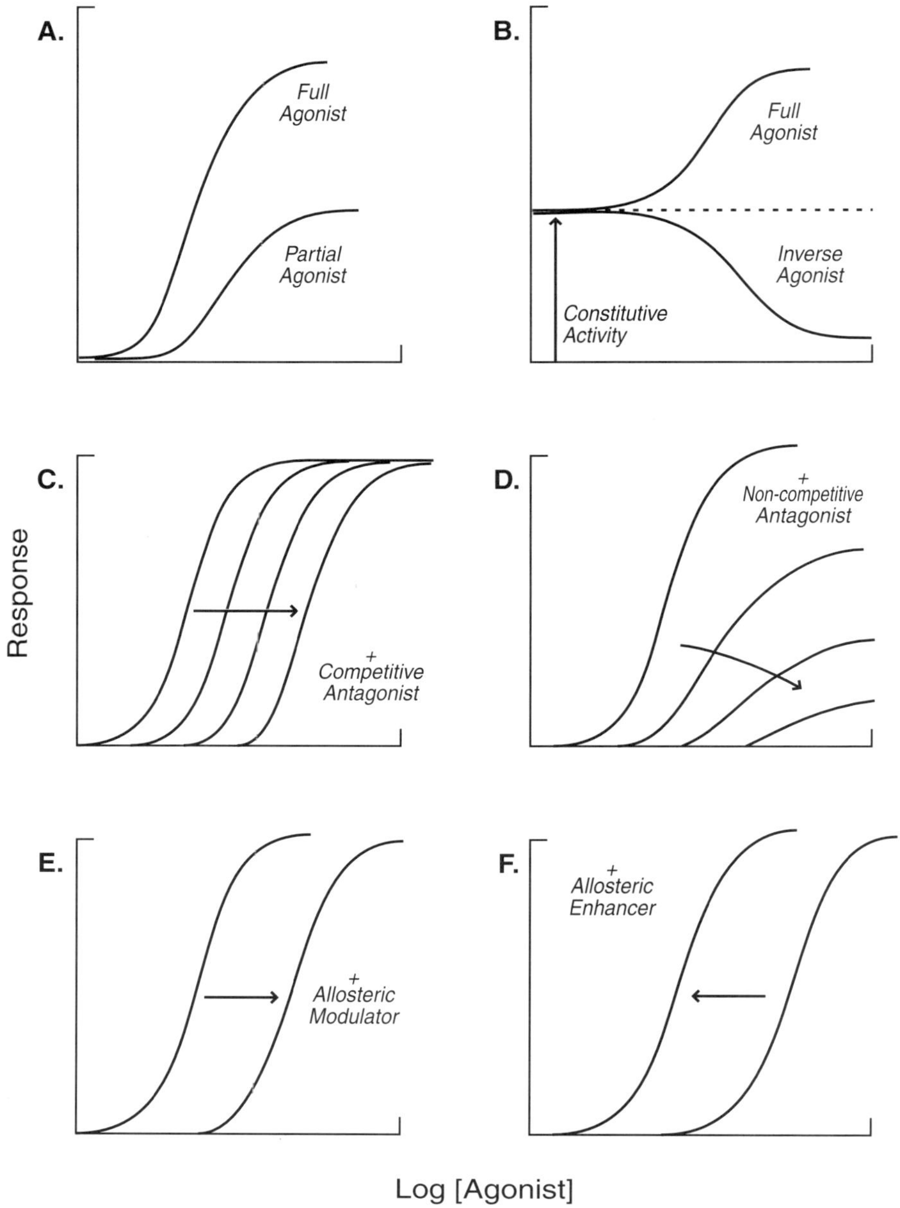
A.
Full
Agonist
Partial
Agonist
B.
Full
Agonist
Inverse
Agonist
Constitutive
Activity
C.
+
Competitive
Antagonist
D.
+
Non-competitive
Antagonist
E.
+
Allosteric
Modulator
F.
+
Allosteric
Enhancer
Response
Log [Agonist]

The observed response in a pharmacological preparation is a translation of the receptor stimulus (imparted by the agonist) by the tissue into a physiological response. The processes that translate the initial receptor stimulus are referred to as the tissue ***stimulus–response mechanisms***. It should be noted that stimulus–response mechanisms, which are saturable biochemical processes within the cell that process the cytosolic messengers produced by the receptor stimulation into a general cellular response, can generally normalize the maximal response observed for strong agonists to cause them to produce the same observed maximal response. Thus, a collection of agonists may all be full agonists in a given system but that does not necessarily mean that they are of the same strength in terms of producing agonism in other systems. This underscores a basic tenet in receptor pharmacology, *namely that the stimulus–response mechanisms of tissues can completely control the observed response to agonists.* Therefore, a weak agonist can produce no response in one tissue if that tissue possesses a weak stimulus–response mechanism, can be a partial agonist in another tissue that has a more efficient processing system and can be a full agonist in yet another tissue that has a highly efficient stimulus–response cascade. These effects will be discussed in detail in Chapter 3.

It is known that receptors themselves can adopt active conformations spontaneously to impart a physiological response to systems. This type of activity is referred to as ***constitutive receptor activity***. Also there are ligands that actively block such activity — these ligands are referred to as ***inverse agonists***. The dose–response curve to an inverse agonist is shown in Fig. 1.17B.

There are different ways in which antagonists can block agonist response; some of these patterns of antagonism are used to characterize antagonists. The production of parallel shifts to the right of an agonist dose–response curve (Fig. 1.17C) is consistent with ***competitive antagonism***. In contrast, some ligands depress the maximal asymptotic response to an agonist (Fig. 1.17D); this may or may not be accompanied by a shift to the right of the curve. This pattern of antagonism is

---

Fig. 1.17. Dose–response curves for different types of pharmacological ligands. (A) A full agonist produces the system maximum response; a partial agonist produces a maximal asymptotic response below that of the system maximum. (B) Constitutive receptor activity is made manifest in the elevated basal response observed in the absence of an agonist. Agonists can augment this response; inverse agonists reverse the response due to constitutive receptor activity. (C) A competitive antagonism produces parallel shifts to the right of the dose–response curve with no diminution of maximum response. (D) A non-competitive antagonist may depress the maximal response; this may or may not be accompanied by shifts to the right of the dose–response curve. (E) An allosteric modulator reduces the activity of an agonist by producing a shift to the right of the dose–response curves to the agonist. (F) An allosteric enhancer shifts the dose–response curve to the agonist to the left, thus potentiating agonist response

Table 1.1. Potency of the β-adrenergic agonist prenalterol in different isolated tissues

Tissue	ED_{50}^{1} (nM)	Affinity[2] (μM)
Guinea pig trachea	0.2	0.05
Cat left atrium	0.35	0.05
Rat left atrium	0.6	0.05
Cat papillary muscle	1.5	0.05
Guinea pig left atrium	3.1	0.04

[1] Concentration producing 50% maximal response as an agonist (system dependent).
[2] Concentration binding to 50% of the receptor population (system independent).
Data from Kenakin and Beek (1980).

referred to as ***non-competitive antagonism.*** It should be stressed that the experimental system containing the receptor can influence the pattern of antagonism as well therefore these observed patterns may not necessarily denote molecular mechanism (this is discussed further in Chapter 2).

Finally, there are ligands that modify the behavior of the receptor towards other ligands. These are referred to as ***allosteric effectors.*** The effects can be inhibitory (a diminution of the effects of other ligands—***allosteric modulators***; see Fig. 1.17E) or cause potentiation (***allosteric enhancers***; see Fig. 1.17F).

The elimination of system dependence is crucial to the improvement of drug activity through chemical synthesis. Table 1.1 shows data for the β-adrenergic partial agonist prenalterol. There is a 16-fold range in the system-dependent parameter of observed potency (concentration producing 50% maximal response, i.e. ED_{50}) but there is only a 1.4-fold range in the system-independent parameter of affinity. System dependence can cause confusion when drugs are tested in one system for discovery and characterization and used therapeutically in another.

As will be seen in this book, the descriptive information about drugs and drug receptors generally will be given within a defined context, i.e. intrinsic efficacy relative to a certain reference drug, etc. This may not allow speculation as to the molecular nature of drug effects, but it does allow for a certain prediction of quality and quantity of drug effect in the therapeutic arena.

1.5 DRUG RECEPTOR THEORY

All of the techniques described in this book for the quantification and characterization of drug properties originate from models of drug receptor theory. The equations used within this theory emanate from Michaelis-Menten enzyme kinetics and descriptions of protein two-state theory and largely are customized to fit existing data on drug and drug–receptor function. The origins and development of these theories will not be discussed specifically in this book except to illustrate the limits of the pharmacological estimates obtained by use of the models. In actuality, specific mechanisms of drug and receptor action often are not required for the definition and quantification of drug effect and prediction of therapeutic effect. Instead, null techniques often circumvent the need for detailed knowledge of complicated physiology, and the interaction of drugs and receptors can be described in simple terms of molecular interaction. The circumventing of complex biological systems to estimate the chemical properties of drugs will be referred to as the operational approach to the classification of drugs and drug receptors.

An important general idea contained within receptor theory is the fact that, unlike enzymes, which interact with substrates and change their nature, receptors interact with molecules repeatedly and read the molecular information encoded within them but do not change the messenger. This extraordinary process is still not understood but presently can be described approximately by the interaction of molecules with multiple forms of proteins and the introduction of bias into complex systems by the differential affinity for protein conformations.

There is a rich history of drug receptor theory that describes various theories of drug–receptor interaction in terms of the available knowledge of the biochemistry of receptors. In general, mathematical formulations of drug–receptor interaction began with the work of A. J. Clark in Cambridge (Clark 1937), in treatises that described receptor occupancy theory (i.e. the receptor responded to a drug when it was occupied by the drug). Important modifications of this view were made by various experimenters and theoreticians (Gaddum 1937, 1957; Schild 1949; Ariens 1954, 1964; Stephenson 1956; Arunlakshana and Schild 1959; Furchgott 1966; MacKay 1966). Each provided the required alterations to the original theory because of new knowledge of receptor function. Concurrently, information about the allosteric nature of proteins (Monod *et al* 1965; Changeux *et al* 1967; Podleski and Changeux 1970) and the discovery of new ways to monitor protein conformation (i.e. ion channels: Del Castillo and Katz 1957; Katz and Thesleff 1957) have led to further modifications of receptor theory and the introduction of more mechanism-based descriptions of drug–receptor effects. Biochemical evidence of protein translocation within membranes (Cuatrecasas 1974) also has added to models of receptor function. These ideas sowed the seeds of mechanistic models of drug action, such as the ternary complex model for receptors (DeLean *et al* 1980) and modifications thereof (Samama *et al*, 1993; Weiss *et al* 1996*a*,*b*). Theoretical models of receptor function, such as the operational model (Black and Leff 1983), also are available as tools to quantify drug activity. These various models will be referred to as required in the discussions of the quantification of drug activity.

1.6 SPECIFICITY AND SELECTIVITY

Usually, a drug is designed to produce a very selective effect on a given pathology, i.e. if possible, to function as a chemical scalpel. Any other drug activity possessed

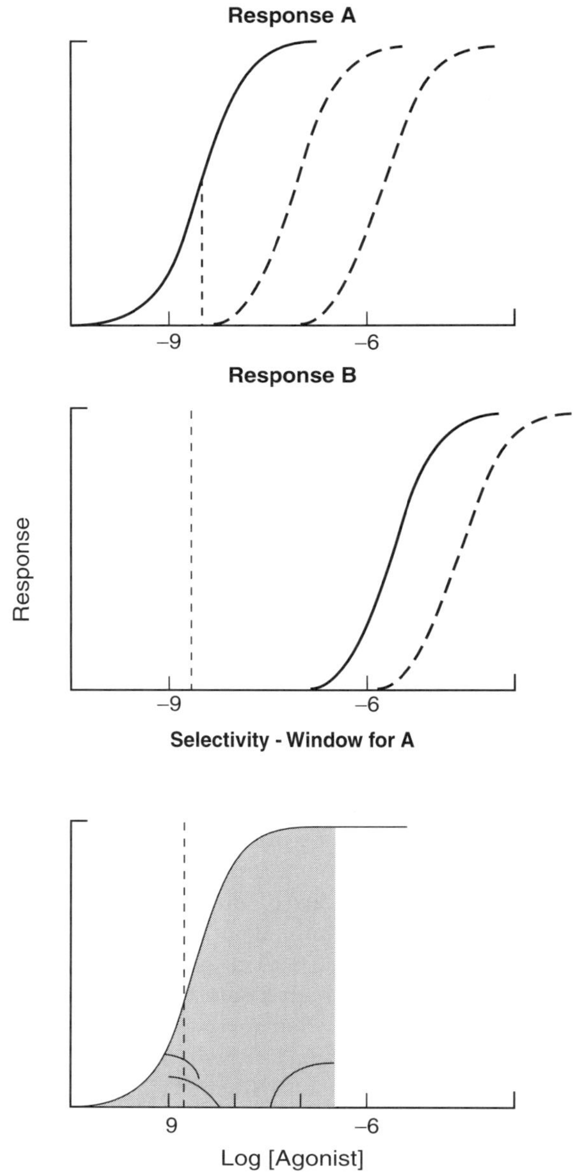

Fig. 1.18. Drug selectivity windows. The agonist for receptor A produces effects at concentrations as low as 1 nM. A secondary effect of the agonist is observed on receptor B at 1 μM. Therefore, the selectivity window for this agonist is 1–500 nM

by that drug may make itself manifest in other organs to produce obfuscating effects; these are referred to as side effects and generally are undesirable. Therefore, the drug discovery process naturally is aimed toward finding drugs with specific activity, i.e. activity of only one sort. Biologically active molecules very seldom are specific (i.e. have only one activity on biological systems) but rather are useful because they are selective. This latter term links biological effect with the concentration of the drug at the receptor. Thus, although a drug may possess two activities, with one occurring at concentrations 1/100 those required for the second activity, an operational type of specificity can be achieved by restricting the concentration of the drug (dosage) to those concentrations that allow only the most prominent activity to be expressed (Fig. 1.18). Thus, the second effect is not permitted to be expressed through the use of a restriction of concentration. The range in which the desired activity can be obtained with no complicating effects of secondary activities is termed the therapeutic window. In general, a basic tenet of pharmacology is that most drugs are selective and not specific and that specific effects can be obtained only be confining the concentration of drugs to the therapeutic window.

1.7 NULL PROCEDURES

Drugs interact with receptors to produce a stimulus, which is then processed by the target organ into a physiological response. Thus, the same stimulus (product of a common drug–receptor pair) could be processed in a number of ways by a number of organs to give varying types and strengths of responses. For this reason, molecular information about drug–receptor interaction cannot be obtained directly from observation of the processed physiological responses. Instead, two drugs are compared in the same system and it is assumed that the post-receptor translation of stimulus that takes place for one drug also takes place for the other (see Fig. 1.19). Under these circumstances, the tissue factors cancel and the relative activity represents molecular drug properties.

1.8 CONCLUSIONS

This chapter is designed to be an introduction to the drug discovery process. The aim of this endeavor is to quantify, in chemical terms, the biological activity of chemicals and also to determine the means to predict the biological activity of these same molecules in all biological systems. Also important to the drug discovery process are the emerging tools available to sort and discover new biological targets and to optimize rapidly those compounds for clinical efficacy in humans.

Drug discovery involves the interaction between many fields of science, including chemistry, biology, genetics, medicine, pharmacology and toxicology. For the process to work, there must be an understanding of the information that

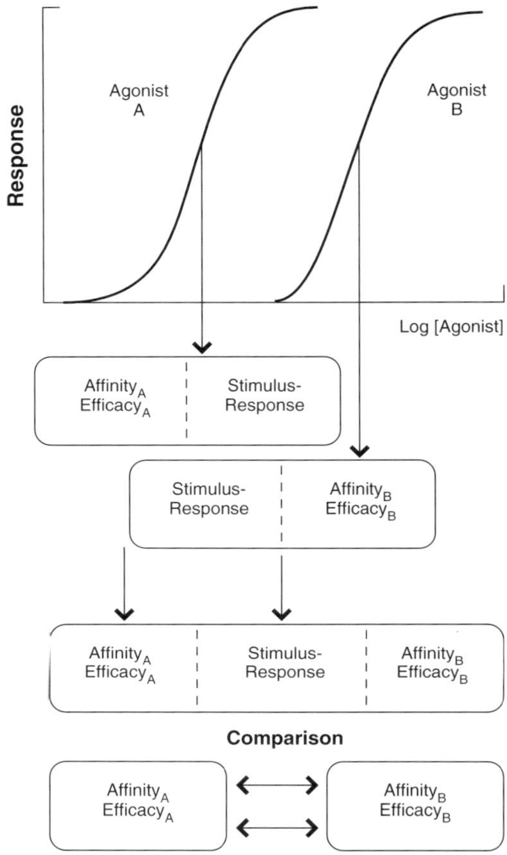

Fig. 1.19. The null method in receptor pharmacology assumes that all stimulus-processing events occurring after receptor activation by two agonists are identical for the two agonists and thus cancel. Under these circumstances, equal responses are assumed to emanate from equal receptor stimulus

each experiment and set of measurements contributes to the profile of information about potential drugs. Problems in design (optimization of a compound or construction of chemical libraries to probe a biochemical target) become more focused as more information becomes available. The remainder of the book provides details on the specific measurement systems, receptor theory, statistical methods and pragmatic issues involved in drug discovery. The theory and methods in these areas continue to evolve and discussion of many of the methods is beyond the scope of this book. Each chapter has a Further Reading section listing both recent and classical literature that is useful as a starting point for readers interested in more detail about a specific topic.

1.9 REFERENCES

Ariens EJ (1954) Affinity and intrinsic activity in the theory of competitive inhibition. *Arch Int Pharmacodyn Ther* **99**: 32–49.
Ariens EJ (1964) *Molecular Pharmacology*, Vol. 1. Academic Press, New York.
Arunlakshana O and Schild HO (1959) Some quantitative uses of drug antagonists. *Br J Pharmacol* **14**: 48–58.
Black JW and Leff P (1983) Operational models of pharmacological agonist. *Proc R Soc London Ser B* **220**: 141–162.
Changeux JP, Thiery J, Tung Y and Kittel C (1967) On the cooperativity of biological membranes. *Proc Natl Acad Sci USA* **57**: 335–341.
Clark AJ (1937) General pharmacology. In *Heffner's Handbuch d-exp. Pharmacol. Erg. band 4*, pp. 1–37. Springer, Berlin.
Cuatrecasas P (1974) Membrane receptors. *Annu Rev Biochem* **43**: 169–214.
Del Castillo J and Katz B (1957) *Proc R Soc London* **146**: 369–381.
DeLean A, Stadel JM and Lefkowitz RJ (1980) A ternary complex model explains the agonist-specific binding propoerties of adenylate cyclase coupled β-adrenergic receptor. *J Biol Chem* **255**: 7108–7117.
Furchgott RF (1966) The use of β-haloalkylamines in the differentiation of receptors and in the determination of dissociation constants of receptor-agonist complexes. In *Advances in Drug Research*, Vol. 3, ed. by NJ Harper and AB Simmonds, pp. 21–55. Academic Press, New York.
Gaddum JH (1937) The quantitative effects of antagonistic drugs. *J Physiol (London)* **89**: 7P–9P.
Gaddum JH (1957) Theories of drug antagonism. *Pharmacol Rev* **9**: 211–218.
Katz B and Thesleff S (1957) A study of 'desensitization' produced by acetylcholine at the motor endplate. *J Physiol (London)* **138**: 63–80.
Kenakin TP and Beek D (1980) Is prenalterol (H133/80) really a selective beta1–adrenoceptor agonist? Tissue selectivity resulting from differences in stimulus–response realtionships. *J Pharmacol Exp Ther* **213**: 406–413.
Kick EK, Roe DC, Skillman AG, *et al* (1997) Structure-based design and combinatorial chemistry yield low nanomolar inhibitors of cathespin D. *Chem Biol* **4**: 297–307.
Lipinski CA, Lombardo F, Dominy BW and Feeney PJ (1997) Experimental and computational approaches to estimate solubility and permeability in drug discovery and development settings. *Adv Drug Deliv Rev* **23**: 3–25.
MacKay D (1966) A new method for the analysis of drug–receptor interactions. In *Advances in Drug Research*, Vol. 3, ed. by NJ Harper and AB Simmonds, pp. 1–19. Academic Press, New York.
Monod J, Wyman J and Changeux JP (1965) On the nature of allosteric transitions. *J Biol Chem* **12**: 88–118.
Podleski TR and Changeux J-P (1970) In *Fundamental Concepts in Drug–Receptor Interaction*, ed. by JF Danielli, JF Moran and DJ Triggle, pp. 93–119, Academic Press, New York.
Samama P, Cotecchia S, Costa T and Lefkowitz RJ (1993) A mutation-induced activated state of the β_2-adrenergic receptor: extending the ternary complex model. *J Biol Chem* **268**: 4625–4636.
Schild HO (1949) pA_x and competitive drug antagonism. *Br J Pharmacol* **4**: 277–280.
Sittampalam G, Iversen PW, Boadt JA, *et al* (1997) Design of signal windows in high-throughput screening assays for drug discovery. *J Biomol Screening* **2**: 159–169.
Stephenson RP (1956) A modification of receptor theory. *Br J Pharmacol* **11**: 379–393.
Walters WP, Stahl MT and Murko MA (1998) Virtual screening—an overview. *Drug Discovery Today* **3**: 160–178.

Weiss JM, Morgan PH, Lutz MW and Kenakin TP (1996*a*) The cubic ternary complex receptor occupancy model. I Model description. *J Theor Biol* **178**: 151–167.
Weiss JM, Morgan PH, Lutz MW and Kenakin TP (1996*b*) The cubic ternary complex receptor occupancy model. II Understanding apparent affinity. *J Theor Biol* **178**: 169–182.
Zuckermann RN, Martin EJ, Spellmeyer DC, *et al* (1994) Discovery of nanomolar ligand for 7-transmembrane G-protein-coupled receptors from a diverse *N*-(substituted) glycine peptoid library. *J Med Chem* **37**: 2678.

1.10 FURTHER READING

Clinical Genetics and Target Selection

Beeley LJ and Duckworth DM (1996) The impact of genomics on drug design. *Drug Discovery Today* **1**: 474.
Bishop MJ (1998) *Guide to Human Genome Computing*. Academic Press, San Deigo.
Lander E (1996) The new genomics: global views of biology. *Science* **274**: 536–539.
Ott J (1991) *Analysis of Human Genetic Linkage*. Johns Hopkins University Press, Baltimore.
Risch N and Merikangas K (1996) The future of genetic studies of complex human diseases. *Science* **273**: 1516–1517.
Spence P (1998) Obtaining value from the human genome: a challenge for the pharmaceutical industry. *Drug Discovery Today* **3**: 179–188.

Compound Selection and Library Design

Blaney JM and Martin EJ (1997) Computational approaches for combinatorial library design and molecular diversity analysis. *Curr Opin Chem Biol* **1**: 54.
Cummins DJ, Andrews CW, Bentley JA and Cory M (1996) Molecular diversity in chemical databases: comparison of medicinal chemistry knowledge bases and databases of commerically available compounds. *J Chem Inf Comput Sci* **36**: 750.
Fecik RA, Frank KE, Gentry EJ, Menon SR, Mitscher LA and Telikepalli H (1998) The search for orally active medications through combinatorial chemistry. *Med Res Rev* **18**: 149.
Higgs RE, Bemis KG, Watson IA and Wikel JH (1997) Experimental designs for selecting molecules from large chemical databases. *J Chem Inf Comput Sci* **36**: 750.
Martin EJ, Blaney JM, Siani MA, Spellmeyer DC, Wong AK and Moos WH (1995) Measuring diversity: experimental design of combinatorial libraries for drug discovery. *J Med Chem* **38**: 1431.

Optimization of Lead Compounds for Use in Humans

Chan OH and Stewart BH (1996) Physicochemical and drug-delivery considerations for oral bioavailability. *Drug Discovery Today* **1**: 461–473.
Navia MA and Chaturvedi PR (1996) Design principles for orally bioactive drugs. *Drug Discovery Today* **1**: 179.
Weber L, Wallbaum S, Broger C and Gubernator K (1995) Optimization of the biological activity of combinatorial libraries by a genetic algorithm. *Angew Chem Int Ed (Engl)* **34**: 2280.

History of Receptor Pharmacology

Parascandola J (1986) The development of receptor theory. In Pharmacological Methods, Receptors and Chemotherapy, Vol. 3, ed. by MJ Parnham and J Bruinvels, pp. 129–158. Elsevier, Amsterdam.

Allosterism in Receptor Theory

Thron CD (1973) On the analysis of pharmacological experiments in terms of an allosteric receptor model. *Mol. Pharmacol* **9**: 1–9.

2 Measurement of Drug Affinity

2.1 INTRODUCTION

One chemical constant that can be measured is the affinity between a ligand and a receptor, i.e. how strong is the attraction between the two. It should be noted that affinity can be a system-dependent parameter that involves the ligand, receptor and other components of the receptor system (this will be discussed later in this chapter). These system-dependent estimates of ligand affinity are not useful for drug discovery and development because they cannot be extrapolated to other systems. This chapter will focus mainly on the elimination of system dependency and the measurement of the chemical constants that define the forces of attraction between the ligand and the receptor. This measurement can be extrapolated across systems and be useful in predictions of activity in humans. As a preface to the discussion of methods to measure the affinity between drugs and receptors, a discussion of the chemical forces involved is useful.

2.2 THE CHEMICAL NATURE OF AFFINITY

Ligands reside at an optimum distance away from the peptide chains in the receptor binding pocket according to the various forces that cause attraction and repulsion between them. The critical factor determining distance between the receptor and the ligand is the energy lost or gained as the ligand moves in the receptor pocket. Thus, the ligand will find a distance of lowest free energy and then move closer and farther away according to the thermal energy it possesses. The salient point of this process is that ligands do not statically reside in the receptor pocket but, rather, take part in a process of movement in and out of the receptor compartment according to thermal energy. Figure 2.1 shows a frequency histogram of the number of molecules of a given ligand that can be found at any one instant in a receptor system at equilibrium as a function of the distance away from the distance of lowest free energy. When the ligand is not in the pocket, another ligand in the vicinity can occupy the space. This concept is essential to the understanding of competitive drug interactions.

2.3 THE LANGMUIR ADSORPTION ISOTHERM: DIRECT MEASUREMENT OF AFFINITY

Most mathematical models used to quantify the affinity of a ligand for a receptor are derived from the simplest model that describes the adsorption of a molecule

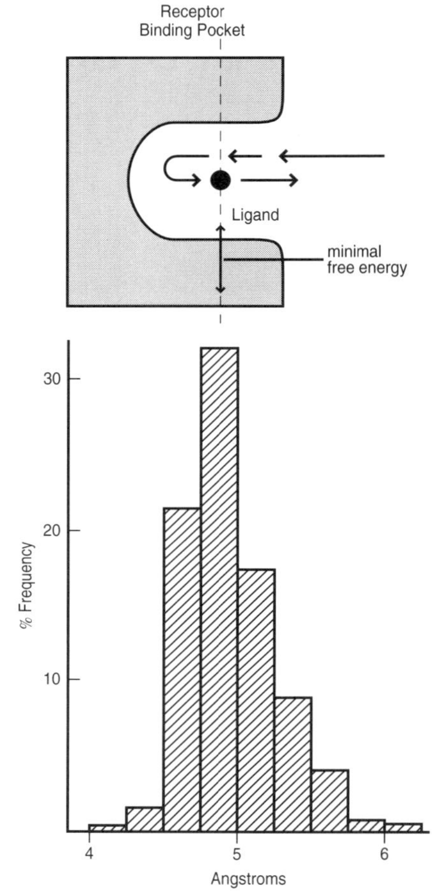

Fig. 2.1. Frequency histogram of the relative number of ligand molecules residing in a receptor protein-binding pocket at any one instant. It is assumed for this molecule that the distance of minimal free energy is 4.7 Å away from the protein backbone of the receptor. The histogram is skewed because of the extremely high energies required to bring the molecule closer than the minimal free energy distance

to a surface, namely the Langmuir adsorption isotherm. It is useful to begin the discussion of the measurement of drug affinity by considering this simple model.

Derived in 1917 by the chemist Irving Langmuir, this model was used to quantify the adsorption of toxic molecules to activated charcoal, an important subject in Langmuir's time because many miners were exposed to toxic substances in mine shafts as they dug for coal. The model is beautiful in its simplicity. The rate of adsorption of a substance to a surface was given as the product of the concentration of the substance (denoted μ), the area of surface available for the substance to bind to (denoted θ) and an intrinsic rate constant for onset of the molecule to the surface (denoted α). Langmuir then assumed that there was a proclivity of bound substance to disassociate away from the surface. This second process was given as the product of the amount of bound substance (denoted as θ_1 and defined as the area of surface occupied by the substance) and an intrinsic rate constant for dissociation away from the surface (denoted V_1). The area of surface bound by the substance, expressed as a fraction of the total area, was given as $1 - \theta$. Langmuir then surmised that the process would proceed until some steady state would ensue, whereby the rate of molecules binding to the surface would equal the rate of molecules leaving the surface. In terms of the previous designations (left in Langmuir's original nomenclature), this is termed as the 'rate of condensation' and given as:

$$\text{Rate of condensation} = \alpha\theta\mu \qquad \ldots (2.1)$$

The rate at which molecules left the surface (termed the 'rate of evaporation') is given as:

$$\text{Rate of evaporation} = \theta_1 V_1 \qquad \ldots (2.2)$$

In terms of conventional chemical nomenclature, the ratio of the rate of dissociation and the rate of association is defined as the equilibrium dissociation constant ($K_d = \theta_1/\alpha$), a chemical term that defines the concentration of substance that half-saturates the total number of binding sites. Replacing Langmuir's nomenclature for concentration of substance as [A], where 'A' is the ligand binding to the receptor, the isotherm can be rewritten as the recognizable equation used to model the binding of drug to receptors:

$$\rho = \frac{[A]}{[A] + K_d} \qquad \ldots (2.3)$$

Thus, if the amount of ligand that is bound to the receptor could be quantified, then the relationship between the amount of ligand added to the receptor preparation and the amount bound could be determined and related to the adsorption isotherm. A graphical depiction of eq. 2.3 is shown in Fig. 2.2A. It can be seen from this figure that as the concentration of A increases, so too does the amount of bound A to the receptor. However, as the number of binding sites become filled with ligand, the amount bound diminishes until a saturation of binding results. Thus, the binding curve can be described by three essential features: a threshold, a slope and a maximal asymptote. This maximal asymptote is difficult to determine experimentally because, in theory, it is the amount bound when the concentration of ligand is infinite. However, it can be estimated in practical terms by carrying out the binding curve until little further binding is observed. This region of the dose–response curve is considerably greater along the concentration axis than where the majority of the binding occurs and the incorporation of this

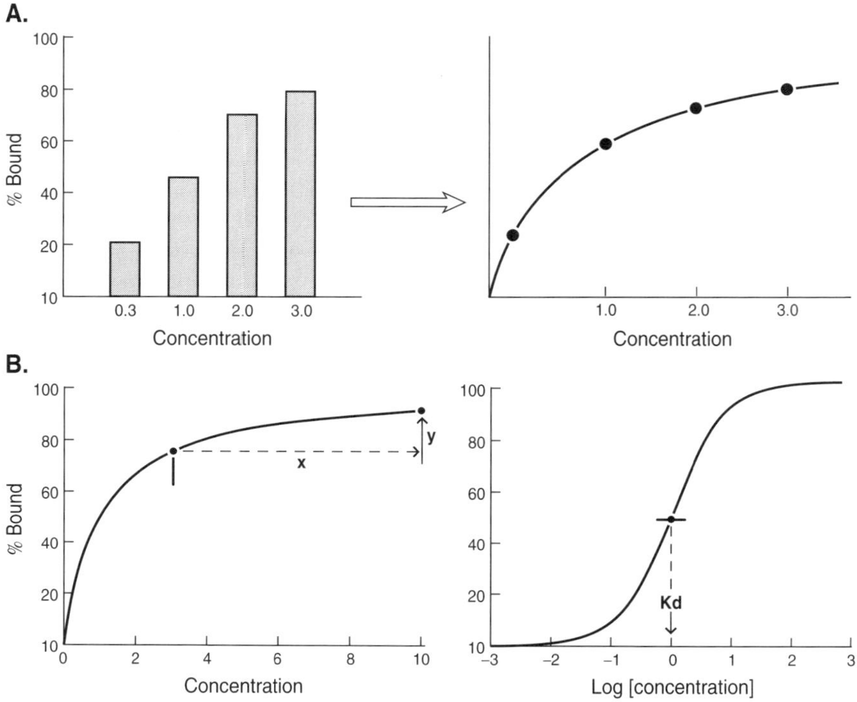

Fig. 2.2. Direct measurement of affinity. (A) The tracer signal produced by increasing concentrations of ligand expressed as the amount of tracer ligand bound to the receptor (percentage of total receptors). A plot of this signal strength on the actual concentration produces a hyperbolic curve. (B) As the maximal asymptote is approached, it takes increasingly large increments of dose to produce measurable changes in response. Note the small change in *y* with a large increase in *x*. This makes visualization of the maximal response difficult on a linear scale. A semi-logarithmic plot of the same curve condenses the concentration increments and shows the maximal asymptote more clearly. The concentration at which the signal is 50% of the maximum corresponds to the equilibrium dissociation constant of the ligand receptor complex if receptor occupancy is the tracer signal

region on a curve can make this latter region difficult to visualize (Fig. 2.2B). For this reason, semi-logarithmic dose–response curves usually are employed because they depict the region of half-maximal and maximal binding with equal facility (Fig. 2.2B).

It can be seen from eq. 2.3 that when the concentration of ligand is equal to K_d the amount of ligand bound will be 50% of the maximal value. Therefore, a practical application of a dose–response curve that directly measures the amount of bound radioligand for given concentrations of free ligands is to determine the half-maximal binding value; by definition, this is the equilibrium dissociation constant of the ligand–receptor complex, i.e. a direct measure of affinity (see Fig. 2.2B).

The obvious constraint of the above approach is the need to have a method of quantifying the amount of ligand–receptor complex formed for any given ligand of interest. One method of doing this is to use a radioactive ligand and measure the amount of radioactivity associated with the ligand-bound receptor; the methods used to do this are described in Chapter 4. Thus, if a radioactive copy of the ligand of interest is available, then its affinity for a receptor can be measured directly by obtaining a saturation binding isotherm according to eq. 2.3. In practice, the synthesis of radioactive analogs of ligands to be screened is impractical, thus indirect methods are needed for the measurement of the affinity of ligands, namely their potency in interfering with the binding of ligands whose presence on the receptor can itself be measured. This will be referred to as indirect measurement of affinity.

2.4 INDIRECT MEASUREMENT OF AFFINITY

This technique depends upon having a tracer ligand, the presence of which on the receptor can be monitored. There are two general modes in which this can be done. The first is, as discussed above, observing the antagonism of binding of radioactive ligands by non-radioactive binding; this is a biochemical binding approach. Thus, a certain equilibrium level of binding is achieved by equilibrating the receptor system with a radioligand and then adding a second non-radioactive ligand, which, when bound to the receptor, will reduce the signal emanating from the portion of receptors previously bound to radioligand (see Fig. 2.3A) The tracer need not be a radioligand but also could be an agonist. Under these circumstances, the strength of response is used as a measure of receptor binding by the ligand (not a direct relationship but rather a dependency) and the reduction in the strength of that response as an indicator of the reduction in tracer ligand binding (see Fig. 2.3B). This is an indirect approach because the actual affinity of the agonist producing the response usually is not known but, through null procedures, models have been devised to estimate the affinity of the antagonist.

Clearly, unless every ligand to be screened as a drug candidate were radioactive, almost all screening is by this indirect method. Under these circumstances, the ligands being screened can be considered as antagonists of the tracer ligand. As a preface to the description of the techniques used in the measurement of affinity, it is useful to consider the general topic of drug antagonism.

2.4.1 DRUG ANTAGONISM

There are pharmacological techniques available to measure the affinity of antagonists if they interfere with the effects of tracer ligands whose binding can be followed either through measurement of radioactivity or physiological effect.

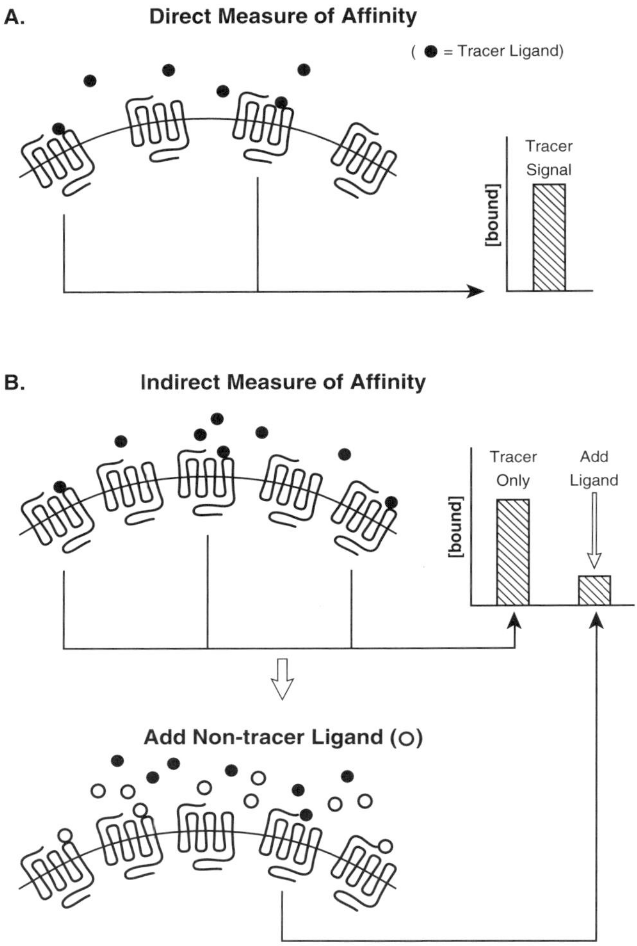

Fig. 2.3. Two general methods of measuring affinity. (A) Direct method. The amount of ligand bound to the receptor is separated from the media; the strength of the tracer signal denotes the amount of ligand bound. (B) Indirect method. A previously equilibrated amount of tracer ligand is bound to the receptor. This amount bound yields a stable tracer signal. A second ligand that does not produce a tracer signal is added. The amount of this second ligand that is bound is inferred from the decrease in tracer signal, denoting displacement of tracer ligand from the receptor

In this latter instance, agonists can be used in a functional receptor system to yield a dose–response curve that can be modeled with eq. 2.3. However, in the case of a functional assay, the K_d in eq. 2.3 will not be the concentration of ligand that binds to 50% of the receptor population but rather some function of that value. This fact does not interfere with the measurement of the antagonist affinity because null techniques are used (*vide infra*).

There are two general effects that an antagonist can have on a dose–response curve showing the effects of a tracer ligand: a change in the location parameter along the concentration axis: and a change (usually a decrease) in the maximal asymptote of the curve. A useful early classification of the effects of antagonists was made by the delineation of surmountable and insurmountable antagonism (Gaddum *et al* 1955). If the maximal response to the initial tracer ligand could be regained in the presence of the antagonist, then the antagonism was termed to be surmountable (Fig. 2.4A). However, if the maximal response was depressed, then the antagonism was termed insurmountable (Fig. 2.4B). This depression of maximal response may or may not be·accompanied by a shift to the right of the dose–response curve. These general categories of antagonism can be produced by a variety of pharmacological mechanisms. For example, drugs that do not bind to the specific receptor of interest but, rather, simply interfere with the physiological function of tissues and cells can produce either surmountable or insurmountable antagonism. Another specific mechanism that produces surmountable antagonism is termed chemical antagonism; this is where the tracer ligand (either radioligand or agonist) and the antagonist form a chemical complex (with no receptor involved), which then cannot interact with the receptor. Because these are not receptor effects, they are not relevant to any discussion of affinity and will not be discussed further in this chapter.

It should be noted that the effects of an antagonist on the tracer dose–response curve cannot be used to define the molecular mode of antagonism because more than one mechanism can cause shifts and/or depression of the curves. The three

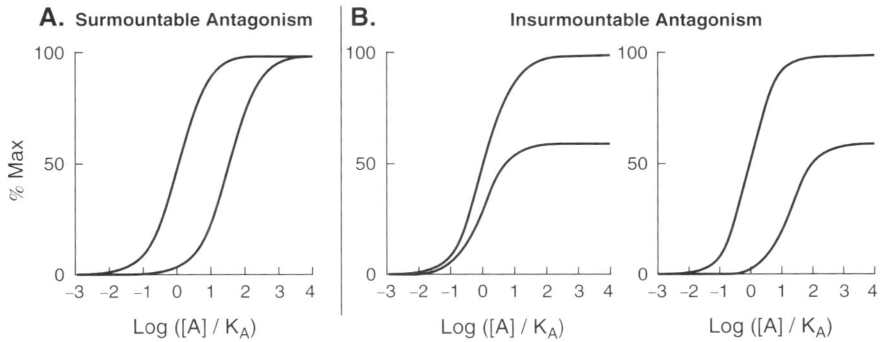

Fig. 2.4. Two general patterns of antagonism of tracer responses. (A) Surmountable antagonism produces a shift to the right of the tracer dose–response curve with no diminution of the maximal asymptote signal. (B) Insurmountable antagonism produces a depression of the maximal asymptote response that may or may not be accompanied by a shift to the right of the dose–response curve

molecular mechanisms of antagonism to be discussed with respect to affinity are competitive, allosteric and irreversible antagonism. Competitive antagonism produces strictly surmountable effects on saturation and/or dose–response curves. The molecular mechanism for this effect is competition of the tracer ligand and antagonist for the same steric site on the receptor. Allosteric and irreversible antagonists can produce either surmountable or insurmountable antagonism, depending on the tracer signal being used to measure the antagonism,the type of receptor system and the experimental conditions (*vide infra*). Allosteric antagonism involves the binding of both tracer and antagonist ligands to separate sites on the receptor, with the receptor perceiving the presence of either and reacting to that presence. Irreversible antagonism involves antagonist ligands that have a very slow rate of offset (when compared to the rate of onset) and thus do not come to equilibrium steady states within experimental periods. The mechanism that has the simplest model and for which there are a number of internal checks of consistency with the model is competitive antagonism.

2.4.1.1 Competitive Antagonism

The scheme shown in Fig. 2.5 illustrates the scenario of competitive antagonism. The tracer ligand, for which we have a method of tracking receptor occupancy either as radioligand–receptor complex or physiological response, is denoted [A] and the antagonist as [B]. The receptor is given as [R] and the equilibrium association constants for the ligand–receptor complexes are denoted K_a and K_b for ligands A and B, respectively.

The equilibrium equations for this scheme are:

$$K_a = \frac{[AR]}{[A][R]} \quad \text{and} \quad K_b = \frac{[BR]}{[B][R]}$$

The conservation equation for total receptor is $[R_{tot}] = [R] + [AR] + [BR]$. Thus, the expression for the amount of tracer ligand bound to the receptor,

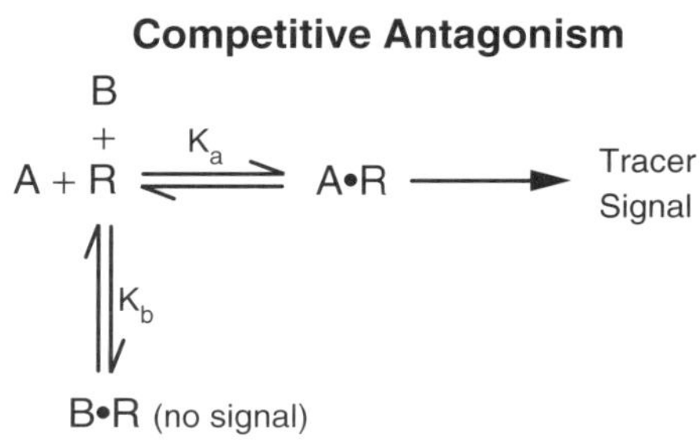

Fig. 2.5. Scheme for competitive antagonism of a signal produced by a tracer ligand A by another ligand B. It is assumed that binding is mutually exclusive, i.e. if A is bound then B cannot be bound, and vice versa

expressed as a fraction of the total receptor ($\rho = [AR]/[R_{tot}]$), can be derived as:

$$\frac{[AR]}{[R_{tot}]} = \rho = \frac{[A]/K_A}{[A]/K_A + [B]/K_B + 1} \qquad \ldots(2.4)$$

where $K_A = 1/K_a$ and $K_B = 1/K_b$.

This equation was first presented by the British pharmacologist Sir John Gaddum (1937) and is often referred to as the Gaddum equation. There are features of the Gaddum equation that are of interest. For example, it can be seen that if $[B] > 0$, then the receptor occupancy by the tracer ligand will be less than it could be, at the same concentration of tracer, in the absence of B. However, this deficit be overcome can be increasing [A]; this is the essence of surmountable antagonism. The effects of a competitive antagonist can be overcome always by the addition of more tracer ligand.

What this translates into over a complete range of tracer concentrations is a parallel shift to the right of the initial dose–response (or saturation binding) curve when a competitive antagonist is present in the receptor compartment. Figure 2.6 shows the curve of the amount of a tracer ligand bound to the receptor in the absence and presence of antagonist (calculated with the Gaddum eq. 2.4). The affinity of an antagonist producing such effects on the dose–response curve to a tracer ligand can be derived with models based on the Gaddum equation.

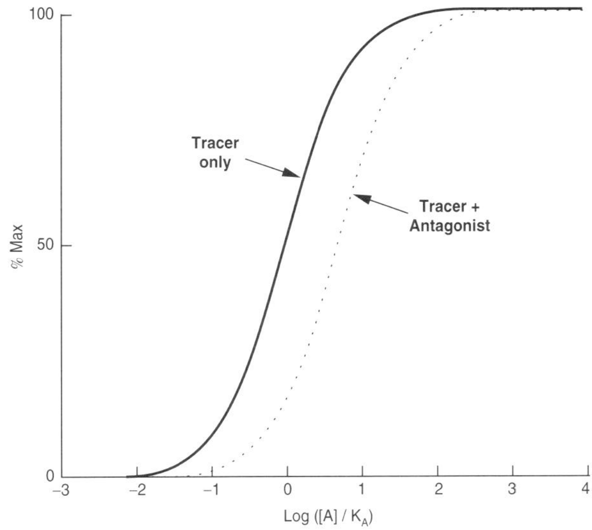

Fig. 2.6. The effect of a competitive antagonist on a tracer dose–response curve. The curve is shifted to the right in a parallel manner with no diminution of maximal response. The dose ratio is defined as the ratio of equiactive doses of tracer

There are two common approaches generally utilized in binding and functional studies, respectively, to estimate the affinity of ligands that produce competitive antagonism. The most straightforward approach to the study of competitive antagonism is to observe the effects of the antagonist on the full dose–response curve to the tracer ligand. Theoretically this can be done in both binding and functional modes, although working with complete radioligand saturation curves can be prohibitively expensive and time consuming. By far the more common approach to the study of full tracer dose–response curves is through the use of physiological responses to agonists in receptor preparations.

2.4.1.1.1 Competitive Antagonism in Experiments of Pharmacological Function: Schild Analysis

The Austrian pharmacologist Heinz Schild derived a tool now used universally, for the measurement of competitive antagonist affinity in functional experiments (Arunlakshana and Schild 1959). Termed 'Schild analysis', this method is based on the concept that the strength of antagonism produced by a competitive antagonist depends upon its concentration at the receptor and the equilibrium dissociation constant (reciprocal of affinity) of the antagonist–receptor complex. Because the antagonism can be observed and the concentration of antagonist is set by the investigator, the equilibrium dissociation constant can be calculated. Thus, a dose–response curve to an agonist is determined in the receptor preparation (i.e. isolated tissue, cell culture) in the absence of the antagonist (Fig. 2.7A). After the agonist has been removed from the preparation by washing, and any possible effects of desensitization have waned, a fixed concentration of antagonist is added and left to equilibrate with the receptors for a certain period of time (usually 30 min–2 h). Then, in the presence of the antagonist, another dose–response curve to the agonist is obtained (Fig. 2.7A). As noted above, the presence of the antagonist produces a shift to the right of the agonist dose–response curve. Comparison of equiactive concentrations of agonist (before and after antagonist) yields a ratio termed the 'dose ratio' (the multiple of the initial concentration of agonist required to achieve the same response in the presence of the antagonist).

These dose ratios are utilized in a derivative of the Gaddum equation, termed the Schild equation, as follows. The response to a concentration of agonist is obtained from the Gaddum equation with the antagonist concentration set at zero ($[B] = 0$):

$$\text{Response}_A = f\left(\frac{[A]/K_A}{[A]/K_A + 1}\right) \qquad \ldots(2.5)$$

where K_A represents the equilibrium dissociation constant of the agonist–receptor complex and f represents a function linking receptor occupancy and system response (see Chapter 3 for further discussion). What is relevant to this discussion is the fact that knowledge of the exact nature of this function is not pertinent to this analysis. This is because equiactive concentrations of agonist are compared

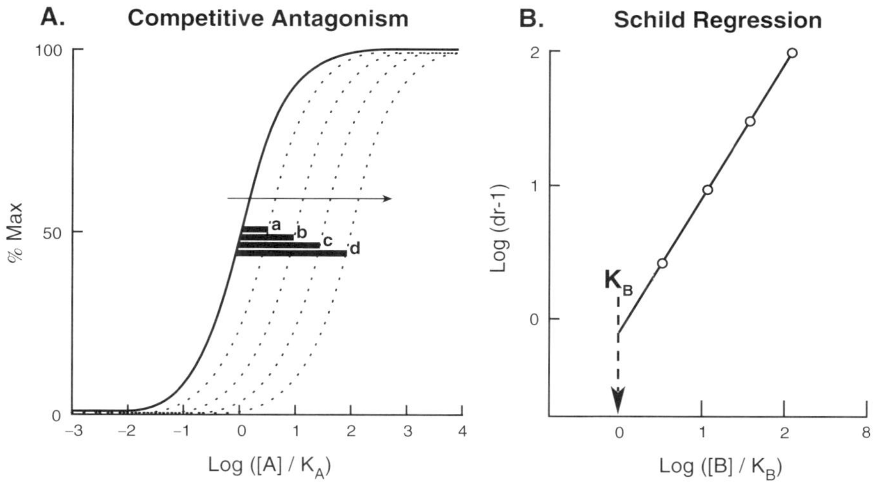

Fig. 2.7. Schild Analysis. (A) Tracer dose–response curves shifted to the right by increasing concentrations of competitive antagonist. Each concentration produces a dose ratio for equiactive concentrations of tracer that are used in the Schild equation to make the regression shown in (B). The straight line regression of log($dr - 1$) upon logarithms of the molar concentration of antagonist should have a slope of unity. The intercept (where log($dr - 1$) = 0) is the concentration of antagonist corresponding to the equilibrium dissociation constant of the antagonist–receptor complex

and the assumption is made that equally active concentrations of agonist emanate from equal receptor occupancies. Thus, in the presence of a concentration of antagonist [B] (with equilibrium dissociation constant of the antagonist–receptor complex denoted K_B) the response is:

$$\text{Response}_B = f\left(\frac{[A']/K_A}{[A']/K_A + [B]/K_B + 1}\right) \qquad \ldots(2.6)$$

The concentration of the agonist has been adjusted to a new level [A'] to produce the same response as that obtained in the absence of antagonist (according to eq. 2.5). The responses are then equated ($\text{Response}_A = \text{Response}_B$). At this point it can be seen that the nature of the function relating receptor occupancy and tissue response can be ignored under the null conditions of the experiment. Rearrangement of the two equations yields the following equation:

$$\frac{[A']}{[A]} = dr = \frac{[B]}{K_B} + 1 \qquad \ldots(2.7)$$

where the ratio of equiactive agonist concentrations is redefined as the 'dose ratio' (denoted dr). The logarithmic metameter of eq. 2.7 is known as the Schild equation:

$$\log(dr - 1) = \log[\mathrm{B}] - \log K_{\mathrm{B}} \qquad \ldots(2.8)$$

The process of Schild analysis consists of obtaining a number of dose–response curves to the agonist in the presence of a number of different concentrations of antagonist and plotting the resulting values of the dose ratios (as $\log(dr - 1)$) as a function of the logarithm of the antagonist molar concentrations ($\log[\mathrm{B}]$). The result should be a straight line with a slope of unity and, importantly, an intercept equal to the negative logarithm of K_{B} (termed $\mathrm{p}K_{\mathrm{B}}$). Figure 2.7B shows the Schild regression for the dose–response curves in Fig. 2.7A.

Schild analysis is extremely useful in that it possesses internal tests for true competitive behavior of antagonists. First, a competitive antagonist should produce parallel displacement of the tracer (agonist) dose–response curve with no diminution of the maximal response; this can be observed directly during the analysis. Second, the slope of the resulting Schild regression should not be significantly different from unity. If it is, then either the antagonist is not a true competitive antagonist or other factors are interfering with the observation of competitive receptor behavior. Table 2.1 shows some factors that are known to

Table 2.1. Experimental factors that can alter slopes of Schild plots

A. Slope is greater than unity

- Insufficient time of equilibration allowed for incubation of antagonist with receptors

  For a first-order rate of onset, lower concentrations require a longer time to reach equilibrium with receptors than higher concentrations. Therefore, the antagonism of concentrations near the $\mathrm{p}K_{\mathrm{B}}$ is underestimated skewing the regression upwards.

- There is a saturable removal mechanism for the antagonist present

  A removal mechanism for the antagonist (i.e. enzymatic degradation, uptake process) reduces the concentration of antagonist in the receptor compartment. Once this process is saturated, there is an increase in the concentration of antagonist at the receptor leading to a corresponding increase in antagonism. This elevates the regression line when it occurs leading to slopes greater than unity.

B. Slope is less than unity

- There is a saturable removal mechanism for the agonist present in the preparation

  A removal mechanism for the agonist diminishes the concentration at the receptor. As the dose–response curve is shifted toward higher concentrations of agonist (due to increasing concentration of antagonist), the removal mechanism is saturated and ceases to decrease the agonist concentration at the receptor. The increased agonist concentration leads to increased response, an apparent decrease in antagonism and a resulting diminution of the Schild plot slope.

- The agonist activates another receptor at higher concentrations

  If the agonist produces augmented responses at the higher concentrations encountered with increasing competitive blockade, and if the antagonist is ineffective at blocking these secondary receptor effects, then a diminution of antagonism at these higher concentrations will result with a concomitant decrease in Schild plot slope.

produce Schild plot slopes of greater or less than unity. If such aberrant Schild regressions are obtained, then the experiment can be adjusted to produce true competitive kinetics.

It should be noted that the simple competitive antagonist model is valid only if the slope of Schild regression is unity (i.e. only under these conditions can the intercept by considered a valid estimate of pK_B). However, experimental variation with a limited number of data points can still result in an experimentally observed slope that is not equal to unity. When this occurs, the data should be tested statistically to determine whether the 95% confidence limits of the slope include unity. If this occurs, then the data should be refited to a regression of slope unity and the intercept used as an estimate of pK_B. However, if the confidence limits of the slope do not include unity, then the intercept is only an empirical measure of the antagonist potency termed pA_2. This value is defined as the negative logarithm of the antagonist that produces a dose ratio for the agonist of 2 (a twofold shift to the right of the agonist dose–response curve). Although this is an empirical measure of potency and cannot be considered a chemical term like the equilibrium dissociation constant of the antagonist–receptor complex, pA_2 is still a useful measure of antagonist potency.

2.4.1.1.2 Affinity of Competitive Antagonists in Binding Studies

As seen from the above discussion, a complete binding curve for a radioactive ligand can be obtained in the absence and presence of a competitive antagonist, and a measure of the affinity of the antagonist gained from the magnitude of the resulting parallel shift to the tracer curve imparted by the antagonist. However, under normal circumstances this would be time consuming and prohibitively expensive in a setting of high-throughput screening, therefore a different approach is used that tracks the observed binding of a single concentration of the radioligand in the presence of a range of concentrations of competitive antagonist. Thus, it can be seen that the increasing concentrations of antagonist produce increasing shifts to the right of the tracer saturation curve (according to the Gaddum eq. 2.4), which result in a diminution of the binding signal from the single concentration of radioligand (see Fig. 2.8A). If the signal from the radioligand is plotted as a function of the concentrations of competitive antagonist producing the shifts of the saturation curve, an inverse sigmoid plot results (see Fig. 2.8A). Usually, the radioligand signal is expressed as a percentage of the initial signal obtained in the absence of the competitive antagonist.

The concept of competitivity dictates that the probability of a receptor being occupied by a tracer ligand or an antagonist ligand at any instant depends on the relative concentration of each ligand and their respective affinities for the receptor. Therefore, the observed potency of the antagonist ligand is inversely proportional to the concentration of the tracer, i.e. more tracer ligand on the receptor requires correspondingly more antagonist ligand to displace it. Figure 2.8B shows the inhibition of the binding of a higher concentration of tracer ligand; the corresponding

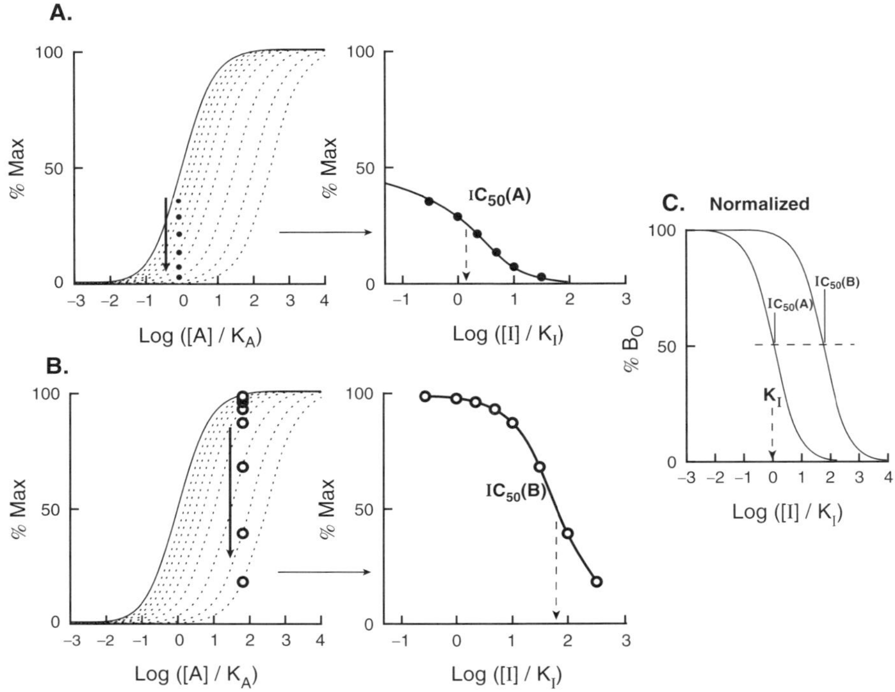

Fig. 2.8. Indirect measurement of affinity in radioligand binding studies. (A) A concentration of tracer is chosen that produces 40% maximal signal. Increasing concentrations of a competitive antagonist shift the tracer dose–response curve to the right, as shown on the left panel of the figure. All that is observed is the diminution of the signal from the single concentration of tracer (closed circles). If this diminishing signal is plotted as a function of the competitive antagonist, then an inverse sigmoidal curve is observed (panel on the right). The concentration of competitive antagonist that produces a 50% diminution of the tracer signal is an operational measure of antagonist potency (termed the IC_{50}). (B) The same effect is observed if a higher concentration of tracer is used, except that more antagonist is needed to produce inhibition of the stronger signal. (C) Both inhibition curves have been normalized (the original tracer signal is 100% and all responses are expressed as a percentage of that value). Note how the higher concentration of tracer (B) yields a higher value for the IC_{50}

inhibition curves are also shown. In Fig. 2.8C the inhibition curves are normalized to show the percentage inhibition of the initial amount of tracer–receptor complex. The point at which the amount of bound tracer is reduced by 50% is defined as the IC_{50} (inhibitory concentration producing 50% inhibition or, conversely, leaving 50% bound tracer). This is an empirical measure of the potency of an antagonist, i.e. if the IC_{50} for a given antagonist is 1/10 of another at the same concentration, then its affinity for the receptor is ten times greater. However, such relative measures are not useful because they vary with the strength of the signal being inhibited and this is an experimental variable that may change from day to day. Therefore, a chemical measure of the affinity of the antagonist is required, because this estimate theoretically is immutable to changes in experimental conditions and is unique for every antagonist–receptor pair.

The receptor system is equilibrated with a fixed concentration of radioligand $[A^*]$. The amount of radioligand bound to receptor is given by:

$$\frac{[A^*R]}{[R_{tot}]} = \frac{[A^*]/K_d}{[A^*]/K_d + 1} \qquad \ldots(2.9)$$

The amount of bound tracer radioligand $[A^*]$ in the presence of a non-radioactive ligand [I] is given by the Gaddum equation as:

$$\frac{[A_I^*R]}{[R_{total}]} = \frac{[A^*]/K_d}{[A^*]/K_d + [I]/K_I + 1} \qquad \ldots(2.10)$$

As noted above, the concentration of antagonist ligand [I] that decreases the amount of radioligand signal by 50% ($[A_I^*R]/[A^*R] = 0.5$) is defined as the IC_{50}. Substituting the equations for $[A_I^*R]$ and $[A^*R]$, the following relationship between the IC_{50} and K_I can be defined:

$$K_I = [IC_{50}]/([A^*]/K_d + 1) \qquad \ldots(2.11)$$

It can be seen from eq. 2.11 that the greater the concentration of radioligand used to obtain the tracer signal, the greater will be the observed IC_{50}. There is a practical reason for utilizing the minimal concentration of radioligand (minimizing $[A^*]/K_d$). As can be seen from eq. 2.11, if $[A^*]/K_d \ll 1$, then essentially $[IC_{50}]$ will equal K_I. As $[A^*]/K_d$ increases, then a correction factor must be used to obtain the K_I. Under these latter circumstances, random variation in the binding of $[A^*]$ (i.e. variation in ($[A^*]/K_d$) will be transferred to a variation in the correction factor relating $[IC_{50}]$ and K_I. Therefore, use of unnecessarily high amounts of tracer radioligand may lead to a higher degree of error in the measurement of antagonist affinity.

2.4.1.1.3 Competitive Antagonism with Ligands that Produce Response: Partial Agonists

As seen in the foregoing discussion, if the ligand simply occupies the receptor and prevents receptor occupancy by the tracer, then the signal emitted by the tracer will diminish in proportion to the occupancy by the antagonizing ligand to a baseline of zero response. However, if the antagonizing ligand itself is able to produce a signal, then what is observed with the competition is a change of signal strength from the tracer to the antagonist ligand. For a diminution of signal to be observed (antagonism), the occupying ligand must have a lower ability to produce response than the tracer ligand. Under these circumstances, the antagonist response-producing ligand is referred to as a partial agonist. Discussion of the inhibition of tracer response by a partial agonist is best done with reference to the dose–response curves for the two ligands. Figure 2.9A shows the dose–response curves to the tracer ligand (full agonist) and the partial agonist. If the dose–response curve to the tracer were to be repeated in the presence of the lowest designated concentration of partial agonist A, then the curve shown to the right of Fig. 2.9A would result. From this figure it can be seen that the baseline response is elevated by the presence of the partial agonist (in the absence of tracer) and that the response to the tracer ligand is additive at low concentrations but then is blocked slightly at the higher concentrations. This latter effect occurs because some of the receptors that might have been occupied by the more powerful tracer are now occupied by the weaker partial agonist; the result is a diminished response from this receptor population. At a higher dose of partial agonist, the baseline is elevated further and the tracer curve is shifted further to the right (Fig. 2.9B). At the maximal concentration of partial agonist (i.e. approximately 95% receptor occupancy, where further increases in response to the partial agonist are difficult to detect), the baseline is elevated and more shift to the right of the tracer curve results (Fig. 2.9B). At concentrations of partial agonist beyond this, no further elevations in baseline are seen but the shift in the tracer curve continues according to mass action (Fig. 2.9C). In keeping with the requirement of Schild analysis that dose ratios be measured under null conditions (equiactive responses), the shifts to the right of the tracer curve can be taken at concentrations that produce greater than 50% maximal response (see Fig. 2.10).

It should also be noted that the EC_{50} (concentration producing 50% maximal response to the agonist) of a partial agonist closely approximates the K_A (concentration of agonist occupying 50% of the receptors). Figure 2.11 shows a dose–response curve to a partial agonist and also the percentage receptor occupancy.

2.4.1.2 Allosteric Antagonism: Studies of Receptor Function

Another mode of antagonism of receptor function can involve separate binding sites on the receptor for the agonist (tracer ligand) and antagonist. Under these circumstances, the occupancy of the receptor by either ligand does not preclude occupancy by the other ligand. For the simplest case, it is assumed that occupancy of the receptor by the antagonist precludes receptor function, i.e. the agonist cannot induce response if the antagonist occupies the receptor. When this occurs, the antagonist is referred to as an allosteric antagonist and the resulting effect on tracer response is termed non-competitive. As with competitive antagonism, it is most instructive to follow the effects of allosteric antagonists on full dose–response curves to a tracer ligand. Although this can be done in either functional or binding mode, the most common usage is with the measurement of agonist response in receptor preparations.

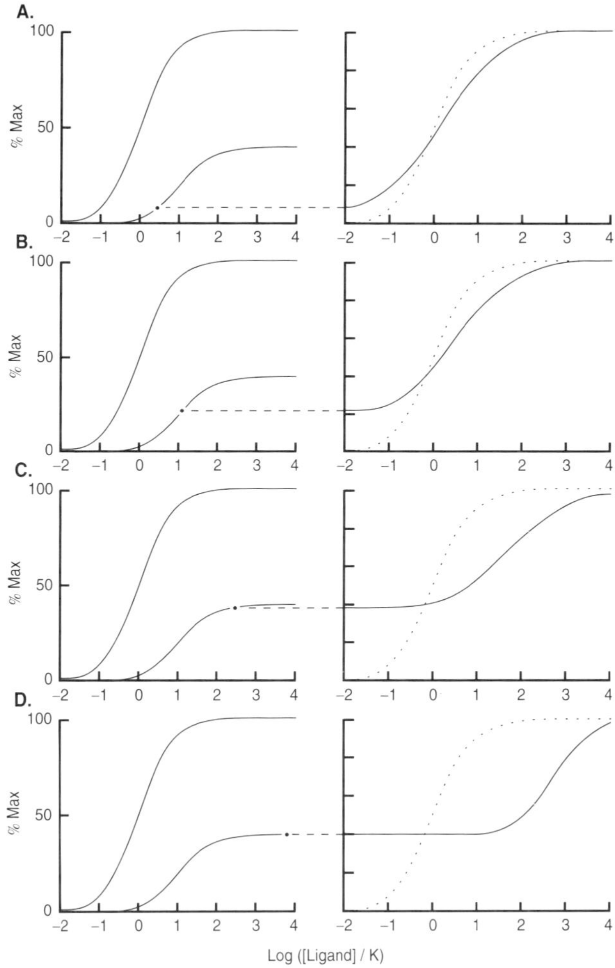

Fig. 2.9. Inhibition of a tracer response by a partial agonist. The dose–response curves on the left show the curve to the tracer (a full agonist producing 100% maximal response) and a partial agonist (producing 40% maximal response). The panels on the right show the response to the full agonist in the presence of designated concentrations of partial agonist (A) A low concentration of partial agonist produces a 10% elevation in baseline and shifts the full agonist dose–response curve partially to the right. (B) A higher concentration produces a further elevation with a concomitant increased shift to the right of the curve. At the concentration of partial agonist that produces a near-maximal response (to the partial agonist), the baseline is elevated nearly to the maximum possible extent and the curve to the full agonist is shifted further to the right. (C) Higher concentrations of partial agonist produce no further increases in the baseline but do produce further shifts to the right of the curve to the full agonist

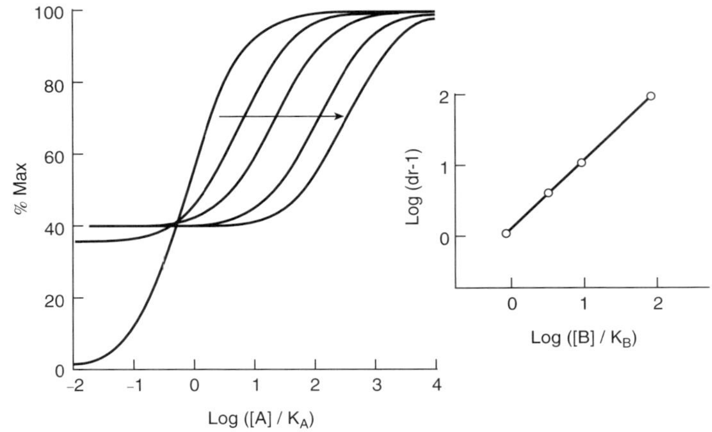

Fig. 2.10. Equiactive dose ratios produced by partial agonists can be obtained from the upper portions of the tracer dose–response curves for use in Schild analysis

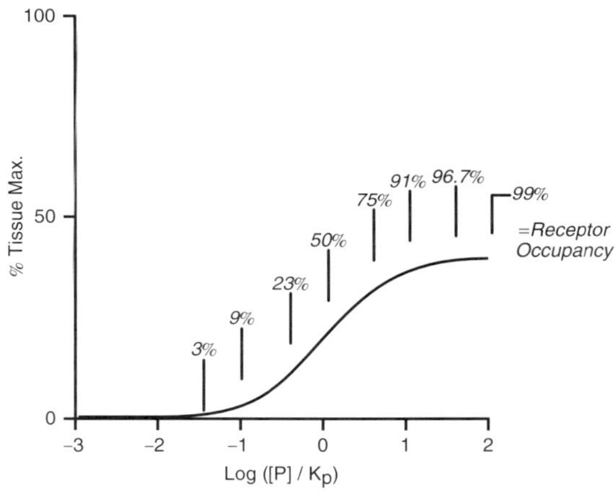

Fig. 2.11. A dose–response curve to a hypothetical partial agonist, also indicating the percentage receptor occupancy at various concentrations along the concentration axis

The model for allosteric antagonism is shown in Fig. 2.12. The parameters K_a and K_b refer to the association constants of the tracer A and allosteric ligand B, respectively, and α is a modifying term that denotes the multiple change in the affinity of one ligand produced by the binding of the other. Thus, if $\alpha = 0.1$, then the binding of the allosteric ligand on the receptor decreases the affinity of

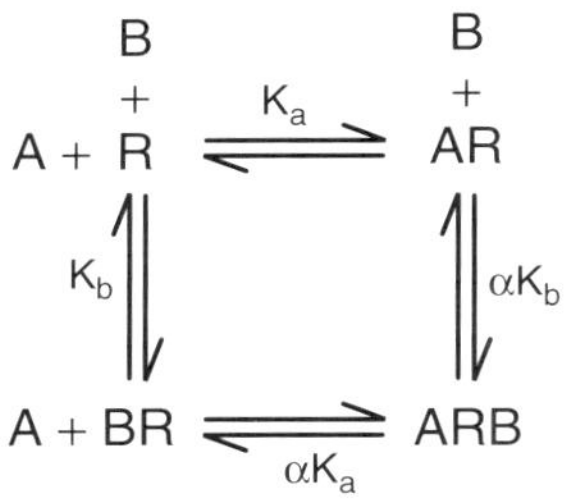

Fig. 2.12. Scheme for allosteric antagonism of receptors. Occupation of the receptor R by the tracer ligand A and allosteric ligand B is not mutually exclusive because each ligand has its own binding site on the receptor. Ligand A binds to the receptor with an equilibrium association constant K_a and B with an equilibrium association constant K_b. When one of the ligands is bound to the receptor already, the affinity of the second ligand is modified by the allosteric factor α

the receptor for tracer A by a factor of 10. Under these circumstances, the ligand is termed an allosteric antagonist. Conversely, if $\alpha = 10$, then the affinity of the receptor for the tracer ligand is increased by a factor of 10. The ligand is then termed an allosteric enhancer.

The equilibrium equations for this model are:

$$K_b = \frac{[BR]}{[B][R]} \qquad \alpha K_b = \frac{[ARB]}{[AR][B]}$$

$$K_a = \frac{[AR]}{[A][R]} \qquad \alpha K_a = \frac{[ARB]}{[BR][A]}$$

Using the conservation equation for receptors as $[R_{tot}] = [R] + [AR] + [BR] + [ARB]$, the equation for the amount of tracer ligand [A] bound to receptor in the presence of an allosteric ligand B is given by:

$$\frac{[AR] + [ARB]}{[R_{tot}]} = \frac{[A]/K_A(1 + \alpha[B]/K_A)}{[A]/K_A(1 + \alpha[B]/K_B) + (1 + [B]/K_B)} \qquad \ldots(2.12)$$

where $K_A = 1/K_a$ and $K_B = 1/K_b$.

Equation 2.12 assumes that the tracer ligand may still bind to the receptor with an affinity modified by factor α. It can be seen from this equation that the maximal amount of tracer bound will not decrease. However, the observed affinity of the tracer ligand will be affected by the presence of the allosteric ligand according to the degree of receptor occupancy by the allosteric ligand (K_B) and the cooperativity between the two binding sites (α). The observed affinity for ligand A is given by:

$$K_{obs} = \frac{K_A(1 + [B]/K_B)}{(1 + \alpha[B]/K_B)} \qquad \ldots(2.13)$$

It can be seen that $K_{obs} < K_A$ if $\alpha > 1$, i.e. the binding of ligand A will be of a higher affinity in the presence of the allosteric ligand (allosteric enhancement). The effects of various concentrations of an allosteric enhancer on the binding curve to ligand A are shown in Fig. 2.13A. Similarly, if $\alpha < 1$, then allotopic inhibition results and $K_{obs} > K_A$. This effect is shown in Fig. 2.13B.

A somewhat different scenario is operative when receptor function, rather than simply ligand binding, is considered. The simplest case is when binding of the

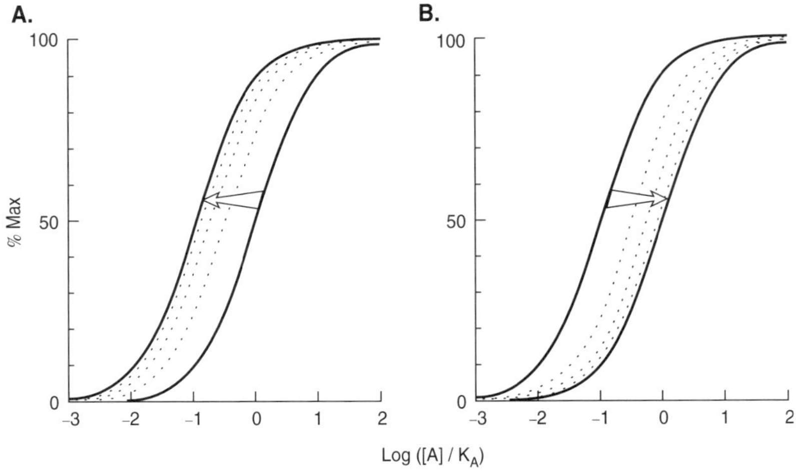

Fig. 2.13. Effects of an allosteric enhancer (A) and an allosteric antagonist (B) on the saturation binding curve to a tracer radioligand. The allosteric enhancer has an allosteric factor $\alpha = 10$ and the allosteric antagonist has an allosteric factor $\alpha = 0.1$ for this tracer ligand. Dotted lines show the increasing effects of allosteric ligand concentrations ($[B]/K_B = 3, 10, 30, 100$. Note how the antagonism in (B) is saturable and no further shifts to the right occur beyond 10-fold (the value of $1/\alpha$)

allosteric antagonist prevents receptor activation by an agonist tracer ligand A. Therefore, the tracer signal emanates only from the complex AR ([ARB] cannot signal) and the expression for response is given by:

$$\frac{[AR]}{[R_{tot}]} = \frac{[A]/K_A}{[A]/K_A(1 + \alpha[B]/K_B) + (1 + [B]/K_B)} \qquad \ldots(2.14)$$

From eq. 2.14 it can be seen that a unique feature of allosteric antagonism is the fact that the maximal binding of ligand A necessarily decreases with binding of the allosteric ligand (see Fig. 2.14A). In contrast, assuming that ligand A saturably binds to the complete receptor population in the absence of allosteric ligand, then an allosteric enhancer cannot increase the maximal AR formed.

Complex effects on dose–response curves can occur if the allosteric ligand simultaneously changes the affinity of the receptor for ligand A (according to eq. 2.13) as well as affecting receptor function. Figure 2.14B shows the effects of a ligand that blocks receptor function but increases the affinity of the receptor for ligand A; here, the curve shifts to the left with depression of maximal response. Figure 2.14C shows the effect of an allosteric antagonist with $\alpha < 1$ (affinity decreased) on a dose–response curve to ligand A. Here, it can be seen that there is a shift to the right of the dose–response curve concomitant with depression of the maximal response.

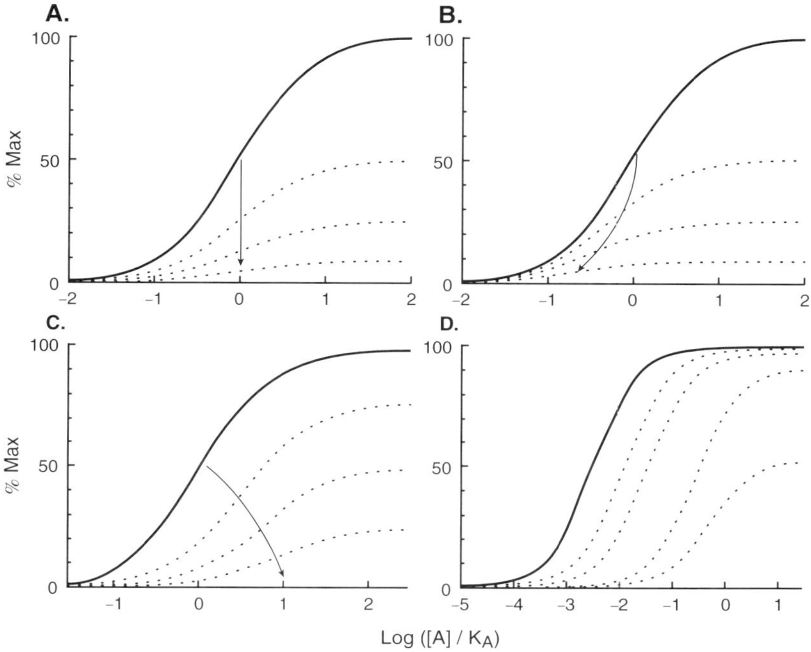

Fig. 2.14. The effects of an allosteric antagonist on the functional response of an agonist. It is assumed that occupation of the receptor by the antagonist precludes the receptor from producing a signal when occupied by the agonist. (A) The allosteric antagonist has no affect on the affinity of the agonist ($\alpha = 1$). Dotted lines show the effects of $[B]/K_B = 1$, 3 and 10. (B) The allosteric antagonist enhances the affinity for the agonist 10-fold ($\alpha = 10$). Dotted lines show the effects of $[B]/K_B = 0.1$, 0.3, and 1. (C) The antagonist decreases the affinity for the agonist by a factor of 10 ($\alpha = 0.1$) Dotted lines show the effects of $[B]/K_B = 10$, 30 and 100. (D) The allosteric antagonist has no effect on the affinity of the receptor for the agonist ($\alpha = 1$) but the response is greatly amplified by the cell such that only 1% of the receptors need be occupied by the agonist to produce maximal tissue response. Under these circumstances, even though the receptor occupancy by the agonist is being decreased by the antagonist (as in A), this does not immediately produce a depression of the maximal response. The curve can be shifted by nearly a factor of 10 before a depression of the maximal response is observed. Further antagonism by the allosteric ligand then produces further shifts to the right, with concomitant depression of the maximal response. Dotted lines show the effects of $[B]/K_B = 3$, 10, 30 and 100

A major shortcoming of the above predictions is the necessary assumption that the physiological tracer signal is linearly related to the amount of AR complex. In practice, this almost certainly is not true (see Chapter 3), therefore the effects of an allosteric antagonist on true physiological tracer response cannot be used to infer molecular mechanism. In most tissues, there are a number of saturable cytosolic reactions that occur between the steps of receptor occupancy by ligand A and the observed physiological response. The net effect of these reactions is to amplify the signal such that low receptor occupancies by agonist result in high levels of response. Under these circumstances the inhibition of receptor function by allosteric antagonists, regardless of allosteric effects on the affinity of the tracer ligand, usually results in a shift to the right of the tracer dose–response curve, concomitant with depression of maximal response (as in Fig. 2.14C). In tissues with an extreme amplification between receptor occupancy and physiological response, there may be concentrations of allosteric antagonist that only shift the dose–response curve to the right, with no depression of the maximal response (Fig. 2.14D).

The complex nature of allosteric effects makes measurement of chemical affinity difficult. However, there is one approach that can be used to measure the affinity of an allosteric antagonist in functional systems. This approach utilizes the null approach (i.e. it is assumed that equal tissue responses result from equal receptor occupancies with the same ligand) and is independent of the effects of the allosteric ligand on the affinity of the tracer ligand (the magnitude of α is not a complicating factor). The assumption is made that occupation of the receptor by the allosteric antagonist blocks receptor activation.

For measurement of non-competitive antagonist affinity by this method, the dose–response curve to the tracer ligand (agonist) is obtained in the absence and presence of a set concentration of allosteric antagonist (Fig. 2.15A). Then, equiactive concentrations of the agonist are compared from each curve according to the following double reciprocal equation:

$$\frac{1}{[A]} = \frac{1}{[A']}(1 + [B]/K_B) + \frac{\alpha[B]}{K_A K_B} \qquad \ldots(2.15)$$

where [A] refers to the concentration of agonist in the absence of antagonist and [A'] to the equiactive concentration in the presence of antagonist B. This equation takes the graphical form of a straight line when values of 1/[A] are plotted as ordinates on the abscissae 1/[A'] (see Fig. 2.15B). Under these circumstances, the equilibrium dissociation constant of the allosteric antagonist is given by the expression:

$$K_B = \frac{[B]}{(\text{Slope} - 1)} \qquad \ldots(2.16)$$

It can be seen from this equation that α cancels out the calculation of K_B. With this expression, the effects of the allosteric antagonist on agonist affinity are not important.

2.4.1.3 Irreversible Antagonism

Another mechanism of antagonism of drug receptor action is through irreversible inactivation of receptor function. Thus, chemical alkylating agents such as

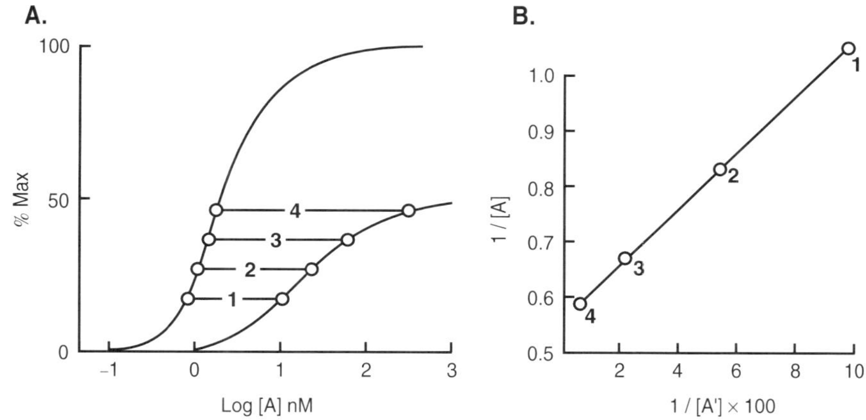

Fig. 2.15. Estimation of the equilibrium dissociation constant of an allosteric antagonist by the method of Gaddum *et al* (1955). (A) Dose–response curves to the full agonist are obtained in the absence and presence of a fixed concentration of antagonist. Then equiactive concentrations (designated 1–4) are obtained from these curves and used in the double reciprocal equation (see text). (B) The straight line regression from the reciprocals of these concentrations (1–4) is used to calculate the equilibrium dissociation constant of the antagonist according to eqs. 2.15 and 2.16

beta-haloalkylamines form chemically reactive species in solution, which then go on to form alkyl bonds with the receptor protein. If the chemical structure of the alkylating agent brings them near sensitive regions of the receptor, then this alkylation can block receptor function and/or agonist binding. Figure 2.16A shows an example of one such chemical reaction and the alkylation of receptor protein. Figure 2.16B shows the resulting effect on receptor occupancy of a tracer ligand. As with allosteric antagonism, if physiological response is the tracer effect being observed, then the amplification processes between receptor occupancy and physiological response may cause dose–response curves to be shifted to the right with no depression of maximum at some treatments of irreversible antagonist (Fig. 2.15C). Further treatment with the irreversible antagonism will produce a depression of maximal response (Fig. 2.16D).

A hallmark of irreversible antagonism is an inability of the system to come to a steady state or equilibrium. Thus, a receptor alkylating agent will continue to block response progressively until response is completely blocked or until the alkylating agent is chemically depleted from the medium. For this reason, all treatments with an alkylating agent must be characterized by both a concentration of the antagonist utilized in the study and the time of exposure to the receptor.

There are numerous experimental conditions whereby the effects of allosteric antagonists, competitive antagonists and irreversible antagonists produce much the same patterns of effects on tracer dose–response curves (i.e. shifts to the right

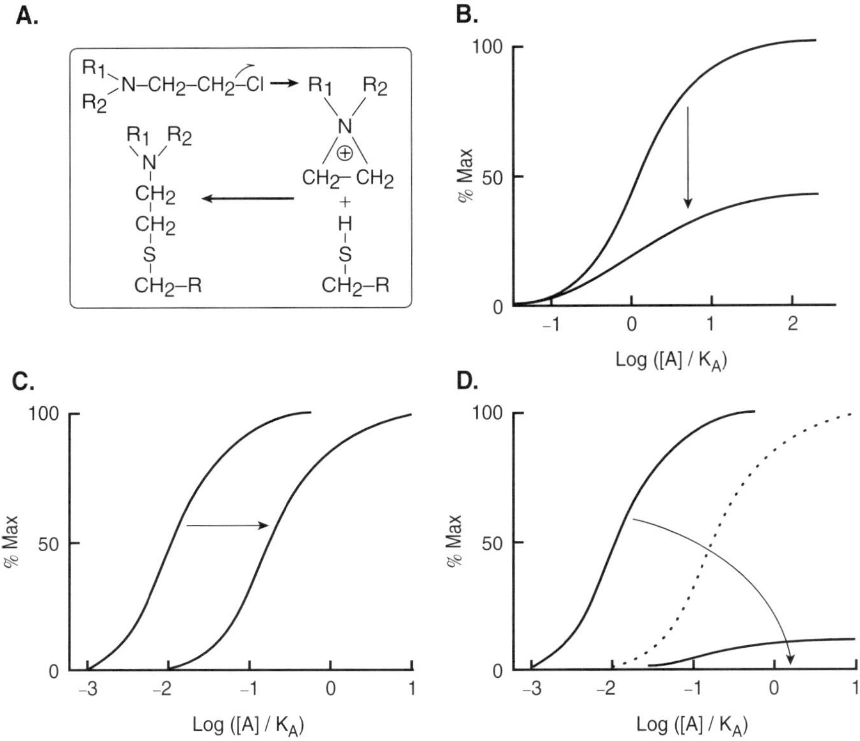

Fig. 2.16. The irreversible alkylation of receptor protein by β-haloalkylamines. (A) The formation of the reactive aziridinium ion by the β-haloalkylamine. This ion alkylates the protein, forming an irreversible attachment of the antagonist to the receptor protein. (B) The effects of receptor occupation by an agonist when the receptor is treated with a fixed concentration of β-haloalkylamine for a fixed time and then washed. (C) If the response to the agonist is amplified by the tissue, such that only a small fraction (i.e. 1%) of the receptors need be activated to produce 100% tissue response, then the response to the agonist may be shifted irreversibly to the right, with no diminution of the maximal response. Washing the preparation will not reverse the antagonism. (D) Further alkylation beyond the point where the agonist can activate 1% of the receptors will result in a depression of the maximal response to the agonist

with concomitant depression of maximal response). However, the kinetics of these effects can be used to differentiate them. Allosteric effects will continue up to some equilibrium steady state and then the system will be stable. For example, Fig. 2.13B shows the shift of a tracer dose–response curve by an allosteric antagonist. In this case the allosteric antagonist has a value of $\alpha = 0.1$, therefore the affinity of the tracer ligand is 1/10 of control when 100% of the allosteric sites are occupied on the receptor. This is a key determinant of allosteric effects, i.e.

they are saturable. Once the allosteric sites are fully occupied, then no further perturbation of the tracer response can be elicited. This is a contrast to the effects of a competitive antagonist. In this case, a further increase in the concentration of competitive antagonist will produce further antagonism. Thus, the dose–response kinetics of the antagonism for allosteric vs. competitive antagonists can be used to differentiate them.

2.5 SYSTEM VS. RECEPTOR-BASED ESTIMATES OF AFFINITY

The microscopic affinities of ligands for receptors are essentially affinities of ligands for tertiary conformations of receptor protein. If these conformations change in the system, either spontaneously or through actual interaction with the ligand, then the ensuing new conformations will affect the observed affinity of the ligand for the entire system, i.e. the observed affinity will be for the concerted spectrum of receptor species and not a single entity. This can cause dissimulation if the proclivity for this to occur varies with the cellular system used to study the receptor-based affinity.

The most common place where this type of artifact is encountered is where the ligand causes isomerization of the receptor protein. Thus, by its very binding to the receptor, the ligand changes the conformation and thus changes the equilibrium of the system controlling the presence of the various receptor conformations. This, in turn, causes a change in the observed affinity of the ligand for the system and introduces a system-based error in the measurement of affinity. This type of effect can be described as 'two-stage' binding of ligands.

2.5.1 THE EFFECT OF TWO-STAGE BINDING ON OBSERVED AFFINITY

The equilibrium equations described up to this point assume that the ligands simply bind to the receptor and do not predispose the receptor to form a new species in the membrane. If this occurs, then a two-stage binding process takes place and this has an effect on the observed affinity of the ligand. Assuming a binding process of ligand A binding to receptor R, the rate of production of the agonist–receptor complex ($d[AR]/dt+$) is equal to the product of the rate of onset of agonist to the receptor (k_1) and the relative concentrations of agonist ([A]) and receptor ([R]) ($d[AR]/dt+ = k_1[A][R]$), and the rate of agonist–receptor degradation is the product of the rate of offset of agonist (k_2) and the concentration of agonist–receptor complex ([AR]) ($d[AR]/dt- = k_2[AR]$). At equilibrium, these rates are equal: $d[AR]/dt+ = d[AR]/dt-$.

$$A + R \underset{k_2}{\overset{k_1}{\rightleftharpoons}} AR \qquad \ldots(2.17)$$

However, if another mechanism removes the agonist–receptor complex from the equilibrium (by isomerization of the receptor to another species AR'), then the forward rate reaction d[AR]/dt+ $= k_1$[A][R] will continue to fill the void (absence of AR complex) and drives the reaction further to the right:

$$\mathrm{A} + \mathrm{R} \underset{k_2}{\overset{k_1}{\rightleftharpoons}} \mathrm{AR} \underset{\theta}{\overset{\omega}{\rightleftharpoons}} \mathrm{AR'} \qquad \ldots(2.18)$$

Thus, if the forward rate of the isomerization reaction is favorable for the formation of the new species (AR') (i.e. if $\omega > \theta$), then the overall reaction will proceed toward the formation of both AR and AR' beyond the limits of k_2. Under these circumstances, as the AR complex is formed according to k_2/k_1, it is removed from that particular equilibrium by the secondary reaction of AR $\rightarrow$ AR'. This can be seen by examining the equilibrium equations for scheme 2.18:

$$\frac{k_1}{k_2} = \frac{[\mathrm{A}]}{[\mathrm{A}][\mathrm{R}]} \qquad \frac{\omega}{\theta} = \frac{[\mathrm{AR'}]}{[\mathrm{AR}]}$$

With the above equations and the conservation equation for receptor (as $[\mathrm{R_{tot}}] = [\mathrm{R}] + [\mathrm{AR}] + [\mathrm{AR'}]$, the equation for the production of AR' can be derived:

$$\frac{[\mathrm{AR'}]}{[\mathrm{R_{tot}}]} = \frac{[\mathrm{A}]/K_\mathrm{A}}{[\mathrm{A}]/K_\mathrm{A}(1 + \theta/\omega) + \theta/\omega} \qquad \ldots(2.19)$$

From eq. 2.19 it can be seen that the observed affinity of the ligand for the receptor system is given by:

$$K_\mathrm{obs} = \frac{K_\mathrm{A}}{(1 + \omega/\theta)} \qquad \ldots(2.20)$$

Therefore for non-zero values of ω/θ, $K_\mathrm{obs} < K_\mathrm{A}$, i.e. the second stage binding will enhance the observed affinity of the ligand. If the isomerization of the receptor by the first binding stage is strong (i.e. a high forward reaction rate ω and a low backward rate θ), then the difference in observed affinity can be considerable. If the isomerization is weak, then $K_\mathrm{obs} \rightarrow K_\mathrm{A}$.

From this it can be seen that if the receptor can convert to another species during the binding process, then there will be a redistribution of the species present in the receptor system, which will concomitantly affect the observed affinity of the ligand for the system. This is important in receptor systems such as G-protein-coupled receptors, where there are membrane components intimately related to the receptor binding process.

2.5.2 THE REDISTRIBUTION OF RECEPTOR BINDING SPECIES

Figure 2.17A shows a schematic diagram of a simple G-protein-coupled receptor system consisting of a receptor that can exist in both an inactive state with respect to G-protein activation (denoted R_i) and an active state (denoted R_a) that can form a complex with a G-protein (active complex denoted RaG; an inactive complex R_iG also may be formed). The equilibrium dissociation for interaction of these species is defined by microscopic equilibria. If a ligand A is added to this system, another array of species corresponding to the ligand-bound counterparts of the species shown in Fig. 2.17A is produced; the complete system is shown in Fig. 2.17B. The microscopic equilibrium dissociation constant of the ligand–inactive receptor complex is denoted K_A; this would be the observed affinity of the ligand for the receptor in isolation if it could not form another conformation (R_a) or interact with G-protein. However, because the receptor can convert to these other species, the observed equilibrium dissociation constant of the ligand for the receptor system is given by:

$$K_{obs} = \frac{K_A[1 + L + [G]/K_G(1 + \beta L)]}{[1 + L + \gamma[G]/K_G(1 + \delta\alpha\beta L)]} \qquad \ldots (2.21)$$

As can be seen from eq. 2.21, not only does the proclivity of the receptor to form that active state affect the affinity (L), but also the relative stoichiometry of the components of the system (namely the ratio of receptor to G-protein). Also, the type of G-protein will affect the observed affinity (K_G). This confers a great deal of system dependence on the observed affinity of the ligand for the receptor. Figure 2.18A shows the variation in the observed affinity of an agonist ligand that promotes interaction of the receptor for the G-protein, as a function of the ratio of receptors to G-protein. This type of variation can be especially problematic in recombinant systems where human receptor genes are used to express human receptor in foreign host cells. Figure 2.18B shows the observed affinity of an agonist ligand (promotes receptor/G-protein coupling) in a sampling of 2000 cell

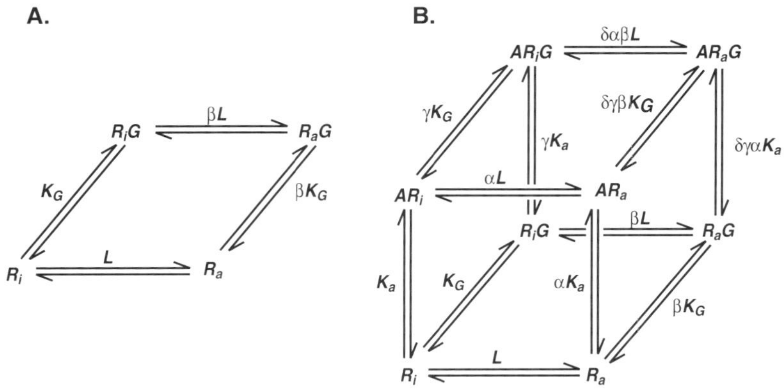

Fig. 2.17. Schematic diagrams of G-protein-coupled receptors. (A) Receptors exist in a active (R_a) and inactive state (R_i) bindings to G-proteins (G). (B) Addition of a ligand A to the scheme shown in (A). A new layer of species is formed

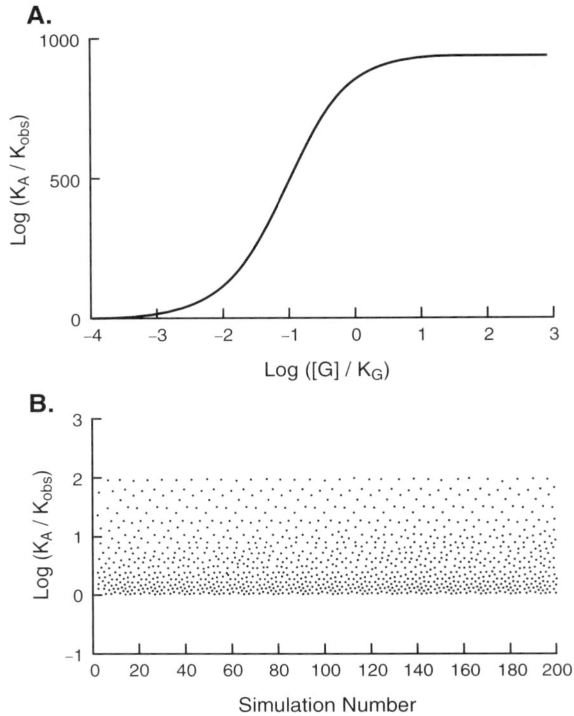

Fig. 2.18. The effect of G-protein coupling on the observed affinity of an agonist ligand to the receptor. (A) Ratio of increased affinity (K_A/K_{obs} – log scale) as a function of the amount of G-protein available for coupling to the receptor. The agonist is assumed to promote G-protein coupling 1000-fold ($\alpha = 10$, $\gamma = 100$). (B) Simulations of observed affinities of an agonist ligand with random changes in the ratio of receptor to G-protein; 2000 'virtual' cellular systems were simulated

lines by computer simulation of random variation in the G-protein content in the cells. The relevant aspect of this simulation is that each of the estimates of observed affinity is correct for that particular cell; this system dependence will be discussed further in Chapter 4.

2.6 CONCLUSIONS

This chapter discusses direct and indirect methods to measure the affinity of ligands for receptors in binding and functional experiments. In practice, indirect measures whereby ligand are used to interfere with the effects of tracer ligands, are nearly exclusively utilized in the screening process for drug discovery. These indirect methods can be used to detect competitive, allosteric and irreversible antagonism.

The system effects of receptor isomerization and/or G-protein binding on the magnitude of affinity estimates also are discussed.

2.7 REFERENCES

Arunlakshana O and Schild HO (1959) Some quantitative uses of drug antagonists. *Br J Pharmacol* **14**: 48–58.

Gaddum JH (1937) The quantitative effects of antagonistic drugs. *J Physiol (London)* **89**: 7P–9P.

Gaddum JH, Hameed K_A, Hathway DE and Stephens FF (1955) Quantitative studies of antagonists for 5-hydroxytryptamine. *Q J Exp Physiol* **40**: 49–74.

2.8 FURTHER READING

Measurement of Affinity in Binding Experiments

Bennet JP, Yamamura HI (1985) Neurotransmitter, hormone, or drug receptor binding methods. In *Neurotransmitter Receptor Binding*, ed. by HI Yamamura, SJ Enna and MJ Kuhar, pp. 61–89. Raven Press, New York.

Burt DR (1986) Receptor binding methodology and analysis. In *Receptor Binding in Drug Research*, ed. by RA OB́rien, pp. 4–29. Marcel Dekker, New York.

Limbird LE (1995) *Cell Surface Receptors: a Short Course on Theory and Methods*. Martinus Nihjoff, Boston.

Complex Binding Phenomena

Birdsall NJ, Hulme MEC, Keen M, Stockton JM, Pedder EK and Wheatley M. (1987) Can complex binding phenomena be resolved to provide a safe basis for receptor classification? In *Perspectives on Receptor Classification, Receptor Biochemistry and Methodology*, Vol. 6, ed. by JW Black, DH Jenkinson and VP Gerskowitch, pp. 61–71. Alan R. Liss, New York.

Black JW and Shankley NP (1990) Interpretation of agonist affinity estimations: The question of distributed receptor states. *Proc R Soc London Set B* **240**: 503–518.

Receptor Models

DeLean A, Stadel JM and Lefkowitz RJ (1980) A ternary complex model explains the agonist-specific binding properties of adenylate cyclase coupled β-adrenergic receptors. *J Biol Chem* **255**: 7108–7117.

Ehlert FJ (1988) Estimation of the affinities of allosteric ligands using radioligand binding and pharmacological null methods. *Mol Pharmacol* **33**: 187–194.

Schild Regressions

Kenakin TP (1980) The effects of equilibration time on the attainment of equilibrium between antagonists and drug receptors. *Eur J Pharmacol* **66**: 295–306.

Kenakin TP (1985). Schild regressions as indicators of nonequilibrium steady-states and heterogeneous receptor populations. *Trends Pharmacol Sci* **6**: 68–71.

Kenakin TP (1992) Tissue response as a functional discriminator of receptor heterogeneity: effects of mixed receptor populations on Schild regressions. *Mol Pharmacol* **41**: 699–707.

Mackay D (1978) How should values of the pA_2 and affinity constants for pharmacological competitive antagonists be estimated? *J Pharm Pharmacol* **30**: 312–313.

3 Efficacy

3.1 INTRODUCTION

By definition, a molecule with affinity for a biological receptor can be regarded as a drug. There is a subset of such molecules that change the behavior of the receptors to which they bind; such drugs have the property of intrinsic efficacy. This is a molecular property of some drugs that is inherent in the chemical structure; the receptor reads this encoded information and accordingly causes the cell to react to it. At present there is no absolute scale for the measurement of intrinsic efficacy and all measurements of this property are made on a relative scale; i.e. the relative efficacy of two agonists is measured in a given receptor system. However, unlike system-dependent measurements such as agonist potency, the measurement of relative efficacy in any one receptor system transcends that system and can be applicable to *all* systems bearing that receptor. It should be noted also that the elements of the chemical structure of a drug that account for affinity are totally separate from those that confer efficacy, i.e. the two properties are entirely independent of each other. For example, in the series of phenylethylamines that activate α-adrenergic receptors in rabbit aorta, the movement of a hydroxy group from the *para* to the *meta* position on the phenyl ring results in a 10-fold increase in affinity of the resulting agonist for the receptor, with no concomitant change in efficacy (Fig 3.1A). The same change in an imidazoline analog causes a sevenfold increase in efficacy but no change in affinity (Fig 3.1B).

There are two general themes in this chapter on efficacy. The first is that the understanding of the biochemical mechanisms of receptor efficacy is not essentially required for the drug discovery process because the strength of the stimulus imparted by an agonist to a receptor can be measured on a relative scale operationally (*vide infra*). The second idea is that there is probably more than one active state conformation of the receptor and therefore different agonists may induce different spectra of active states. This could translate into different 'qualities' of efficacy that may be taken advantage of for greater organ and receptor selectivity. Both of these ideas will be developed further in this chapter.

As a preface to a discussion on the operational quantification of efficacy and the accompanying advantages of doing so, a discussion of the molecular nature of agonism is useful.

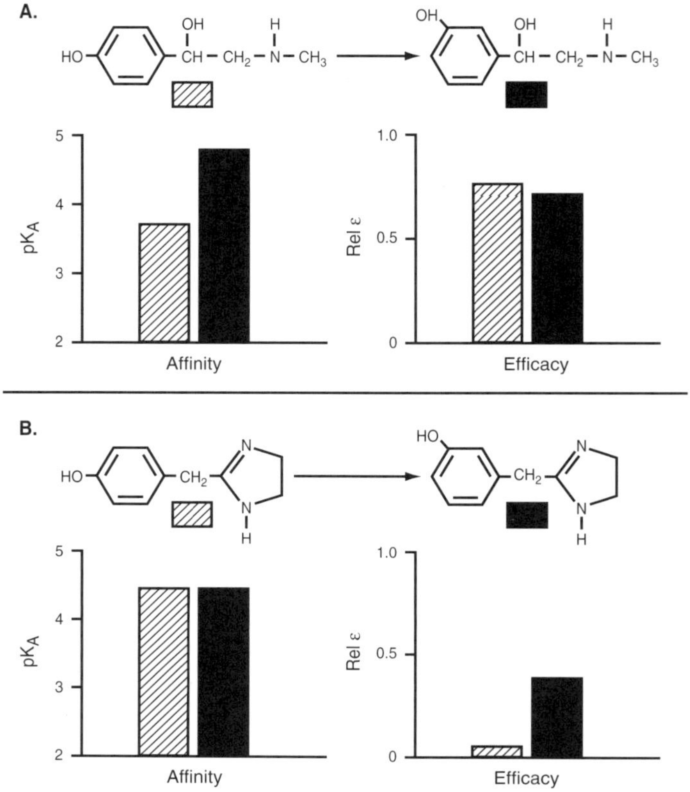

Fig. 3.1. Changes in affinity and efficacy of α-adrenergic agonists with changes in chemical structure. Ordinates for affinity: negative logarithm of the equilibrium dissociation constant of the agonist–receptor complex. Ordinates for efficacy: relative intrinsic efficacy as a fraction of efficacy of norepinephrine. (A) A change of the ring hydroxyl position in the phenethylamine shown produces a powerful change in affinity with no change in efficacy. (B) The same change in the hydroxyl position of the imidazoline produces no change in affinity but a change in the efficacy. (Data from Ruffolo and Waddell (1983).)

3.2 THE MOLECULAR NATURE OF EFFICACY

The mechanisms of enzyme catalysis are rooted in chemical concepts and are more clearly understood than the mechanisms controlling receptor action. Whereas chemical concepts can be used to describe the way in which enzymes bind and orient substrates to cause energetically strained transition states, the repeated production of cellular responses by molecules reversibly interacting with proteins that do not change the messenger is more difficult to describe. The understanding of the way in which molecules cause receptors to impart cellular signals is still a matter of intense research in pharmacology. In early formulations of receptor theory, efficacy was simply treated as a proportionality constant. Thus, the fraction of the receptor population that was occupied by an agonist was then multiplied by a factor set to account for the observed physiological response. There are advantages to such a simple scheme in terms of the operational relative methods that were derived subsequently to estimate the relative efficacy of agonists. In fact, separate lines of research have yielded the methods that will be described to quantify agonist relative efficacy and the nature of the molecular mechanisms of efficacy. This latter field also can yield tremendous fruits in terms of the emerging frontiers of drug discovery, namely the design of more selective agonists.

The major premise of receptor efficacy is the notion that receptors exist in a so-called inactive conformation when not bound to agonist and adopt an 'activated' conformation when bound to agonist. This 'activated conformation' interacts with other elements of the cell membrane to induce a physiological response. Therefore, discussion of the molecular mechanisms of agonist efficacy center on the mechanisms in which molecules can bias protein conformations.

An early and very useful delineation of receptor efficacy mechanisms was given by Sir Arnold Burgen (1981), who defined two biochemical mechanisms of efficacy in terms of protein thermodynamics and conformation. The first he termed 'conformational induction', which simply meant that the agonist bound to the receptor and, by doing so, deformed the receptor into the 'activated' conformation.

A second mechanism was described by Burgen as 'conformational selection', whereby an agonist could selectivity bind to one of a number of spontaneously formed receptor conformations and, by this binding, stabilize it to become the dominant conformation in the membrane. This would then lead to an enrichment of this conformation. If this conformation were a spontaneously 'active' one, then physiological response would result from the enrichment of this conformation. The molecular nature of this effect is rooted in the thermodynamics of equilibrium kinetics.

3.2.1 ENRICHMENT OF RECEPTOR CONFORMATION BY CONFORMATIONAL SELECTION

Assume that the agonist has a choice of binding to two conformations of the receptor, R and R* (R* being the 'active' conformation that induces physiological response). These receptor conformations coexist in the system according to an allosteric constant denoted L:

$$\mathrm{R} \overset{\mathrm{L}}{=\!=\!=} \mathrm{R}^* \qquad \ldots(3.1)$$

The agonist binds to the inactive receptor with an equilibrium association constant K_a and to the activated receptor by an equilibrium association constant αK_a. The factor α denotes the differential affinity of the agonist for the activated receptor form, i.e. if $\alpha = 10$, then the agonist has a 10-fold greater affinity for the R* form. The complete scheme with agonist involved is then:

$$\begin{array}{ccc} & \alpha L & \\ AR & \overline{\overline{\qquad\qquad}} & AR^* \\ K_a \parallel & L & \parallel \alpha K_a \\ R & \overline{\overline{\qquad\qquad}} & R^* \\ + & & + \\ A & & A \end{array}$$

The question then is how can selective affinity of A for R* (i.e. $\alpha > 1$) enrich the R* species? This can be calculated by examining the amount of R* species (both as R* and AR*) present in the system in the absence of agonist and in the presence of agonist. The equilibrium expression for $([R^*] + [AR^*])/[R_{tot}]$ (where $[R_{tot}]$ is the total receptor concentration given by the conservation equation $[R_{tot}] = [R] + [AR] + [R^*] + [AR^*]$) is:

$$\rho = \frac{L(1 + \alpha[A]/K_A)}{([A]/K_A)(1 + \alpha L) + 1 + L} \qquad \ldots(3.2)$$

where the concentration of agonist is [A], L is the allosteric constant, K_A is the equilibrium dissociation constant of the agonist–receptor complex ($K_A = 1/K_a$) and α is the differential affinity of the agonist for the R* state. It can be seen that in the absence of agonist ($[A] = 0$) $\rho_0 = L/(1 + L)$ and in the presence of a maximal concentration of agonist (saturating the receptors; $[A] \to \infty$) $\rho_\infty = (\alpha(1 + L))/(1 + \alpha L)$. Therefore the effect of an agonist on enriching the R* state is given by the ratio ρ_∞/ρ_0; when this ratio is >1 then the presence of the agonist enriches the R* state. This ratio is given by:

$$\frac{\rho_\infty}{\rho_0} = \frac{\alpha(1 + L)}{(1 + \alpha L)} \qquad \ldots(3.3)$$

It can be seen from the above equation that if the agonist has an equal affinity for both the R and R* states ($\alpha = 1$) then ρ_∞/ρ_0 will equal unity and no enrichment of R* will result from maximal agonist binding. However, if $\alpha > 1$ then the presence of the conformationally selective agonist will cause the ratio ρ_∞/ρ_0 to be >1. For example, if the affinity of the agonist is 10-fold greater for the R* state, then in a system where 10% of the receptors are spontaneously in this state ($L = 0.1$) the saturation of the receptors with this agonist will increase the amount of R* by a factor of 1.8 (10–18%).

3.2.2 CONFORMATIONAL INDUCTION vs. SELECTION AND ENERGY LANDSCAPES

A major problem with considering that agonists 'deform' receptor proteins to produce an active state is the extremely high energy barrier to the production of protein conformation that exists only in the presence of a dominant binding of foreign chemical and nowhere else. Some reconciliation of this extreme can be found in theories that describe the spontaneous formation of multiple conformations of protein. In fact, the thermodynamics of proteins demand that receptors spontaneously adopt any number of conformations and at any one instant in time a spectrum of conformations exists in the cell membrane. As described by Frauenfelder and colleagues, proteins can exist in a myriad of conformations, depending upon the energy they possess (Frauenfelder *et al* 1988, 1991) Thus, a receptor protein can be viewed as rolling on a surface filled with energy wells of varying depth. Thermal energy sends the receptors into and out of these wells and at any one instant the number of receptors in each energy well (i.e. of each conformation) can be described with frequency histograms. It should be stressed that these various conformations form a fluid and ever-changing system, i.e. not all of the receptors will stay at any one conformation, and that bias can be introduced into this system. Thus, a ligand could convert receptors into a given conformation by selective binding to pre-existing conformations (enriching a given conformation by selective affinity; see above). Given this scenario, it is possible to see the two apparently diverse ideas of conformational induction and conformational selection (as defined by Burgen (1981)) as simply extremes of the same mechanism. Specifically, the enrichment of a particularly rare conformation of the receptor (from the standpoint of it not often appearing spontaneously in nature in the absence of the agonist) by selective affinity (conformational selection) would formally resemble conformational induction. Under these circumstances, the presence of the agonist would appear to 'create' a new conformation. In reality, it would simply enrich a rare spontaneously forming conformation. The attractive aspect of this idea, from a thermodynamic point of view, is that the receptor would not be made to form a conformation that it could not make spontaneously (albeit reluctantly). The production of a receptor conformation by agonist binding (true conformational induction) clearly would be energetically less favorable and unstable. Therefore, although Burgen considered conformational induction and selection to be mutually exclusive mechanisms, an examination of the thermodynamics of protein conformational states suggests that they are simply extremes of the same continuum (Kenakin, 1996).

The exact nature of receptor efficacy is still not completely understood. It is reasonable to assume that the various conformations of receptors expose regions of peptide sequences that activate other cellular components. For example, G-protein-coupled receptors (GPCR) activate membrane proteins referred to as G-proteins (so called because they catalyze the transition of guanosine triphosphate to guanosine diphosphate upon activation). Small oligopeptides are known to activate G-proteins, and similar sequences in the cytosolic loops of GPCRs are known to interact with G-proteins to activate them (Fig. 3.2). Point mutation studies also indicate that different regions of GPCRs activate different G-proteins, allowing for a complex system of mixed signaling.

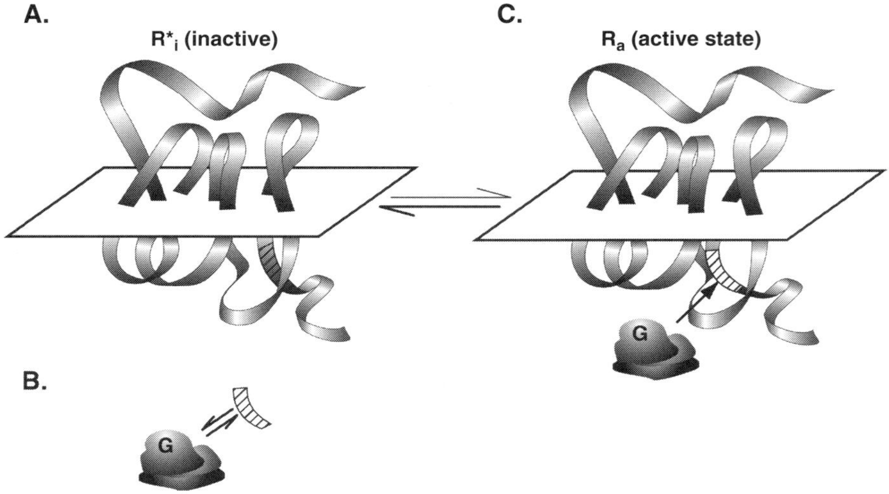

Fig. 3.2. G-Protein-coupled receptors: (A) in the inactive state of the receptor, the regions of the intracellular loops responsible for activation of various G-proteins are inaccessible to G-protein; (B) these regions can be isolated and shown to activate G-proteins directly; (C) active receptor state conformations expose these critical amino acid sequences to produce activation of G-proteins

If it is assumed that active receptor states expose certain key sequences in the intracellular loops to G-proteins to initiate activation, then it is possible that any number of conformations could expose these regions to varying degrees; i.e. there may be a number of conformations capable of causing G-protein activation. Support for this point of view also comes from receptor point mutations where the tertiary structure of receptors is disrupted by introducing different amino acids in certain regions of the receptor. In some cases, as in the α_1-adrenergic receptor, the resulting mutants spontaneously activate G-proteins irrespective of which amino acid is substituted in the position. This might suggest that the inactive form of the receptor is the special one (hiding key sequences from the G-protein) and that a range of other conformations may, to various extents, activate G-proteins. Given this scenario, the possibility of numerous receptor active states is possible (Fig. 3.2C).

3.3 THE CONCEPT OF EFFICACY IN RECEPTOR THEORY

A major impetus for the inclusion of efficacy as a concept in receptor pharmacology was the observation that some agonists were incapable of producing the maximal tissue response but others were, i.e. the discovery of full and partial agonists. Ariens (1954) was the first to introduce a proportionality term (termed intrinsic activity and denoted α) to scale the response-producing capability of an agonist below the maximum. Thus, if an agonist had an intrinsic activity of 0.4, it would only produce 40% of the maximal tissue response (see Fig. 3.3). A full agonist would have $\alpha = 1$ and an antagonist would have $\alpha = 0$. This approach was severely limited because the magnitude of the intrinsic activity is completely system dependent. A drug could be a full agonist in one tissue but a partial agonist in another. The reason for this is that different tissues process receptor-based agonism in different ways and with different efficiencies. The general mechanism used to do this is referred to as the stimulus–response cascade of receptor systems.

3.3.1 STIMULUS–RESPONSE

The stimulus imparted to a receptor by an agonist is an initial step in what is probably a multistep process within cells to produce physiological response. For example, activation of β-adrenergic receptors causes activation of Gs protein, which in turn activates adenylate cyclase, which then produces the intracellular messenger cyclic AMP. This chemical then goes on to cause other biochemical processes in the cell to occur. The total cascade of reactions from the stimulus at the receptor to the final response observed pharmacologically is termed the stimulus–response mechanism of the cell. Note that it is not a single mechanism but probably a succession of many biochemical processes.

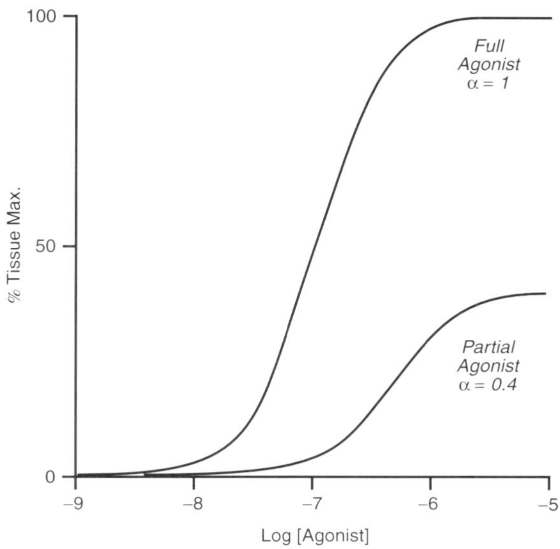

Fig. 3.3. Dose–response curves to a full agonist producing the maximal tissue response and a partial agonist with an intrinsic activity of 0.4 (producing 40% of the maximal tissue response)

There are two features of the stimulus–response mechanism that are relevant to their effect on observed agonist response. The first is that they usually, if not always, confer high amplification to the initial signal. This can be shown mathematically if it is assumed that the successive saturable processes can be modeled by standard Michaelis-Menten-type kinetics (i.e. a hyperbolic relationship between input and output). Thus, one step of the stimulus–response chain can be modeled by a function (see Fig. 3.4):

$$\text{Output}_1 = \frac{\text{Input}_1}{\text{Input}_1 + \theta_1} \qquad \ldots (3.4)$$

The location parameter describing the sensitivity of the Output_1 function with respect to the amount of input it receives is θ_1 (see Fig. 3.4) — this quantifies how sensitive the system is to Input_1. For example, 50% maximal output signal from this reaction occurs when $\text{Input}_1 = \theta_1$. Now, let the output of this first function become the input for a second similar function:

$$\text{Output}_2 = \frac{\text{Output}_1}{\text{Output}_1 + \theta_2} \qquad \ldots (3.5)$$

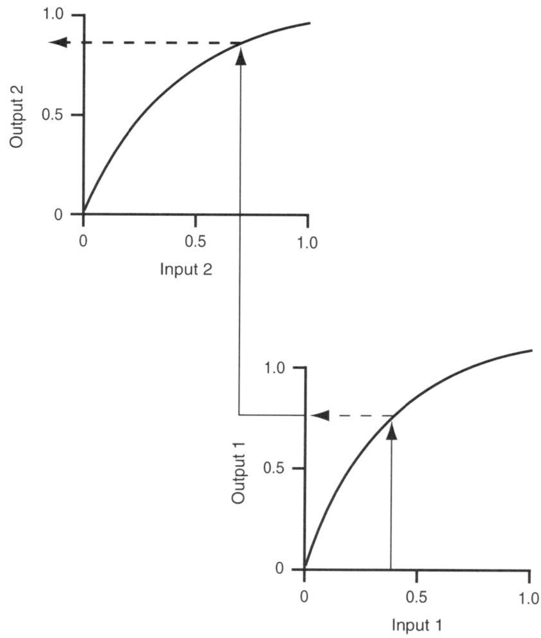

Fig. 3.4. Two saturable stimulus–response coupling functions in series. The output from the first function becomes the input to the second function. Because of the non-linear nature of these functions, note how amplification occurs (an input of 0.4 units becomes an output of 0.9 units)

The output from the two functions in series can be expressed now as a function of the first input (Input_1):

$$\text{Output}_2 = \frac{\text{Input}_1}{\text{Input}_1(1 + \theta_2) + \theta_1\theta_2} \qquad \ldots(3.6)$$

The location parameter for this total output function is given by:

$$\text{Location parameter of two-stage reaction} = \frac{\theta_1\theta_2}{(1 + \theta_1\theta_2)} \qquad \ldots(3.7)$$

It can be seen from this equation that the sensitivity of the two-stage process is equal to the location parameter of the first stage (θ_1) multiplied by the factor $\theta_2/(1 + \theta_2)$. Because this latter factor necessarily is <1, it can be seen that the sensitivity of the multistage process will always be greater than any single part of the stimulus–response pathway (the lower value for the location parameter along the input axis indicates greater sensitivity).

This phenomenon was demonstrated in a striking manner using the irreversible alkylating β-haloalkylamine, dibenamine (Nickerson, 1956). In a classic experiment, a dose–response curve to histamine was obtained in guinea pig ileum. Then a proportion of the histamine receptors was irreversibly blocked with dibenamine, and a second dose–response curve to histamine was obtained (see Fig. 3.5). The relevant feature of this second dose–response curve was that the maximal response was not diminished, i.e. even with a proportion of the receptor inactivated, histamine still produced the maximal response. This was interpreted to mean that the tissue maximal response required activation of only a portion of the complete histamine receptor population in the tissue. This phenomenon was termed 'spare receptors' referring to the portion of receptors that were not needed to produce the maximal tissue response. The amplification of receptor stimulus by tissues is shown by these experiments — i.e. only a small fraction of the receptors need to be activated to produce the maximum tissue response. It should be noted that the phenomenon of spare receptors (also referred to as receptor reserve) depends upon the strength of the agonist used to produce a response and is not a tissue property, i.e. one agonist's spare receptor may be a weaker agonist's essential one.

The second feature of stimulus–response mechanisms is that they are saturable, therefore once a receptor signal has exceeded the limit of one of the processes in the chain, then no increases in the maximal asymptotic response can be expected from agonists (i.e. the tissue maximum may be reached by a number of agonists of differing intrinsic efficacy). The ability of receptor systems to truncate maximal responses to agonists can lead to dissimulation. It should not be assumed that because a number of agonists all produce the same maximal response they are of equal efficacy. The maximal response to an agonist is solely dependent upon the intrinsic efficacy of the agonist only, until the stimulus exceeds the capability of the receptor system. Once the response-producing capability of the

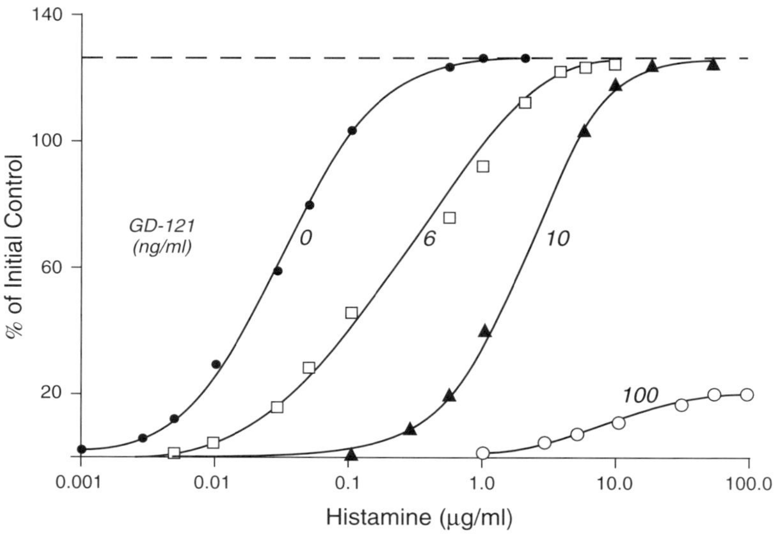

Fig. 3.5. Historical data from the classic paper by Mark Nickerson showing that the irreversible alkylation of histamine receptors in guinea pig ileum by GD-121 results in a rightward shift in the dose–response curve to histamine but no diminution of the maximal response until 100 ng/ml GD-121. This showed that a considerable proportion of the receptors could be removed from the system yet histamine could still produce the tissue maximum, i.e. only a small fraction of the receptors was required for response. (Redrawn from Nickerson (1956).)

receptor system is exceeded, a uniform system-dependent maximal response is observed. This truncation of apparent maximal stimulus production by agonists is shown in Fig. 3.6. Thus, although the initial receptor stimulus to the two agonists shows that agonist A is of higher efficacy than agonist B, when both stimuli are processed through a saturable non-linear stimulus-response cascade the differences in the maxima are not evident and both agonists appear to be of equal efficacy.

The efficiency of a given stimulus–response mechanism is a characteristic of a given tissue (or recombinant receptor system). Thus, there are tissues that are more efficiently coupled than others in translating a receptor-based stimulus into a tissue response. This can be illustrated by considering β-adrenergic receptor-mediated relaxation of guinea pig trachea to isoproterenol. Figure 3.7A shows the dose–response curve to isoproterenol in this tissue and Fig. 3.7B shows the corresponding receptor occupancy vs. response curve in the same tissue. It can be seen from this curve that a 50% response can be produced by an isoproterenol occupancy of only 0.2% of the receptors. Nearly maximal response is obtained with 1% occupancy of the receptors by this agonist, i.e. 99% of the receptors for isoproterenol in this preparation are 'spare'. Figure 3.7C shows corresponding

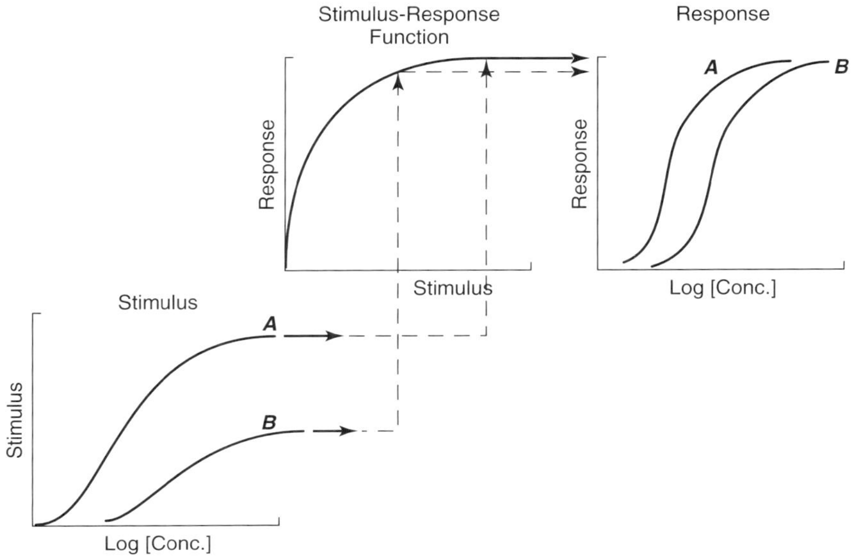

Fig. 3.6. Two unequal stimuli are produced by agonists A and B. The stimuli are processed by a saturable hyperbolic stimulus–response function in the tissue to produce the observed response. Note how the unequal maximal stimuli yield equal maximal observable responses because of the saturation of the stimulus–response capacity

receptor occupancy–response curves for isoproterenol in a range of isolated tissues containing β-adrenergic receptors. It can be seen that although a 50% response can be produced by occupation of 0.2% of the receptors in guinea pig trachea, to achieve half the maximal response in the guinea pig digitorum longus 10% occupancy of the receptors is needed. These curves illustrate the differences in translation efficiencies of different systems in processing a receptor-based stimulus to a cellular response. These differences in receptor coupling efficiencies produce differences in the location parameters of dose–response curves in these tissues. Table 3.1 shows the concentrations of isoproterenol required for 50% maximal response in these tissues. It can be seen that the highly efficiently coupled guinea pig trachea is the most sensitive to isoproterenol whereas the less efficiently coupled extensor digitorum longus is the least sensitive.

These differences also can completely override the receptor stimulation of agonists. Figure 3.8 shows an extraordinary array of profiles for the β-adrenergic agonist prenalterol: from nearly full agonist in thyroxine pre-treated guinea pig right atria, to partial agonist in guinea pig left atrium, to full antagonist in canine coronary artery. These type of data underscore the shortcoming of the intrinsic activity scale as a measure of the power of agonists to produce response in various systems and make evident a need for a less

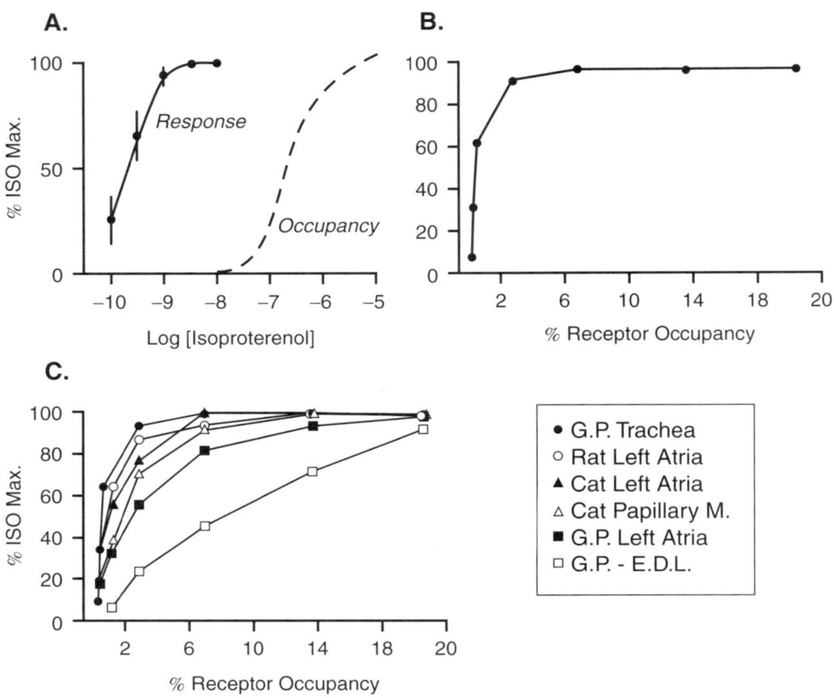

Fig. 3.7. Stimulation of β-adrenoceptors by the agonist isoproterenol. (A) The receptor occupancy curve for isoproterenol is far to the right of the observed response curve in guinea pig (g.p) trachea. This is because of the high degree of amplification of receptor stimulus produced in this tissue by the stimulus–response mechanisms. (B) Relationship between the receptor occupancy by isoproterenol and the tissue response in guinea pig trachea. Note the high amplification. (C) The stimulus–response functions for six isolated tissues for processing isoproterenol stimulus. Note the differences in stimulus–response amplification properties of the different tissues. M = muscle; E.D.L. = extensor digitorum longus). (Redrawn from Kenakin and Beek (1980).)

Table 3.1. Sensitivity of isolated tissues to the β-adrenergic receptor agonist isoproterenol

Tissue	EC_{50} (nM)	Relative potency	% Receptor occupancy for 50% response
Guinea pig trachea	0.2	50	0.2%
Cat left atria	0.4	25	0.5%
Rat left atria	0.6	16.7	0.6%
Cat papillary	2.0	5	2%
Guinea pig left atria	3.0	3.3	3%
Guinea pig extensor digitorum longus	10	1	10%

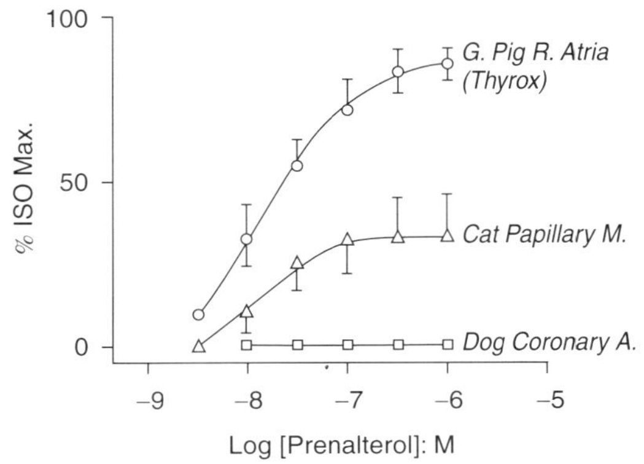

Fig. 3.8. The expression of efficacy by different tissues. Prenalterol is a low efficacy agonist for β-adrenoceptors. In a very efficiently receptor-coupled tissue such as right atria from thyroxine-treated guinea pigs, prenalterol is nearly a full agonist. In cat papillary muscle (less efficiently receptor coupled) prenalterol is a partial agonist, whereas in a poorly coupled tissue (dog coronary artery) prenalterol appears as a simple competitive antagonist with no efficacy. (Redrawn from Kenakin and Beek (1980).)

system-dependent measure of agonism. As a preface to further discussion of efficacy as a useful drug discovery tool, it is useful to describe its mathematical description in drug receptor theory.

3.3.2 EFFICACY AS A DIMENSIONLESS PROPORTIONALITY FACTOR

Efficacy both as a concept and as a pharmacological term was introduced by Stephenson (1956) in a classic paper describing responses of guinea pig ileum to a series of muscarinic receptor agonists. These agonists were considered to produce stimulus to the receptor in proportion to the amount of receptor occupancy achieved by a given concentration (as given by the adsorption isotherm) and a proportionality factor denoting the power of the ligand to elicit physiological response (denoted e and termed efficacy):

$$\text{Stimulus} = \frac{e[\text{A}]/K_\text{A}}{[\text{A}]/K_\text{A} + 1} \qquad \ldots(3.8)$$

The tissue response then is described as being a monotonic function of the stimulus (denoted f, such that Response $= f(S)$). It can be seen that this function then formally becomes the stimulus–response function discussed above. The equation for tissue response within receptor occupation theory then becomes:

$$\text{Response} = f\left(\frac{e[\text{A}]/K_\text{A}}{[\text{A}]/K_\text{A} + 1}\right) \qquad \ldots(3.9)$$

As described by Stephenson, efficacy is both a drug- and system-related parameter. Furchgott (1966) amended this approach by describing the term intrinsic efficacy (denoted ε, where $e = \varepsilon\ [R_T]$). Thus, the intrinsic efficacy describes the power imparted to a single receptor by the agonist to produce stimulus. Thus, the equation for tissue response becomes:

$$\text{Response} = f\left(\frac{\varepsilon[R_T][A]/K_A}{[A]/K_A + 1}\right) \qquad \ldots(3.10)$$

In this scheme, intrinsic efficacy is a dimensionless proportionality factor defining the power of a molecule to induce a change in the receptor that is observed as a physiological response.

Until the advent of molecular cloning of receptor genes and subsequent production of recombinant receptor systems, efficacy was expressed as a physiological response in a naturally quiescent system. Under these circumstances, efficacy generally was thought to be excitatory in nature. With the production of recombinant constitutively active receptor systems (*vide infra*), it was observed that efficacy also could be negative in nature.

3.4 NEGATIVE EFFICACY AND INVERSE AGONISTS

G-protein-coupled receptors activate G-proteins to initiate physiological responses. Under normal conditions, agonists are required to produce an amount of the receptor species (to be referred to as the active state receptor) sufficient to produce measurable cytosolic responses. However, when recombinant systems were produced that had large quantities of receptor, an elevated basal cytosolic response could be measured. These data suggested that receptors themselves were capable of spontaneously forming active states but that this phenomenon, under normal conditions, led to extremely low levels of active state receptor incapable of producing detectable basal responses. However, with the increased numbers of receptors in recombinant systems, enough of the activated receptor was produced to elevate basal responses. The elevated baseline responses produced by receptors spontaneously is referred to as constitutive receptor activity. This phenomenon will be discussed further in Chapter 4 in the discussion of recombinant receptor systems.

One of the most interesting findings to come from the study of constitutive receptor systems is the observation of ligand-induced inhibition of basal receptor-mediated physiological response. Thus, in contrast to agonists that elevate cellular response through receptors, these ligands reduce constitutive receptor activity; these ligands are referred to as ***inverse agonists*** and are considered to have ***negative efficacy***. Interestingly, many apparently pure competitive antagonists, when studied in constitutively active systems, have been found to possess negative efficacy. Previous systems did not possess the necessary elevated basal activity

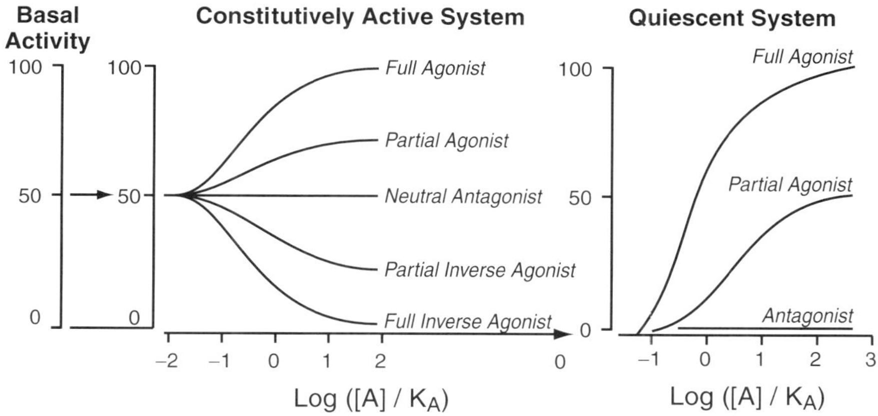

Fig. 3.9. Different views of efficacy by quiescent vs. constitutively active receptor systems. In quiescent systems, only stimulation is observed (full and partial agonist). When the system is constitutively active, negative efficacy is observed as a decrease in the constitutive receptor activity. This effect appears as simple competitive antagonism in quiescent systems

to detect the negative efficacy. Figure 3.9 shows the relationship between the setpoint of the receptor system and the resulting observed activity of ligands.

Up to this point, the operational definitions of efficacy have been discussed as well as the dependence of observed agonism on the efficiency of the stimulus–response function of the system. This sets the stage to discuss the unique utility of efficacy as a drug-dependent factor and tool for the prediction of effects in therapeutic systems.

3.5 WHY MEASURE EFFICACY?

As discussed in the first chapter, a primary tenet of this book is the use of pharmacological and statistical tools to glean chemical knowledge about biological systems that transcends the particular test system and can be applied across all systems, including humans. The concept of efficacy as a scaling factor for agonism (both negative and positive) is extremely useful in this regard.

As shown in Fig. 3.8, stimulus–response functions in different tissues can completely override the molecular information contained in agonists, making classifications such as full and partial agonist, antagonist, and inverse agonist useless as system-independent descriptions. The main utility of the efficacy concept is the elimination of such system dependency. This can be shown for the previously discussed example of prenalterol agonism in β-adrenergic receptor systems. Table 3.2 shows data describing agonism to prenalterol in terms of the concentration producing a 50% maximal effect (to prenalterol; denoted EC_{50})

Table 3.2. System-dependent and- independent properties of prenalterol as a β-adrenergic agonist

Tissue	Observed potency[1] (nm)	Maximal agonist response[2] (%)	Affinity[3] (nm)	Relative intrinsic efficacy[4]
Guinea pig trachea	0.2	79	31.6	0.005
Cat left atrium	0.37	64	32	0.005
Rat left atrium	0.6	65	32	0.0045
Cat papillary	1.4	40	46	0.0045
Guinea pig left atrium	3.2	24	32	0.005
Guinea pig E.D.L.[5]	NA[6]	0	39.8	NA[6]

[1] Molar concentration producing 50% maximal response.
[2] Fraction of the maximal tissue response as measured by the response to the full β-adrenergic agonist isoproterenol.
[3] Molar concentration of prenalterol that occupies 50% of the receptor population at equilibrium (a chemical term — see Chapter 2).
[4] The relative power of prenalterol to produce a response relative to that of the full agonist isoproterenol. This is a measurement of agonism that transcends the receptor preparation (see Chapter 3).
[5] Extensor digitorum longus.
[6] Not applicable. Because no agonist response was observed in the E.D.L., no potency as an agonist or no estimate of relative efficacy could be obtained. Data from Kenakin and Beek (1980).

and the maximal response of prenalterol (as a percentage of the response to the powerful agonist isoproterenol in the same system). Both of these descriptors are system dependent and completely reliant on the efficiency of stimulus–response coupling. Shown also in Table 3.2 is the system-independent property of relative intrinsic efficacy measured in the same preparations; note how the system-independent properties vary little, whereas the system-dependent property of observed agonism varies greatly. For example, there is a range of system-dependent maximal responses to prenalterol of 79–24% but only a corresponding 1.25-fold range in the system-independent parameter of relative intrinsic efficacy. Therefore, the terms full and partial agonist (and, in fact, agonist in general) are system dependent and not necessarily transferable across tissues and into the therapeutic arena. This can cause confusion when drugs are tested in one system for discovery and characterization and used therapeutically in another.

Efficacy can be used also to extend the observed effects of agonists in test receptor systems. In the drug discovery process, it is essential to obtain system-independent measures of the power of molecules that produce physiological response. A first step to such a measure is the agonist potency ratio. This relative measure is independent of the power of the tissue amplification of stimulus because it is made under null conditions, i.e. the stimulus–response process is the same for both agonists and thus cancels itself (see also Fig. 1.19). It can be shown that the ratio of equiactive concentrations of each agonist (the potency ratio) is defined by:

$$\frac{[A_2]}{[A_1]} = \frac{[A_2](\varepsilon_1 - \varepsilon_2)}{K_{A2}\varepsilon_2} + \frac{K_{A2}\varepsilon_1}{K_{A1}\varepsilon_2} \qquad \ldots(3.11)$$

where the relative equilibrium dissociation constants of the agonist–receptor complexes are denoted K_A and the efficacy is ε. If the agonists are of comparable efficacy, the potency ratio can be estimated with a simple ratio of affinity and efficacy:

$$\frac{[A_2]}{[A_1]} = \frac{K_{A2}\varepsilon_1}{K_{A1}\varepsilon_2} \qquad \ldots(3.12)$$

As can be seen from eq. 3.12, the ratio is defined by only the drug-related terms K_A and ε. Such potency ratios can be useful but their general applicability in measuring the response potential of molecules is limited. This is because potency ratios are, to a large extent, system dependent, i.e. they change with the cellular host system of the receptor. Therefore, potency ratios cannot be applied directly from the receptor test system to the therapeutic conditions in humans.

The reason for this system dependence is that the potency of an agonist depends upon both its efficacy and affinity and the ability to produce responses mainly depends on efficacy. This can be seen from a reconfiguration of eq. 3.10:

$$\text{Response} = f\left(\frac{\varepsilon[R_T]}{1 + K_A/[A]}\right) \qquad \ldots(3.13)$$

Changes in the affinity (K_A) can produce only changes in the location parameter of the dose–response curve along the concentration axis (see Fig. 3.10A). In contrast, changes in the efficacy of an agonist can produce both changes in the

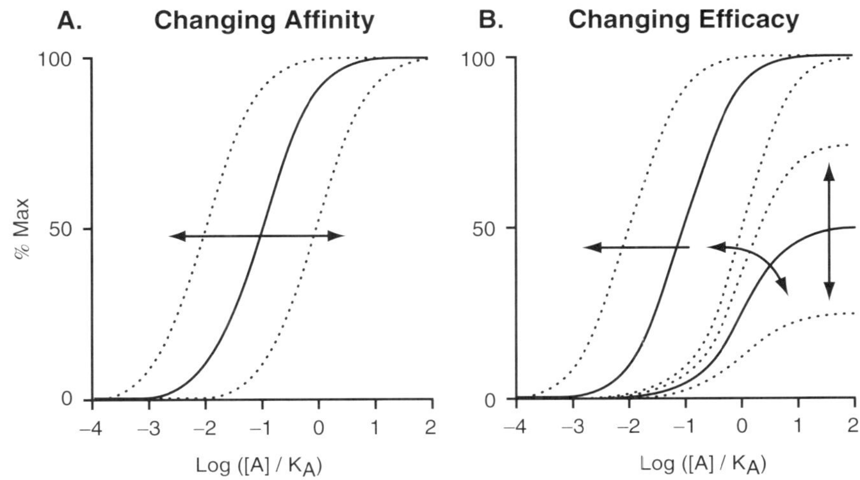

Fig. 3.10. The effects of changing affinity and efficacy on dose–response curves to agonists. (A) Changes in affinity produce changes in the location parameter of the dose–response curve along the concentration axis (potency). (B) Changing efficacy can change the location parameter and the maximal asymptote

maximal response and the location parameter of the dose–response curve along the concentration axis (Fig. 3.10B). The effect on curve location parameter occurs when the receptor density and/or the receptor efficiency coupling is such that the agonist produces the maximal tissue response (when the ligand is a full agonist). Therefore, the location of the agonist curve (the potency) is dependent upon the two tissue-dependent factors of receptor density: $[R_T]$ and the efficiency of stimulus–response coupling (function f). When the potency ratio of two agonists is measured in the same tissue, these factors cancel and the ratio is solely determined by the drug-related parameters of efficacy and affinity. However, both the efficacy and the affinity (actually the ratio of ε/K_A) control the potency and two agonists could be equipotent for different reasons, i.e. one agonist could have a high efficacy but low affinity (to be referred to as an efficacy-driven agonist) whereas another could be as active because of high affinity but low efficacy (referred to as an affinity-driven agonist). It is important to distinguish between these because they behave very differently in different receptor systems.

In general, because efficacy controls the maximal responses to agonists, high-efficacy agonists are more resistant to decreases in receptor density and/or decreases in receptor coupling efficiency. Figure 3.11 shows the relative potency

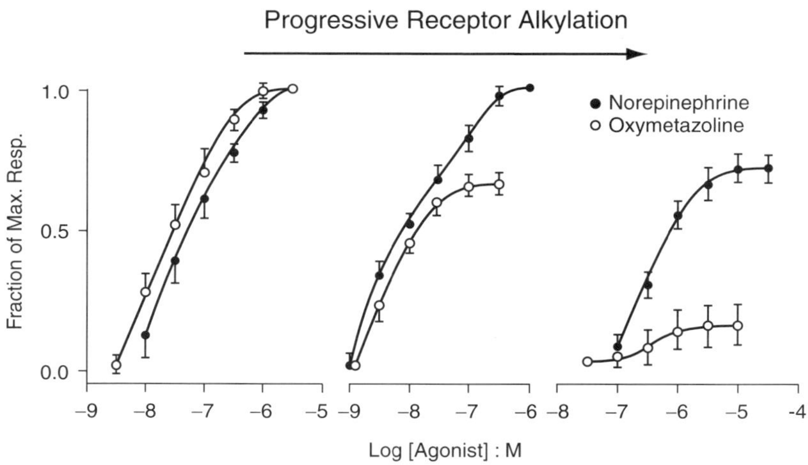

Fig. 3.11. The differential effects of efficacy on the resilience of agonist response to decreasing receptor coupling transduction. Dose–response curves for α-adrenoceptor agonists oxymetazoline (open circles) and norepinephrine (filled circles) in rat anococcygeus muscle. Oxymetazoline has twice the affinity but 30% the efficacy of norepinephrine. Left panel shows that both are full agonists and that oxymetazoline is twofold more potent. As the α-adrenoceptors are reduced by alkylation (treatment with dibenamine), the response to the low efficacy agonist selectively decreases (middle and right panels). (Redrawn from Kenakin: Pharm. Analysis Drug Recept 3E, spec Mat: (1997). Reproduced with permission of Lippincott, Williams & Wilkins (LWW).)

of two α-adrenergic receptor agonists norepinephrine and oxymetazoline in the rat anococcygeus muscle. As can be seen from the first panel to the left, both are full agonists and oxymetazoline is slightly more potent than norepinephrine. The panels to the right of this figure show the same tissue after increasing proportions of the α-adrenergic receptors are irreversibly inactivated with the alkylating agent dibenamine. As can be seen from these figures, the maximal response to oxymetazoline begins to decline whereas that to norepinephrine persists. In the furthest right panel a robust response to norepinephrine remains whereas the response to oxymetazoline is nearly obliterated. Thus, oxymetazoline went from being a more potent full agonist to a weak partial agonist, whereas the norepinephrine response was less affected by the diminution in α-adrenergic receptor density.

These data illustrate the differential sensitivity to efficacy-driven vs. affinity-driven agonists. The potency of norepinephrine emanates from high efficacy whereas the potency of oxymetazoline stems from high affinity. When the receptor density and/or stimulus–response capability of the tissue becomes compromised, the responses to affinity-driven agonists are selectively depressed. Figure 3.12 shows the effects of diminishing receptor density and/or decreasing receptor-coupling efficiency on an efficacy-driven agonist. As can be seen from this figure, the dose–response curve is shifted to the right with no depression in the maximal response until the concentrations approach those that saturate receptor occupancy (K_A). Then the maximal response diminishes. The same is true for affinity-driven agonists except, because the concentrations producing a response are nearer to those saturating the receptor occupancy (low efficacy), decreases in receptor density and/or decreases in stimulus–response efficiency lead to a much greater proclivity for depression of the maximal response. Thus, the dose–response curve shows depression of the maximum.

Figure 3.12 demonstrates the general idea that affinity-driven agonists are much more prone to desensitization than are efficacy-driven agonists. It can be seen from Fig. 3.13A that if desensitization of the system occurs, an increase in dose of the agonist can compensate and the original response can be regained. For affinity-driven agonists, once desensitization occurs, no increase in the dosage of the agonist can overcome it and the response cannot be regained. There are therapeutic areas where resistance to desensitization may be important, therefore an efficacy-driven agonist would be of greater utility. There are also other areas where organ selectivity can be obtained by a sensitivity to receptor density and/or receptor coupling efficiency (i.e. the agonist would produce responses in only the most efficiently coupled organs). Therefore, knowledge of the relative dependency of an agonist's potency on affinity and efficacy can be useful in the choice of agonist to develop for therapeutic utility. In this regard, knowing the relative dependence of potency of a range of agonists on efficacy and affinity can assist in defining the texture of agonism in all systems.

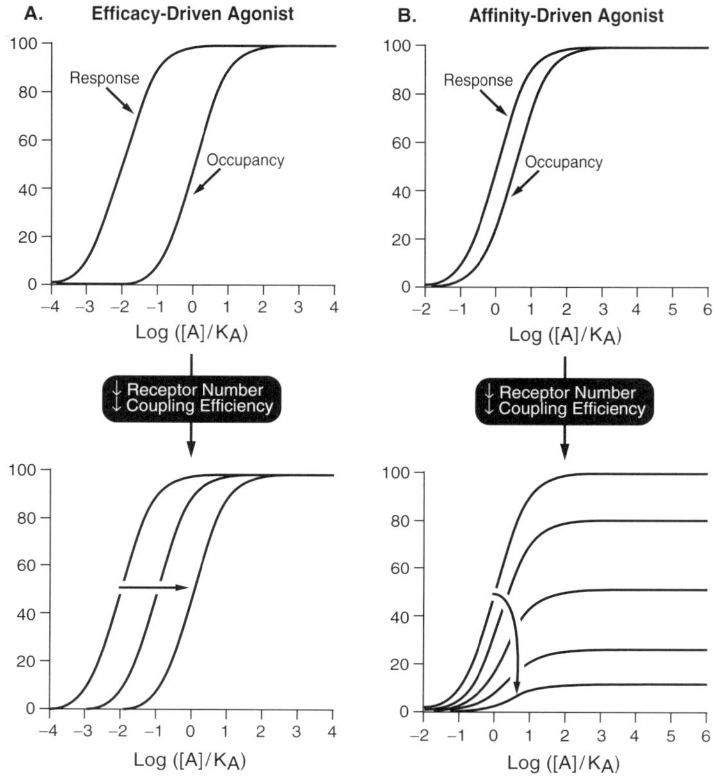

Fig. 3.12. The effect of reduction in receptor density or stimulus–response coupling on response to an agonist that is potent because of high efficacy (left panels) vs. an agonist that is potent because of high affinity (right panels). The graphs show the relationship between the concentrations of agonist that occupy receptors and produce a response. The bottom panels show the effects of reducing stimulus–response capability. The curves to efficacy-driven agonists shift to the right but retain the maximal response. In contrast, the curves to affinity-driven (low-efficacy) agonists do not shift to the right but, rather, show a depressed maximal asymptote

The dependence of agonism on efficacy also can be important in interpreting data obtained from agonist selectivity screens. Figure 3.13 shows the relative effects of two hypothetical agonists in four tissues, with receptor coupling efficiencies ranging from very well to very poorly coupled receptors. The potency ratio is these systems ranges from 100 to infinity, raising the question as to which system would best predict the relative effects of these agonists in humans. A real example of this effect is shown in Fig. 3.14, where the relative agonism produced by the α-adrenergic agonists norepinephrine and oxymetazoline are shown in the

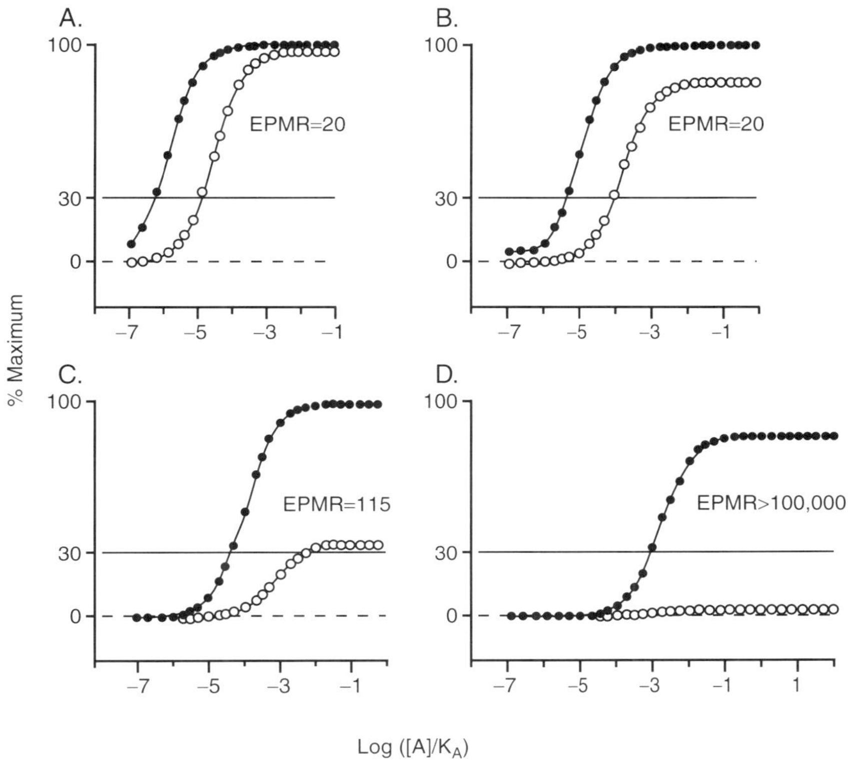

Fig. 3.13. Values for the relative potency of two agonists of differing efficacy in four receptor systems of decreasing efficiency of receptor coupling. Ratios refer to equipotent molar ratios (EPMR). (A) In well-coupled systems where both are full agonists, the ratio is constant at 20. (B) As the maximal response to the lower efficacy decreases with decreasing efficiency of receptor coupling, the EPMR values change. (C) As receptor coupling declines further, the effect becomes exacerbated. (D) Conceivably, some systems can be of such inefficient receptor coupling efficiency as to make low-efficacy agonism undetectable. Dissimulations occur if the screening system for selectivity is at setpoint D and side effects occur in humans at setpoint A. Alternatively, if the screening system is at setpoint A and the therapeutic condition at setpoint D, then a therapeutic effect predicted by the screen will not be seen in humans

rat anococcygeus muscle and the rat vas deferens. The profile in the anococcygeus muscle indicates that oxymetazoline is more potent than norepinephrine, whereas in the vas deferens it is quite the opposite. As discussed above, estimates of the relative efficacy of these agonists would resolve this apparent dilemma but, in terms of which is the most appropriate system to screen for relative agonism, it can be seen that a false sense of security (if the agonism is testing

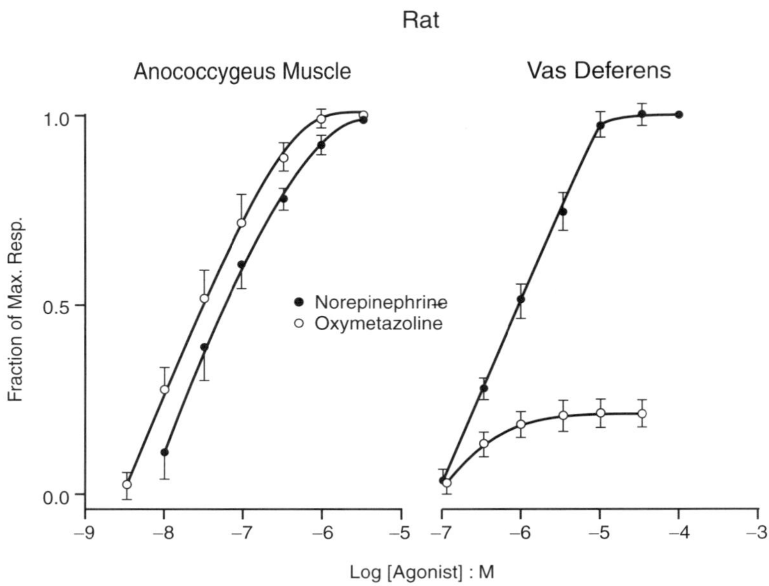

Fig. 3.14. The relative potency of a high-affinity/low-efficacy α-adrenoceptor agonist (oxymetazoline) and a high-efficacy/low-affinity agonist (norpinephrine) in two isolated tissues. The rat anococcygeus muscle is highly efficiently receptor coupled (similar to (A) in Fig. 3.13) whereas the rat vas deferens is less efficiently receptor coupled (similar to (C) in Fig. 3.13). (Redrawn from Kenakin (1997).)

for side effects) can be obtained from using systems of low efficiency of receptor coupling. If system C or D in Fig. 3.13 is the test system used for testing the relative propensity of an agonist to produce agonism that is unwanted, then an unpleasant surprise could occur if the drugs encounter tissues A and B in humans under therapeutic conditions.

Another potential benefit of quantifying efficacy is the determination of possible negative efficacy in apparent neutral antagonists. As discussed above, unless a receptor system is constitutively active, inverse agonism will not be detected and inverse agonists will appear as competitive antagonists. The secondary testing of antagonists in constitutively active receptor systems can detect negative efficacy and this, in turn, can be useful in predictions of possible tolerance to antagonism. For example, some histamine H_2 receptor antagonists demonstrate clinical tolerance upon prolonged use. A study of the apparent simple competitive histamine H_2 receptor antagonists ranitidine and cimetidine shows that, in a constitutively active recombinant histamine H_2 receptor system (transfected CHO cells), these are in fact inverse agonists (see Table 3.3). The H_2 antagonist burimamide shows no negative efficacy in the same study. Receptor binding studies with these three

Table 3.3. Inverse agonism and receptor upregulation with histamine H_2 receptor antagonists

Antagonist	K_i[1]	Cyclic AMP		% Upregulation
		IC_{50}	% Reduction	
Cimetidine	0.6 μM	1.2 μM	72 ± 6	68 ± 9
Ranitidine	0.14 μM	0.09 μM	88 ± 3	55 ± 3
Burimamide	5.5 μM	—	2 ± 4	−4 ± 1

[1] Equilibrium dissociation constant of antagonist–receptor complex. From Smit *et al* (1996).

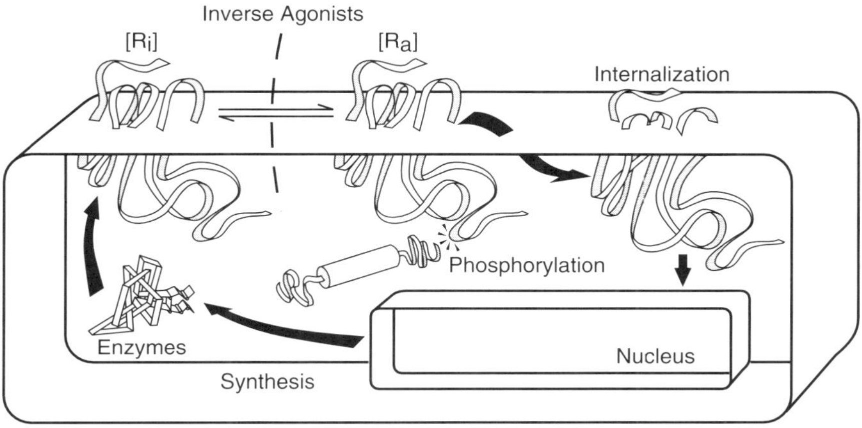

Fig. 3.15. Steady-state synthesis and internalization of membrane surface receptors. The natural progression of the spontaneous formation of the active state of receptors with subsequent internalization can be interrupted by inverse agonists that prevent the transition from R_i to R_a. If receptor synthesis continues during this interruption, then an increase in the receptor density on the cell surface could result

ligands show that the inverse agonists induce an upregulation in the number of histamine H_2 receptors in the cell whereas burimamide (neutral antagonist) does not (Smit *et al* 1996; Table 3.3). These data are consistent with the idea that cells are dynamic in their control of membrane surface receptor populations and that receptors are continually synthesized in the cytosol, transported to the surface and internalized again (see Fig. 3.15). Receptor activation is a known trigger for internalization because the formation of active receptor conformations can expose phosphorylation sites on the inner loops of the receptor. The spontaneous formation of active conformations is an integral part of the receptor cycle and interruption of this process by, for example, an inverse agonist selectively stabilizing the inactive state would arrest the cycle. Because receptors are continually synthesized, it would be expected that inverse agonists would lead to an increased accumulation of surface receptors (MacEwan and Milligan, 1996). For antagonists, increases in receptor density will cause an increase in hormonal signaling and, therefore, translate to a loss of effectiveness of the antagonist. These data suggest that it would be useful to test all antagonists in constitutively active receptor systems to detect possible negative efficacy and, thus, a propensity to cause tolerance upon chronic use.

3.6 METHODS TO QUANTIFY EFFICACY

There are three basic methods used to measure the relative efficacy of agonists. It should be noted that there is no absolute scale for efficacy and that the relative measure of the ability of agonist to produce response is derived as a ratio to the same quantity for another agonist. Two of these methods either assume a base affinity of the agonist for the inactive receptor (method of Furchgott) or create a condition where this base affinity presumably can be measured (GTP shift). The third method obviates the need for knowledge of affinity by comparing drug effects at near-maximal receptor occupancy. This method is the most theoretically sound and calculates efficacy from the relative magnitude of agonist maximal responses.

3.6.1 THE METHOD OF FURCHGOTT

Efficacy is a measure of the ability of a ligand to produce response at a given receptor occupancy. For example, if one ligand produces a 50% maximal physiological response by occupying 5% of the tissue receptors and another ligand produces the same response but needs to occupy 50% of the receptors, then this latter ligand produces less stimulus per receptor activation than the former, i.e. it has a lower intrinsic efficacy. This idea forms the basis for a method of estimating agonist efficacy, devised by the pharmacologist Robert Furchgott. In terms of occupancy theory, the expressions for two equiactive concentrations of agonist A_1 and A_2 are:

$$\text{Response} = f\left(\frac{\varepsilon_1[\mathrm{R_t}][\mathrm{A_1}]/K_{\mathrm{A1}}}{[\mathrm{A_1}]/K_{\mathrm{A1}}+1}\right) = f\left(\frac{\varepsilon_2[\mathrm{R_t}][\mathrm{A_2}]/K_{\mathrm{A2}}}{[\mathrm{A_2}]/K_{\mathrm{A2}}+1}\right) \qquad \ldots(3.14)$$

If the comparison is made in the same tissue, then the receptor density $[R_t]$ and the stimulus–response function f are common to both agonists and their effects cancel. Under these circumstances:

$$\frac{\varepsilon_1[\mathrm{A_1}]/K_{\mathrm{A1}}}{[\mathrm{A_1}]/K_{\mathrm{A1}}+1} = \frac{\varepsilon_2[\mathrm{A_2}]/K_{\mathrm{A2}}}{[\mathrm{A_2}]/K_{\mathrm{A2}}+1} \qquad \ldots(3.15)$$

It can be seen that each side of the equation is the product of the intrinsic efficacy and the receptor occupancy (Langmuir adsorption isotherm, Chapter 2) and thus can be rewritten as:

$$\frac{\rho_1}{\rho_2} = \frac{\varepsilon_2}{\varepsilon_1} \qquad \ldots(3.16)$$

For the example described previously, the ratio of the intrinsic efficacies of the two agonists is the ratio of their receptor occupancies required to produce the same magnitude of response, in that case being 5% and 50%. Thus, the more powerful agonist has ten times the intrinsic efficacy of the weaker agonist. Therefore, according to the method of Furchgott, the dose–response curve for each agonist is expressed as a function of the receptor occupancy required for response (to produce an occupancy response curve) and the ratio of equiactive receptor occupancies yields the relative efficacy of the agonists. This procedure is illustrated in Fig. 3.16. A pre-requisite to the utilization of the method of Furchgott is a reliable estimate of the equilibrium dissociation constant of the agonist (K_A). This method was originated in isolated tissue work, where it was assumed that the high intracellular levels of GTP precluded the formation of ternary complexes between agonist, receptor and G-protein, therefore receptor isomerization would not corrupt the measurement of base affinity of the ligand for the receptor. It should be noted that this is an assumption in this method and may not be valid. Under these latter circumstances, the estimates of efficacy by this method would underestimate true efficacy. Also, if binding estimates of affinity are used in this method, then the assumption of base affinity made for isolated tissues may not be valid either.

A short method of doing this for an estimate of relative efficacy can be useful in the early screening process for agonists to determine which are efficacy driven and which are affinity driven. Figure 3.17 shows the dose–response curves to a number of hypothetical agonists. The data describing these curves as well as the affinities (measured as K_A) are shown in Table 3.4. For this calculation, the receptor occupancy required for each agonist to produce a defined percentage of the maximal response (denoted R) is calculated; this is designated ρ_{Resp}. This receptor occupancy is calculated by determining the concentration of agonist that produces this effect and then calculating the receptor occupancy produced by this concentration with the adsorption isotherm. The following equation can be used:

$$\rho_{\mathrm{Resp}} = \frac{R \bullet \mathrm{EC}_{50}}{K_A(Max - R) + R \bullet \mathrm{EC}_{50}} \qquad \ldots(3.17)$$

where EC_{50} refers to the concentration of agonist that produces 50% of the maximal response to that agonist, and *Max* refers to the maximal response to that agonist expressed as a fraction of the tissue maximal response.

For the example shown in Fig. 3.17, a 20% response level was used ($R = 0.2$) to yield the receptor occupancies shown in Table 3.4. The ratios of these occupancies then are direct estimates of the relative intrinsic efficacies of the agonists (Table 3.4). Efficacy, as a measure of the relative power of an agonist to produce response, is illustrated by the values of ρ_{Resp} shown in Table 3.4.

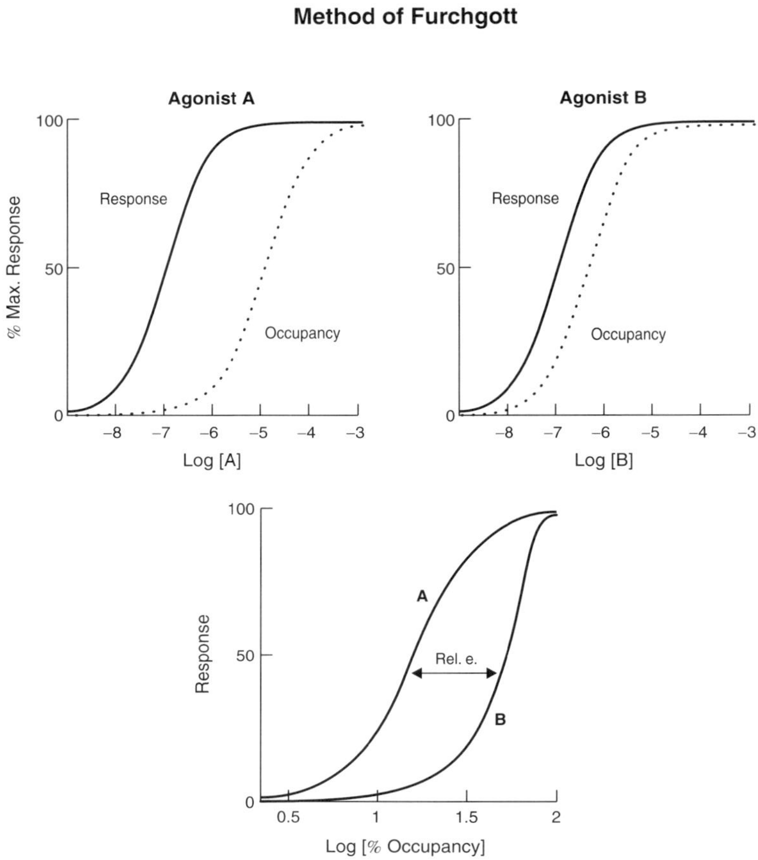

Fig. 3.16. Estimation of the relative efficacy of two agonists by the method of Furchgott. The magnitude of the separation between the response curves and the receptor occupancy curves depends upon the efficacy of the agonist and the efficiency of stimulus–response coupling in the system. If two agonists are compared in the same system, this latter factor cancels. Re-expression of the dose–response curves as a function of the receptor occupancy (bottom panel) results in a measure of the relative efficacy (Rel. e.) of the agonists

For example, although agonist F must active 44.5% of the receptors to elicit a response of 20% maximum, agonist A can do the same by activating only 0.62% of the receptors.

3.6.2 RELATIVE MAXIMAL RESPONSES

It can be seen from the results of the previous calculation that the relative efficacies of the partial agonists (producing submaximal responses) parallels their

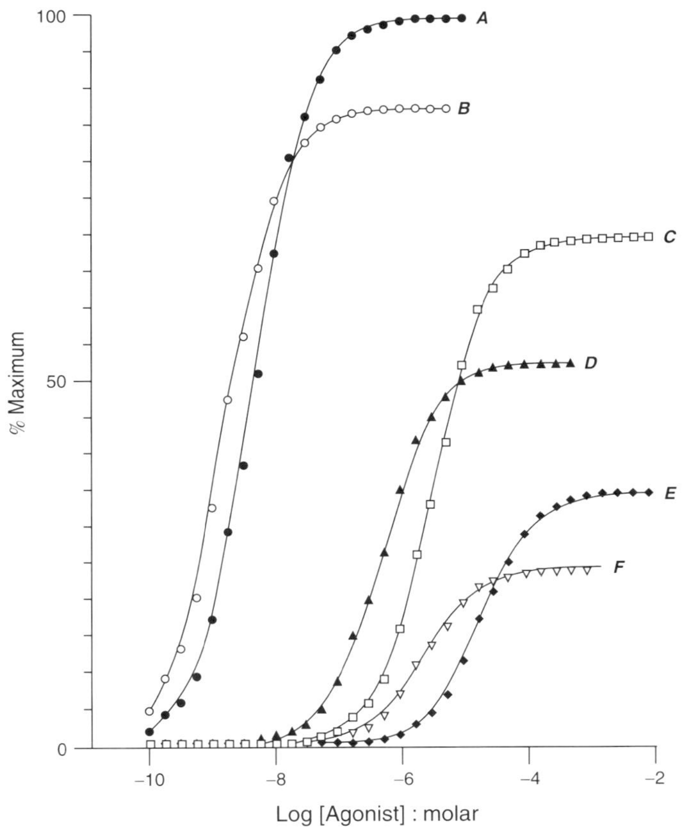

Fig. 3.17. Dose–response curves to six hypothetical agonists of varying efficacy and affinity (see Table 3.4 for parameters). It can be seen that the relative potency of the agonists does not reflect relative efficacy.

Table 3.4. Estimation of relative efficacy of agonists

Agonist	EC_{50} (μM)	Maximal response *Max*	K_A (μM)	Receptor occupancy ρ_{Resp} (%)	Relative efficacy ε
A	0.012	1.0	0.1	0.62	1
B	0.003	0.73	0.01	2.16	0.29
C	5.4	0.46	10	6.3	0.1
D	0.67	0.29	1.0	12.25	0.05
E	26	0.16	30	28	0.022
F	2.9	0.11	3.0	44.5	0.014

relative maximal responses. This is because efficacy, not affinity, controls the magnitude of the maximal response. Provided that the stimulus–response mechanism of the tissue is not saturated (and therefore the agonist does not produce the maximal tissue response), the ratio of the maximal responses of the agonists yields the ratio of the intrinsic efficacies of the agonists. Therefore, one experimental approach to determining relative efficacy is to compromise the stimulus–response capabilities of the tissue, or deplete the receptor density (through chemical alkylation with irreversible antagonist; see Chapter 2) to the point where the agonists all produce a maximal response that is below that of the tissue maximum (all are partial agonists). There is strong theoretical rationale for this method because the affinity of the agonist is not required. In terms of conventional receptor occupancy theory, the relative response of two agonists denoted A and B is given by a ratio defined by eq. 3.18.

$$\frac{\text{Response}_B}{\text{Response}_A} = \frac{\varepsilon_B(1 + K_B/[B])}{\varepsilon_A(1 + K_A/[A])} \qquad \ldots(3.18)$$

If this response is the maximal response to the agonist (and is of lower magnitude than the system's maximum response) then it will be measured at near-maximal receptor occupancies (i.e. $[A]/K_A$ and $[B]/K_B \to \infty$). Under these circumstances, eq. 3.18 predicts that the ratio of these two maximal responses yields the term $\varepsilon_B/\varepsilon_A$.

3.6.3 EFFICACY MEASUREMENTS WITH BINDING STUDIES

Another method of estimating the relative efficacy of agonists utilizes biochemical binding of the agonist ligand to the receptor. This technique depends upon the ability of agonists to isomerize receptors into active states and thus further bind

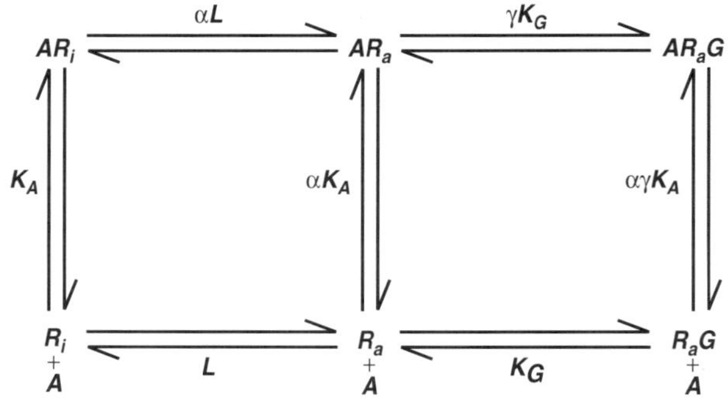

Fig. 3.18. The extended ternary complex model of G-protein-coupled receptors. The receptor exists in an inactive form R_i and an activated form R_a. The activated form R_a can activate G-proteins to produce a physiological response. (Model from Samama *et al* (1993).)

to G-proteins. As discussed in Chapter 2, receptor isomerization confers an added apparent affinity of the ligand for the receptor. Thus, ligand efficacy produces a high-affinity binding for agonists in receptor systems containing G-protein. The greater the efficacy of the ligand, the greater is the promotion of binding to G-protein. Therefore, the magnitude of the high-affinity equilibrium dissociation constant is dependent upon the equilibrium affinity of the ligand for the receptor (K_A) and the magnitude of the intrinsic efficacy. If this latter parameter could be estimated (this is referred to as the low affinity state), then the ratio of the two affinities would be a measure of the intrinsic efficacy. Considering the ternary complex model whereby the receptor can exist in an inactive (R_i) and active (R_a) state and whereby response is produced by activation of G-protein by the active state (species R_aG and AR_aG in Fig. 3.18), the observed affinity of a ligand is given by:

$$K_{obs} = \frac{K_A[1 + L(1 + [G]/K_G)]}{1 + \alpha L(1 + \gamma[G]/K_G)} \qquad \ldots (3.19)$$

Intrinsic efficacy is the ability of the ligand to induce the active state (the multiple higher affinity that the ligand has for R_a as measured by factor α) or the added propensity that the ligand-bound receptor has for binding to G-protein (factor γ). It can be seen from eq. 3.19 that high values for α and/or γ will lead to the condition that $K_{obs} < K_A$, i.e. the observed equilibrium dissociation constant of the system will be smaller than the equilibrium dissociation constant of the ligand for bare receptor. By definition, this is the high affinity state.

An experimental method to obviate the effects of G-protein coupling on affinity is to produce a condition whereby no steady-state accumulation of the AR_aG species can occur. Activation of G-protein by receptor causes the exchange of bound GDP on the G-protein for GTP; the subsequent hydrolysis of GTP by G-protein produces energy and leads to a physiological effect. If the experimental receptor system is incubated with an excess of GTP, this process occurs so quickly that the receptor-activated G-protein recycles back to free G-protein and the AR_aG species does not form in appreciable quantities. Under these conditions, the observed affinity for the ligand (in the presence of an excess of GTP or non-hydrolysable analog GTPγS) is equal to K_A. Therefore, the ratio of the affinity of the agonist in the absence and presence of GTP is a measure of intrinsic efficacy, specifically $= (K_{obs}/K_A)$. Experimentally, the agonist is used to displace a radioligand antagonist (see Chapter 2) in the absence and presence of an excess of GTPγS, and the ratio of observed affinities is used as the measure of relative efficacy (see Fig. 3.19).

3.7 AGONIST-SELECTIVE RECEPTOR STATES: A POSSIBLE TOOL FOR FURTHER SELECTIVITY

As noted in the discussion of the molecular nature of efficacy, there is evidence to suggest that the inactive form of the receptor is the most unique (providing

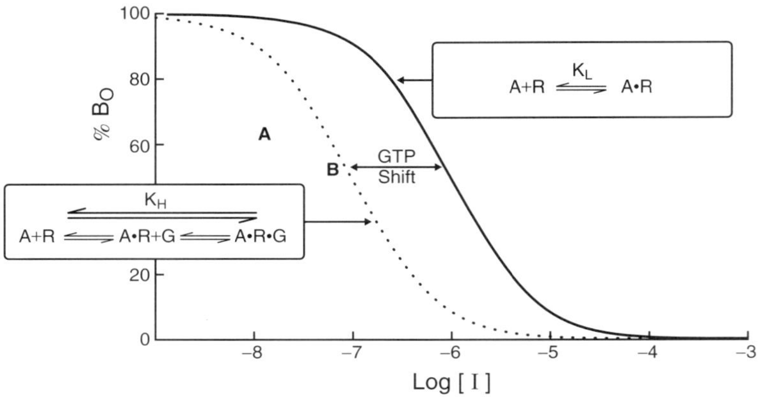

Fig. 3.19. Estimation of efficacy by observation of the 'GTP shift'. The high-affinity state is produced by complexation of the activated receptor to G-protein to form the ARG species. The avidity with which this occurs depends upon the ability of the ligand to induce the active state (i.e. efficacy). The low-affinity state ostensibly measures only the affinity of the receptor for that inactive state and is produced by recycling of the ARG species back to free G and R by high concentrations of GTP (or stable analog). The difference between these affinities reflects efficacy

sufficient shielding of critical amino acid sequences that interact with G-proteins to produce activation) and that there may be a number of conformations of the receptor that allow these sequences to interact with G-proteins and thus produce a response. Point mutations of receptors have shown that different regions of the cytosolic loops of receptors activate different G-proteins. In view of this idea, it would be expected that as the receptor adopts different tertiary conformations, these would not expose all of the various activating sequences in the intracellular loops equally, i.e. some conformations would favor different G-proteins (Fig. 3.20). This concept, coupled with the notion that different agonists may stabilize different receptor active states, leads to the possibility that different agonists may favor activation of different G-proteins. This would lead to agonist-selective signaling to cells and offer new vistas for the design of selective therapeutic agonists. The quest for agonist selectivity in therapy began with the search for receptor subtypes. Thus, although a superfamily of receptors exists for a given hormone or neurotransmitter (i.e. β-adrenergic receptors for epinephrine and norepinephrine), further selectivity can be found in nature by the evolution of three subtypes of β-adrenergic receptor (designated β_1-, β_2-, β_3-) differing in amino acid composition, sequence and tertiary conformation. Thus, although the same endogenous agonists activate all of the subtypes, they have varying affinities and efficacies for them and therefore a diverse system of signaling can be obtained from the same agonist through differential receptor localization. This diversity of subtypes can be taken advantage of therapeutically; for example, β_2-adrenergic receptor-selective

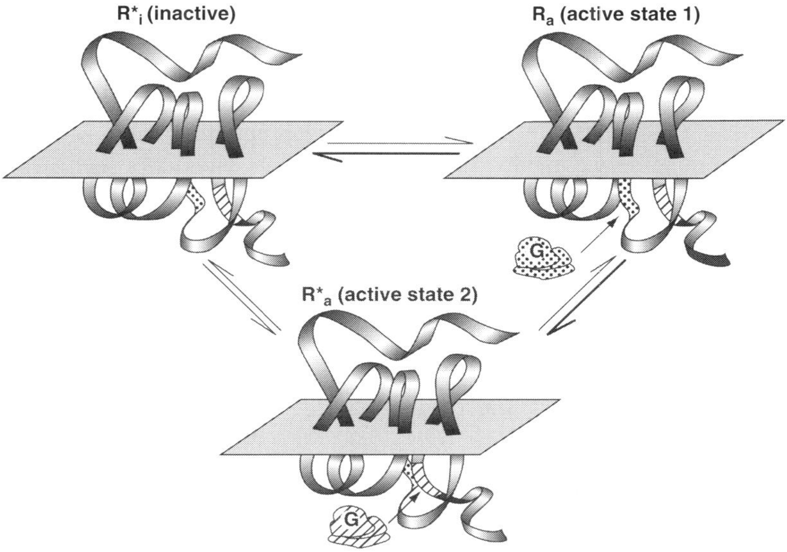

Fig. 3.20. Different G-proteins can be activated by different regions of the cytosolic loops of G-protein-coupled receptors (denoted by bars and dots). Different tertiary receptor conformations are likely to expose these different regions to varying extents, thereby conferring bias as to which G-protein will be activated by which conformation

agonists relax bronchiole smooth muscle and are of use in asthma because they lack activity at β_1-receptors that mediate damaging tachycardia.

There is increasing evidence that different agonists produce different (agonist-specific) active receptor states. Viewed in terms of selective affinity for the existing spectrum of receptor conformations found in nature, it could be expected that different agonists, through variable affinity for the array of conformations, produce a whole new population of receptor conformations (different relative abundance of the natural array of conformations). This could, in turn, lead to a further diversity of response if multiple G-protein activation mediates cellular response (i.e. different agonists could traffic receptor stimulus toward different G-proteins, producing different types and strengths of response). Thus, a further level of selectivity may be gained by designing agonists that selectively form a single active state. Specialized assay systems to identify such activity are discussed in Chapter 4.

3.8 LIMITATIONS TO EFFICACY AS A DRUG PARAMETER

Although it is clear that efficacy is a molecular property of the ligand/receptor pair, it is not always clear that it can be measured independently. Problems can occur in a practical sense when a receptor demonstrates multiple behaviors in different systems. The specific condition where this makes measurement of efficacy as a molecular property difficult is when ligands form different receptor active states where receptors interact with more than one G-protein. Under these circumstances, the efficacy measured is a specific property of the ligand/receptor-state/G-protein ternary complex. However, because the amount of such a complex probably will vary with the relative stoichiometry of the G-proteins available to the receptor, then the observed 'efficacy' in any one system will be a unique value assigned to that system. Under these circumstances, the efficacy parameter will not be transferable across to other systems and it ceases to become a molecular property able to characterize the ligand and the receptor. For example, Fig. 3.21 shows the relative potency of porcine calcitonin and rat amylin in wild-type HEK cells transiently transfected with human calcitonin receptor type II. This receptor is known to interact with multiple G-proteins. As can be seen from this figure, the relative potency of the agonists is 4.6, a value supposedly dependent only upon the relative affinity and efficacy of these agonists. However, transfection of the same receptor into the same cells enriched in Gs-protein (by stable transfection with cDNA for Gs-protein) shows a striking near-20-fold increase in the relative potency (potency ratio = 84). It should be stressed that nothing has changed except for the fact that the receptor is placed in a slightly altered host cell.

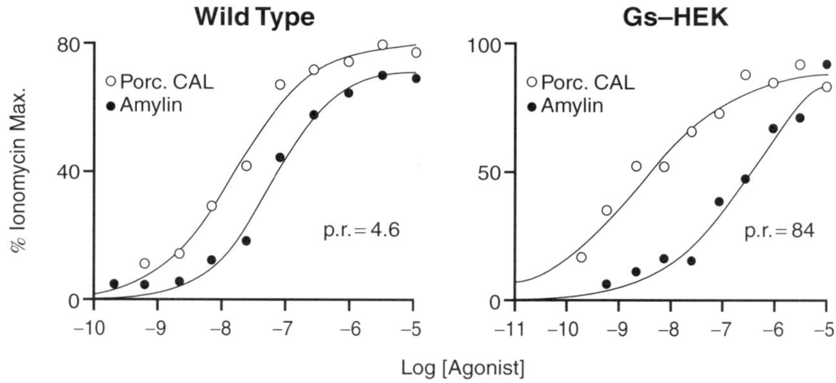

Fig. 3.21. Differing potency ratios (p.r.) for calcitonin agonists produced by the presence of the human calcitonin receptor (Type 2) in HEK 293 host cells of differing composition of G-protein. Gs-HEK are wild-type HEK cells stably co-transfected with cDNA for Gs protein and are thus enriched in this G-protein compared to wild-type HEK cells. The presence of the added complement of Gs produces nearly a 20-fold increase in the relative potency of porcine calcitonin (Porc. CAL) over rat amylin, indicating the selective reliance of this agonist on this G-protein. These data suggest that the receptor active state made by porcine calcitonin interacts more avidly with Gs-protein than the one made by amylin

Separate experiments show that the affinity of the ligands for the receptor did not change, therefore the observed relative efficacy of the two agonists appears to have changed with the change in host system. Such data underscore the difficulties of obtaining system-independent estimates of the molecular properties of agonists for some receptors.

3.9 CONCLUSIONS

This chapter discusses the molecular property of a drug that causes it to change the behavior of a biological receptor and, in turn, to cause a physiological response. The mechanism of such change may involve the selection and stabilization of different conformations of receptors and may result in a positive increase in response or a negative inhibition of constitutive receptor activity. This chapter also discusses the importance of tissue stimulus–response mechanisms in the translation of receptor stimulus to tissue response. The use of efficacy in eliminating the system dependence of agonism and the prediction of agonism in other systems is outlined, as well as the common methods used to measure the relative efficacy of agonists. Finally, the complex nature of receptor efficacy is discussed in terms of the production of multiple active states of receptors by ligands and the resulting possible increase in physiological selectivity and the problems created in assigning unique values of efficacy to ligands.

3.10 REFERENCES

Ariens EJ (1954) Affinity and intrinsic activity in the theory of competitive inhibition. Part I. Problems and theory. *Arch Int Pharmacodyn Ther* **99**: 32–49.

Burgen ASV (1981) Conformational changes and drug action. *Fed Proc* **40**: 2723–2728.

Frauenfelder H, Parak F, Young RD (1988) Conformational substates in proteins. *Annu Rev Biophys Biophys Chem* **17**: 451–479.

Frauenfelder H, Sligar SG and Wolynes PG (1991) The energy landscapes and motions of proteins. *Science* **254**: 1598–1603.

Furchgott RF (1966) The use of β-haloalkylamines in the differentiation of receptors and in the determination of dissociation constants of receptor–agonist complexes. In *Advances in Drug Research*, Vol. 3, ed. by N J Harper and A B Simmonds, pp. 21–55. Academic Press, New York.

Kenakin TP (1996) Receptor conformational induction vs. selection: all part of the same energy landscape. *Trends Pharmacol Sci* **17**: 190–191.

Kenakin TP and Beek D (1984) Relative efficacy of prenalterol and pirbuterol for β_1-adrenoceptors: measurement of agonist affinity by alteration of receptor number. *J Pharmacol Exp Ther* **229**: 340–345.

MacEwan DJ and Milligan G (1996) Inverse agonist-induced up-regulation of the human β_2-adrenoceptor in transfected neuroblastoma X glioma hybrid cells. *Mol Pharmacol* **50**: 1479–1486.

Nickerson M (1956) Receptor occupancy and tissue response. *Nature* **178**: 697–698.

Ruffolo RR Jr and Waddell JE (1983) Aromatic and benzylic -hydroxyl substitution of imidazolines and phenethylamines. Differences in activity at alpha-1 and alpha-2 adrenergic receptors. *J Phamacol Exp Ther* **224**: 559–565.

Samama P, Cotecchia S, Costa T and Lefkowitz RJ (1993) A mutation-induced activated state of the β_2-adrenergic receptor: extending the ternary complex model. *J Biol Chem* **268**: 4625–4636.

Smit MJ, Leurs R, Alewijnse AE, *et al* (1996) Inverse agonism of histamine H-2 antagonists accounts for upregulation of spontaneously active histamine H-2 receptors. *Proc Natl Acad Sci USA* **93**: 6802–6807.

Stephenson RP (1956) A modification of receptor theory. *Br J Pharmacol* **11**: 379–393

3.11 FURTHER READING

Limitations of Efficacy Measurement

Colquhoun D (1985) Imprecision in presentation of binding studies. *Trends Pharmacol Sci* **6**: 197.

Negative Efficacy

Costa T, Ogino Y, Munson PJ, Onaran HO and Rodbard D (1992) Drug efficacy at guanine nucleotide-binding regulatory protein-linked receptors: thermodynamic interpretation of negative antagonism and of receptor activity in the absence of ligand. *Mol Pharmacol* **41**: 549–560.

Agonist-specific Active Receptor Conformations

Kenakin TP (1995*a*) Agonist–receptor efficacy I: mechanisms of efficacy and receptor promiscuity. *Trends Pharmacol Sci* **16**: 188–192.

Kenakin TP (1995*b*) Agonist–receptor efficacy II: agonist trafficking of receptor signals. *Trends Pharmacol Sci* **16**: 232–238.

Receptor Active States

Lefkowitz RJ, Cotecchia S, Samama P and Costa T (1994) Constituitive activity of receptors coupled to guanine nucleotide regulatory proteins. *Trends Pharmacol Sci* **14**: 303–306.

4 Pharmacological Assays used in Screening for Therapeutic Ligands

4.1 INTRODUCTION

As noted in the previous three chapters, there are fundamentally two properties of drugs: affinity (presence on the receptor) and efficacy (the functional change in the receptor and its host system that results from drug–receptor interaction). There also are two corresponding approaches to the detection and study of drug–receptor interaction. One is referred to as binding, which measures the presence of a drug on the receptor. The second is referred to as function, and measures the pharmacological consequence of the drug–receptor interaction. The specific instances where it is advantageous to use either one of these approaches over the other are discussed in detail in the next chapter and these depend upon the type of chemical target sought in the discovery process. This chapter will discuss the practical use of both binding and function in the screening of drugs.

In order for either a binding or functional assay to be developed, certain prerequisites must be met. Clearly there must be a cell line available containing the receptor of interest. This can be found naturally or created with molecular biology. The genetic engineering of surrogate cells containing transfected receptors will be discussed specifically later in this chapter. Second, there must be some means of detecting when a ligand is bound to the receptor. In binding studies this is achieved by using radioactive ligands where the bound ligand can be separated from the free ligand. Alternatively, a change of state of the ligand that occurs when it is bound can be detected. For example, when a fluorescent ligand is bound it may emit different wavelengths of light. In functional studies, a physiological response must be detected from the binding of an agonist to a receptor.

4.2 BINDING ASSAYS

In general, binding techniques measure the proportion of a given concentration of ligand bound to the receptor. This usually can be achieved if the kinetics of binding allow for an equilibrium between the ligand and the receptor to be attained and there is a sufficient means of separating the signal emanating from the receptor-bound ligand from the non-receptor-bound ligand (free in solution and that bound to non-specific sites on the membrane). There must be a method to trace and quantify the ligand; usually the ligand is radioactive. Recently fluorescent labels have been introduced that can be distinguished from free solution vs. bound states. These traceable ligands are equilibrated with preparations of receptor (either contained in membrane fragments or present on whole cells in culture) in an appropriate solution for a given time. After this procedure, the bound ligand either is physically separated from the unbound ligand (by filtration; see Fig. 4.1A) or detected via spacial discrimination (scintillation proximity assay (SPA); see Fig. 4.1B) or physical chemical state (fluorescence polarization; see Fig. 4.1C). This procedure is repeated with a concentration of non-radioactive ligand that is known to bind selectively to the receptors at a concentration that almost completely protects the receptor population from interaction with the radioactive ligand. This latter procedure defines the amount of radioligand that is bound to non-receptor sites on the biological preparation (defined as non-specific binding and denoted NSB).

It is optimal if the chemical structure of the ligand that defines NSB is different from the radioligand. Although theoretically a protecting non-radioactive concentration of the same ligand as the radioligand can be used to define NSB, this could lead to the inclusion of non-specific sites in the apparent receptor population because both ligands have the same affinity for receptors and non-specific sites (see Fig. 4.2). If a receptor-selective non-radioactive ligand with a different chemical structure is used to define NSB then, under ideal circumstances, the only common binding between the radioligand and non-radioactive ligand will be the receptor population (see Fig. 4.2). This should yield a more correct estimate of the specific binding to receptors.

Once the basic system is in place for the production and detection of receptor-bound traceable ligand, certain other criteria must be fulfilled before the assay is suitable for the screening of receptor-active compounds. The first prerequisite to be met is that the signal to noise ratio be sufficient to detect receptor-bound ligand from background binding. One method of increasing this is to increase the amount of biological material (i.e. either membrane or cells). However, this may contravene another prerequisite of a useful assay, namely that the binding of the radioligand to receptor should not appreciably affect the amount of free radioligand in solution available for binding. If the amount of bound ligand is large enough to decrease the free ligand concentration, then the Langmuir adsorption isotherm for radioligand binding is not valid. If the amount of radioligand bound to the receptor is insignificant with respect to the free concentration of ligand, then there will be a linear relationship between the bound ligand and the amount of biological material (i.e. membrane, number of cells) added to the assay medium. As the concentration of biological material is increased and the receptor binding begins to deplete the free concentration of radioligand, then the amount of tracer ligand bound to receptor will reach an asymptotic limit, i.e. the initially linear

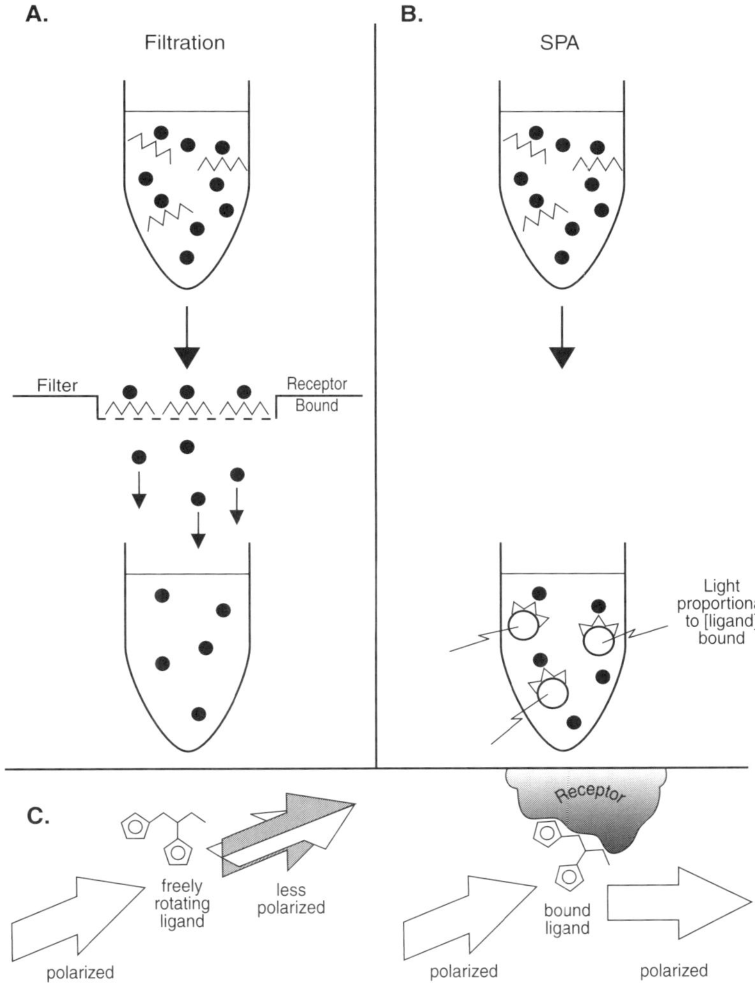

Fig. 4.1. Methods of determining amount of ligand that is bound to a receptor: separating receptor-bound from non-receptor-bound ligand. (A) After equilibration between membrane fragments containing receptors and ligand, the mixture can be filtered, thus washing away unbound free ligand. (B) Scintillation proximity assays (SPA): binding of ligand to the receptor brings the radioligand close enough to a bead containing radioscintillant to cause the emission of light. Free ligand is too far away from the scillintant to cause a signal. (C) Fluorescence polarization: fluorescently-labeled ligand in the medium, through free rotation in solution, reduces the degree of polarization of polarized light passed through the medium. Bound ligand is fixed and preserves polarization, therefore the amount of binding is proportional to the polarization of the light emerging from the assay

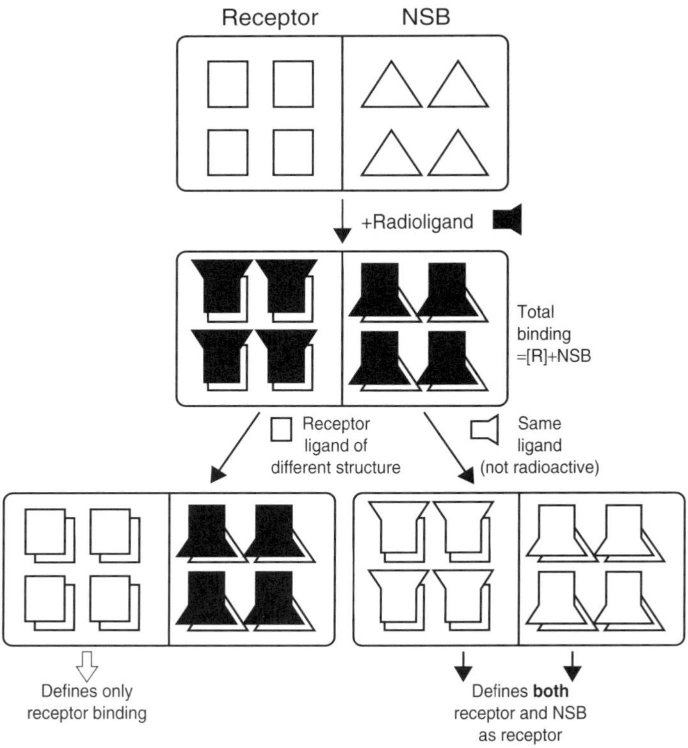

Fig. 4.2. Differentiation of specific receptor-bound ligand from ligand bound to non-receptor sites (non specific binding). Tracer ligand binds to receptors and other non-receptor sites (designated NSB). If a high concentration of a non-radioactive version of the same ligand is used to protect receptors and determine the magnitude of NSB, it is possible that this concentration also will protect some high-affinity non-receptor sites. However, if a receptor ligand of different structure is used to determine specific receptor binding, then there is a greater probability that this ligand also will not protect the non-receptor sites binding the radioligand, and a more accurate estimate of the receptor binding will be made

relationship will become hyperbolic. Fig. 4.3 shows the relationship between the amount of receptor added to the medium and amount of radioligand bound to the receptor for a given concentration of radioligand. For a binding assay to be optimum, the amount of receptor should be kept below the concentration that begins to deplete free ligand; this is in the linear region of Fig 4.3A. If this condition is not met, then the resulting saturation curve ostensibly showing the affinity of the radioligand for the receptor will be shifted to the right of the correct one (see Fig. 4.3B). Demonstrating a linear relationship between the

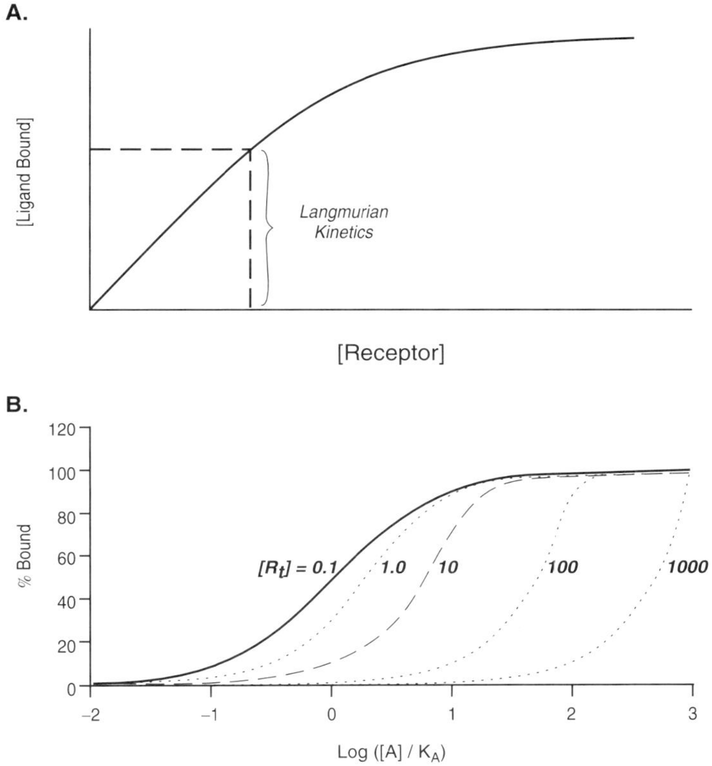

Fig. 4.3. Effect of receptor (membrane or cell) concentration on specific binding to receptors. (A) As the amount of membrane becomes large enough to deplete the concentration of radioligand, the binding will diminish. An asymptote in binding will be reached when the amount of receptor takes up all of the radioligand (and large errors in the binding estimates will result). In the linear region of this curve, the amount of ligand bound does not appreciably alter the free concentration of ligand; this is a prerequisite for the Langmuir model of binding to receptors. All binding experiments should demonstrate a linear relationship between the amount of membrane in the samples and radioligand binding. (B) The effect of depletion of radioligand by receptor binding on the saturation binding curve of the ligand. As the concentration of receptor ($[R_t]$) begins to deplete the ligand, the saturation curves become steeper and shifts to the right

receptor concentration and bound ligand is a good way to ensure that the receptor binding is not depleting the free ligand concentration.

Another prerequisite for a useful binding assay is that the radioligand be displaceable and that the kinetics of binding be rapid enough to have the assay come to equilibrium within a reasonable amount of time for the experiment. The first-order rate of onset for binding is given by:

$$\rho_t = \rho_e[1 - e^{-(k_1[\mathrm{A}]+\mathrm{k}_2)\mathrm{t}}] \qquad \ldots(4.1)$$

where ρ_t and ρ_e refer to the fractional amount of ligand concentration [A] bound at time t and equilibrium e, respectively and k_1 and k_2 refer to respective rates of onset and offset of the ligand. Upon washing with drug-free medium, the ligand should dissociate from the receptor according to the equation:

$$\rho_t = \rho_e e^{-k_2 t} \qquad \ldots(4.2)$$

Thus, according to eqs. 4.1 and 4.2, the amount of ligand bound with time upon addition to the receptor-containing medium is shown in Fig. 4.4A. Upon washing with drug-free medium, the temporal relationship for the binding also is shown in Fig. 4.4A. Binding experiments should only be done when equilibrium has been achieved. Typically, a timecourse for radioligand binding is carried out and a minimal equilibration time is determined (i.e. reagents must be incubated for a period no less than 60 min for equilibrium to be attained). However, these estimations can still be in error because binding experiments aimed at screening ligands observe a displacement of the radioligand by non-radioactive ligands, and the time required for two ligands to come to equilibrium with the receptor can be considerably longer than the time required for only the radioligand. The equation for the equilibration of two ligands ([A] and [B]) with a receptor is given by (Motulsky and Mahan 1984):

$$\rho_t = \frac{k_1[\mathrm{R_t}][\mathrm{A}]}{K_\mathrm{F} - K_\mathrm{S}}\left[\frac{k_4(K_\mathrm{F} - K_\mathrm{S})}{K_\mathrm{F}K_\mathrm{S}} + \frac{(k_4 - K_\mathrm{F})e^{-K_\mathrm{F}t}}{K_\mathrm{F}} - \frac{(k_4 - K_\mathrm{S})e^{-K_\mathrm{S}t}}{K_\mathrm{S}}\right] \qquad \ldots(4.3)$$

where

$$K_\mathrm{A} = k_1[\mathrm{A}] + K_2$$

$$K_\mathrm{B} = k_3[\mathrm{B}] + k_4$$

$$K_\mathrm{F} = 0.5\left\{([K_\mathrm{A}] + K_\mathrm{B}) + ((K_\mathrm{A} - K_\mathrm{B})^2 + 4k_1k_3[\mathrm{A}][\mathrm{B}])^{1/2}\right\} k_4$$

$$K_\mathrm{S} = 0.5\left\{([K_\mathrm{A}] + K_\mathrm{B}) - ((K_\mathrm{A} - K_\mathrm{B})^2 + 4k_1k_3[\mathrm{A}][\mathrm{B}])^{1/2}\right\} k_4$$

and where k_1 and k_2 refer to the respective rates of onset and offset for ligand A and k_3 and k_4 the same for ligand B. It can be seen from eq. 4.3 that, depending on the magnitudes of k_3 and k_4, the overall rate of equilibration for the two ligands can be considerably greater than that determined by just k_1 and k_2. Figure 4.4B shows the time required for equilibration between receptor and ligand A (also shown by itself) and another ligand B with a rate of offset 1/10 that of the radioligand. As shown in this figure, the time determined for the binding of just ligand A considerably underestimates the time required for the two ligands to come to equilibrium.

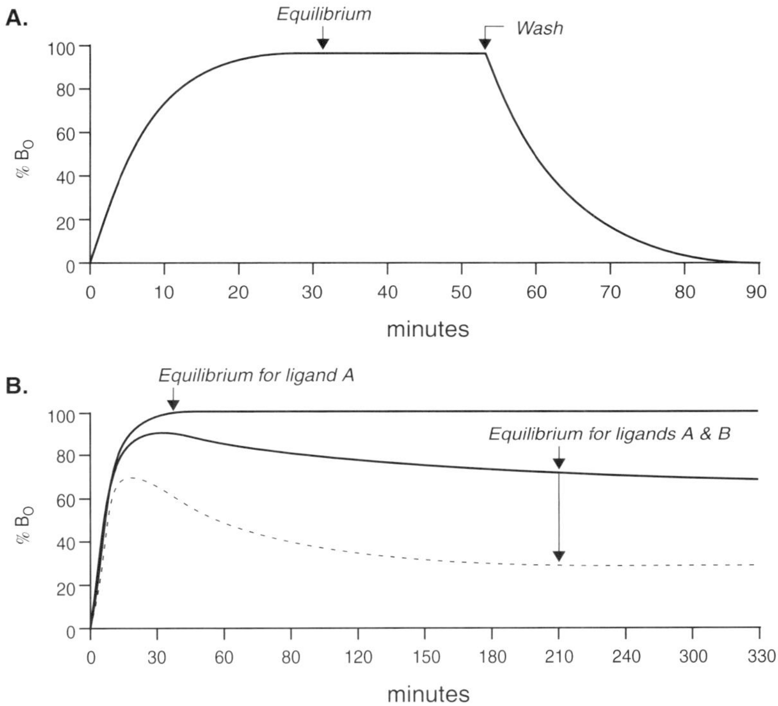

Fig. 4.4. Temporal approach to equilibrium with a single ligand and two ligands in a binding assay. (A) First-order binding of ligand to receptor results in a hyperbolically shaped onset curve. Equilibrium is denoted by the asymptote where no further increase in time leads to increased binding. Reversibility is demonstrated by observation of the decline in binding upon washing with drug-free medium. (B) The time required for attainment of equilibrium with one ligand (curve marked A only) can be lengthened dramatically in the presence of another ligand with a slower rate of onset (curves A + B): top curve is for ligand A alone; the next curve is for equal concentrations of A and B, the bottom curve is [A] + 2[B]

It is essential that this be considered in binding experiments because robotic and high-throughput format assays premix the radioligand and non-radioactive ligand in the well and then begin the binding reaction with addition of the receptor. Thus, all reagents must come to temporal equilibrium together. It can be seen that this timespan for two ligands can be quite different from the timespan required for only one (Fig. 4.4B).

Another optimal condition for a binding assay is that the amount of specific binding be as high as possible and much greater than the NSB. Although a window of specific binding of 1200 counts per minute (cpm) is sufficient, there is a difference if this window resides in a preparation with a high vs. a low degree

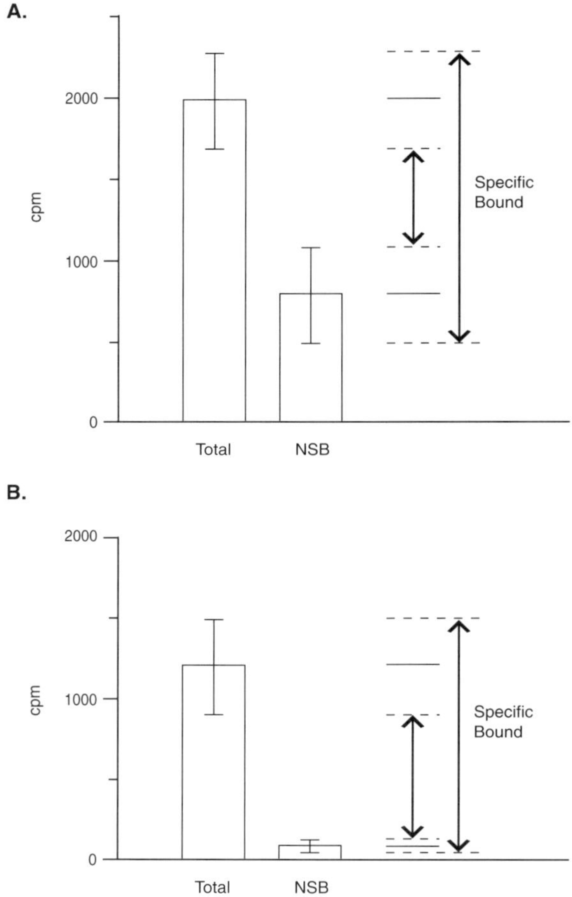

Fig. 4.5. Effects of random error on the specific binding window in binding assays. The window for specific binding is 1200 cpm under two different experimental conditions. (A) Total binding of 2000 cpm and a non-specific binding (NSB) of 800 cpm causes a large error in the specific binding window. (B) In contrast, when the total binding is 1280 cpm and the NSB is 80 cpm, the 15% random error in binding causes a lower degree of uncertainty in the specific binding window

of NSB. For example, if the total binding is 2000 cpm and NSB is 800 cpm, and there is a 15% noise level on the NSB, then the corresponding error on the specific binding window is +120 cpm (see Fig. 4.5A). However, if the total binding is 1280 cpm and NSB is 80 cpm, then the corresponding error on the specific binding window is only 12 cpm (see Fig. 4.5B). In general, the major prerequisite and optimization factors for a useful binding assay are reviewed in Table 4.1.

Table 4.1. Prerequisite and optimization factors for binding assays

Prerequisites
- There is an available radioligand or sufficient material to custom radiolabel (the labeling procedure does not alter biological activity)
- The binding is saturable and reversible (and displaceable)
- There is a ligand available to determine non-specific binding
- The amount of biological material does not significantly depelete the radioligand

Optimal
- The kinetics of binding are rapid
- There is high specific binding and low non-specific binding
- An agonist and antagonist radioligand are available
- The ligand used for determining non-specific binding is of a different chemical structure than the radioligand

Binding is ideal for high-throughput screening if there is a sufficiently high window of specific receptor binding and the kinetics of binding are reasonable. A concentration of tracer ligand is chosen and compounds are added to determine if the binding of the ligand in the presence of the compound is less than in the absence of compound (presumably due to competition at the receptor). The characteristics of the binding of the radioligand are determined by a saturation curve. Thus, the total binding (comprised of receptor plus non-specific sites) is measured for a concentration range of the radioligand. This can be described by the equation:

$$\text{Total binding} = \frac{[\mathrm{A}]B_{\max}}{[\mathrm{A}] + K_{\mathrm{d}}} + c[\mathrm{A}] \qquad \ldots(4.4)$$

where [A] is the concentation of radioligand, K_d is the equilibrium dissociation constant of the radioligand–receptor complex (see Chapter 2) and c is a constant determining the extent of NSB. It can be seen from eq. 4.4 that total binding is the sum of the specific binding to receptors (first half of the equation) and a linear unsaturable binding to a host of other non-specific sites.

This process is repeated in the presence of a non-radioactive ligand known to bind to the receptor (to protect the receptors from binding radioactive ligand) to determine the extent of NSB at each concentration of radioligand. The observed binding is linear and described by the equation:

$$\mathrm{NSB} = c[\mathrm{A}] \qquad \ldots(4.5)$$

One method of determining the specific binding is to subtract the NSB from the total binding. However, this does not make maximum use of all of the datapoints, and a more efficient method is to fit the total binding and NSB simultaneously to eqs. 4.4 and 4.5; this utilizes the complete dataset. An example of the type of data generated for a saturation curve is shown in Fig. 4.6.

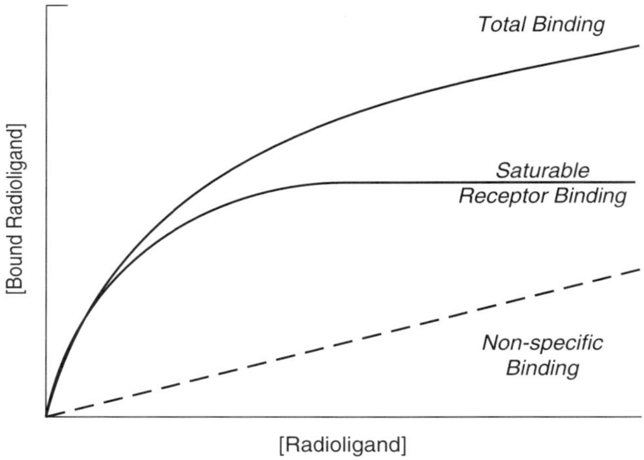

Fig. 4.6. The total binding is the sum of the non-specific binding (NSB) and receptor binding; the function is an amalgam of a hyperbolic function and a straight line of positive slope (eq. 4.4). In the presence of a protecting concentration of ligand to prevent the radioligand from binding to receptors, the NSB is measured as a straight line of positive slope. Subtraction of the NSB from total binding (or simultaneous fitting of the data points to the two equations for total binding and NSB) reveals the saturable specific binding to the receptor population

Once the binding characteristics of the radioligand are determined, the interference of the binding of radioligand by foreign non-radioactive ligands is determined. Thus, a suitable concentration of radioligand is chosen (preferably of lower magnitude than the K_d; see Chapter 2 for discussion) and the effect of test ligands on tracer ligand is observed. The interpretation of the resulting data is discussed more fully in Chapter 2.

4.3 FUNCTIONAL ASSAYS

Another approach to the screening of biologically active molecules is to use a functional assay, i.e. an assay that either seeks molecules that produce physiological response or molecules that block the effects of agonists that produce physiological response. There are numerous possibilities for functional assay preparations. The earliest *in vitro* functional assays were isolated tissues, usually from animals. Thus, isolated atria, ileum, seminal vesicle, blood vessels and numerous other organs were placed in environments that preserved biological function (i.e. organ baths filled with warmed, oxygenated medium containing essential nutrients at physiological pH). These tissues could survive for the periods required for *in vitro* experiments, usually carried out according to null techniques (see Chapter 1).

The high-volume screening required for the testing of large numbers of molecules for possible biological activity precludes the use of isolated tissues. Also, the advanced technology that allows the expression of human receptors in surrogate cells (*vide infra*) obviates the potential problems of testing molecules for use in human therapeutic situations on animal biological targets. Therefore, the cell-based testing of molecules for biological function has now supplanted the use of isolated animal tissues. The creation of recombinant functional screening systems is discussed specifically at the end of this chapter. This present discussion is concerned with the general pharmacological requirements of a functional assay for screening.

In general, a functional assay must have a cell line that produces a measurable response to a reference agonist in a realistic timeframe for screening. The agonist response should be reversible with washing (with drug-free medium) and, ideally, a reference antagonist should be available to test for specificity of receptor effect. Also, a secondary agonist for the testing of biological viability of the cell response system should be in place, i.e. if an antagonist of the target agonist is found, there should be another agonist for a different receptor available to show that the screened potentially active antagonist is not simply toxic to the host cell (see Table 4.2).

There are numerous vantage points available to view the functional effects of agonists in receptor systems. Fig. 4.7 shows some possible points along the stimulus–response chain that can be used to observe physiological response. These vantage points can be described in four general regions. There are advantages and disadvantages to viewing response closer and farther from the point of receptor stimulation. The advantage of viewing response nearer the receptor stimulus is that there is less amplification and a greater texture in efficacy is obtained, i.e. differences in maximal response show differences in the efficacy of agonists (see Chapter 3, Fig. 3.6). However, that very advantage can also be a disadvantage because this region is least amplified and therefore the assay will not be nearly as sensitive to weakly efficacious agonists. Similarly, these arguments apply in reverse to viewing response at the end of stimulus–response, namely that assays

Table 4.2. Prerequisite and optimization factors for functional assays

Prerequisities

- An agonist and selective antagonist are available
- The agonist is reversible

Optimal

- The response should be sustained, not transient
- The response should be rapid
- There are methods available to potentiate cellular response
- There are cells available with low (with respect to G-protein content) receptor density

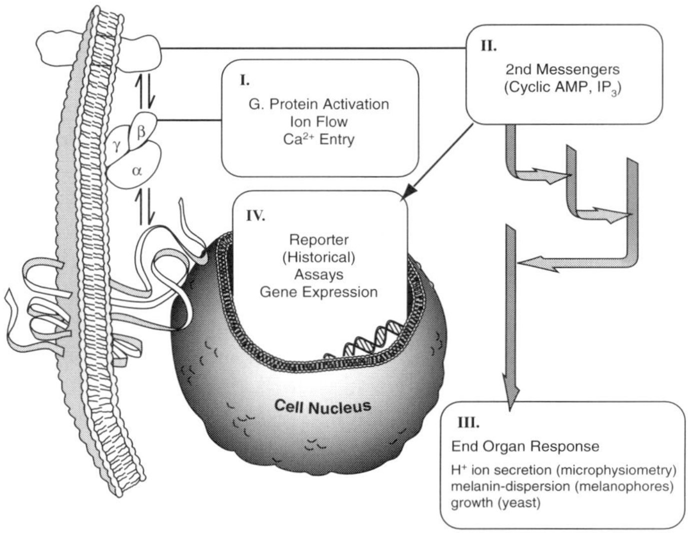

Fig. 4.7. Membrane receptors mediate cytosolic functions. The results of receptor activation can be measured at various points along the stimulus–response chain of reactions from receptor activation to end organ response. (I) The first point of measurement is the extent of activation of the first protein to interact with the activated receptor. In the case of G-protein-coupled receptors, this is the G-protein. The escalated exchange of GDP for GTP on the G-protein, in response to interaction with the activated receptor, can be measured. Ion flow also can be measured at this point. (II) Another point of measurement is the magnitude of second messenger produced by the receptor interaction (i.e. cyclic AMP, IP_3, Ca^{2+} mobilization, etc.). (III) The effect of these cascades on cellular function can be measured. Examples of such processes are the rate of hydrogen ion secretion (through microphysiometry), melanin dispersion in melanophores or yeast growth. (IV) A fourth mode of response measurement historically records the results the receptor interaction (i.e. through logging of the effects of second messenger elevation) through specialized reporter genes responsive to the second messenger

with this response readout are capable of detecting weak agonists that the former type of assay would show to be antagonists (see Fig. 4.8).

The first region of response detection is the first interaction of the activated receptor with the next component of the stimulus-response cascade. If G-protein-coupled receptors are the biological target, then the exchange of GDP for GTP can be utilized as the first step in the activation of G-protein coupled receptors (see Fig. 4.7). Other vantage points can be found along the stimulus–response chain of cytosolic reactions. The result of G-protein and effector, activation namely the production of second messengers such as cyclic AMP or IP_3, can be monitored. Another vantage point is the end organ response. For isolated tissues, this is usually the only method of determining response. With the advent of

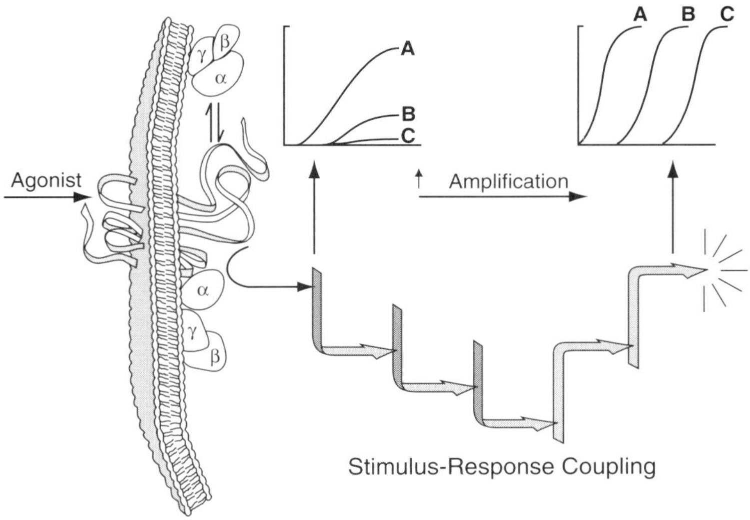

Fig. 4.8. Amplification associated with vantage points. If the receptor-mediated response is viewed near the point of ligand–receptor interaction, then little amplification will be observed and the efficacy of the ligand may determine the maximal response. However, the ligand may not demonstrate high potency. If the end organ response to the ligand is observed, then a series of amplification processes may increase the potency of the ligand but there will be a concomitant loss in texture of efficacy in terms of maximal response (i.e. the maximal response may not reflect the efficacy and all ligands may appear to be full agonists). Note how the least potent ligand (agonist C) produces no measurable response near the receptor interaction but does produce response after stimulus–response amplification

technological advance comes the ability to measure responses in single cells either by measuring electrophysiological events or other types of responses such as respiration. For example, microphysiometry allows the measurement of minute differences in the rate of hydrogen ion secretion by cells, which is tightly linked to the cellular rate of metabolism. A major advantage of this technique is that basically any perturbation of cellular metabolism will result in either an increase or a decrease in hydrogen ion secretion, therefore it is nearly a universal detection system for receptor-mediated responses. The coupling of this technique to recombinant cell technology enables the transient or stable transfection of any type of receptor into surrogate host cell lines for the production of a functional assay. Microphysiometry has numerous applications, including the detection of receptor expression. Figure 4.9A shows the lack of response to calcitonin in wild-type HEK 293 and the subsequent observation of response to human calcitonin in cells transiently transfected with human calcitonin receptor. Figure 4.9B shows a complete dose–response curve to human calcitonin in the same cells.

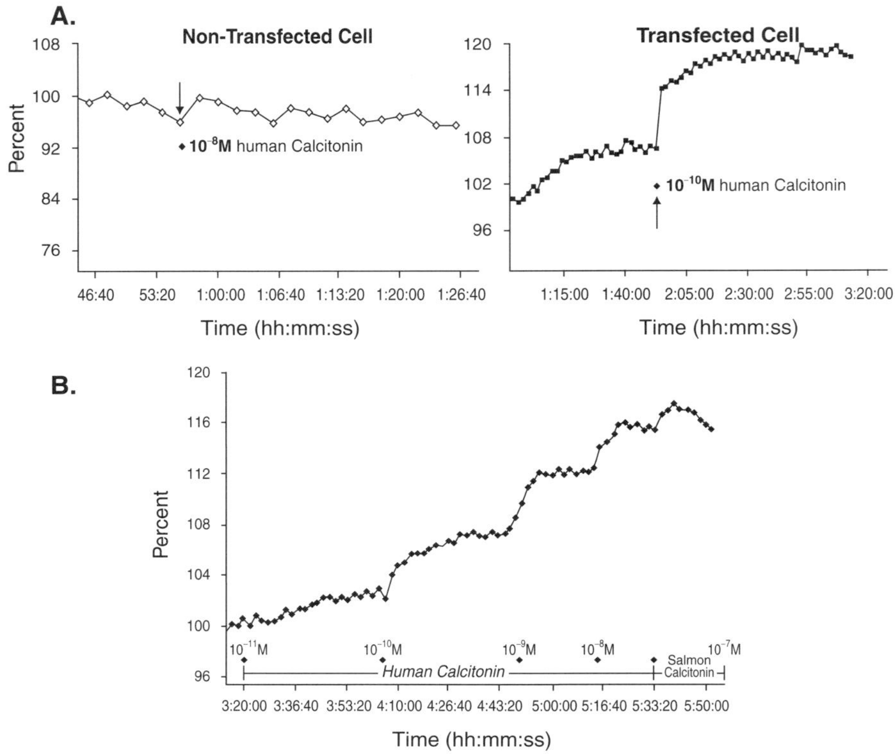

Fig. 4.9. Microphysiometry. Perturbation of cellular function causes a change in the hydrogen ion secretion of cells in culture. (A) In HEK 293 cells, calcitonin produces no agonist response. However, after transient transfection with human calcitonin cDNA, calcitonin produces a change in the metabolic rate of transfected HEK 293 cells, indicating the expression of human calcitonin receptors on the cell membrane. (B) A dose–response curve in real-time is produced by increasing concentrations of calcitonin in HEK cells transfected with human calcitonin receptors

Although microphysiometry is an excellent cell-based assay system for secondary testing, it does not yet have the capability of processing the large volumes of samples required for screening. This latter process requires high-volume sampling, usually accomplished through miniaturization and automation. Although binding is amenable to such screening, there are also functional systems adaptable; some are naturally suited for screening, such as *Xenopus laevi* melanophores, and others can be adapted for screening through recombinant genetics (i.e. yeast cells, mammalian reporters).

End organ response can be observed in specialized cells to advantage. For example, *Xenopus laevi* melanophores have a natural ability to translocate melanin in response to second messengers such as cyclic AMP. Thus, increases in cyclic AMP cause the dispersion of melanin, whereas the lowering of cytosolic cyclic AMP levels causes a granulation of melanin (see Fig. 4.10A). The dispersion of melanin in these cells causes them to become pigmented or transparent, a response that is readily quantified by measurement of light transmittance through a layer of cells. Melanophores also are easily transfected with cDNA for receptors, thus recombinant response systems can be obtained. These effects can be seen in real time; Fig. 4.10B shows the response to human calcitonin in melanophores transfected with human calcitonin receptor. The complete dose–response curve to calcitonin is shown in Fig. 4.10C. A major advantage of this approach is that it is readily amenable to high-volume screening, because plates with numerous wells containing cell monolayers can be used for measurement of light transmittance (Karne *et al* 1993; Potenza *et al* 1992, 1994; Graminski *et al* 1993, 1994; McClintock *et al* 1993; Jayawickreme *et al* 1994*a*, *b*; Lerner 1994).

Another system amenable to screening are yeast cells containing genetic modification that allows the induction of growth in response to transfected G-protein-coupled receptors (Price *et al* 1995, 1996; Pausch 1997). Yeast cells have the advantage that they grow rapidly, have a sparse receptor background (low noise) and low cost of maintainance. Yeast cells can be altered genetically such that they do not grow in medium not containing histidine unless the receptor, previously transfected, is activated. Growth zones of agonist-stimulated growth can be detected either in an agar matrix or by changes in optical density in liquid media. Cotransfection of different G-proteins further increases the flexibility of this approach.

4.3.1 REAL-TIME, STOP-TIME AND HISTORICAL ASSAYS

There are basically three temporal modes for receptor assays and all of them can apply to functional studies. Real-time assays show the response as it happens. An example of this was shown in Fig. 4.10B; the response is produced and observed in the same timeframe by the onset of the agonist on the receptor. Real-time assays have the advantage of allowing the observation of equilibrium instead of assuming that it has been reached. A disadvantage, for screening purposes, is the difficulty of viewing the approach to equilibrium of multiple reaction wells in functional studies.

Stop-time assays allow the reagents (agonists, receptors, response elements, antagonists, etc.) to come to equilibrium (or steady state) in some time period and then the results can be viewed. Binding studies utilize this approach (see Fig. 4.4). This mode is acceptable as long as sufficient time is allowed for equilibrium to occur.

The third type of response measurement available for screening in functional assays is the historical (or reporter) assay. Historical assays record the amount of reaction product from the beginning of the experiment (mixture of reagents) and accumulate this record throughout the course of the experiment. Reporter

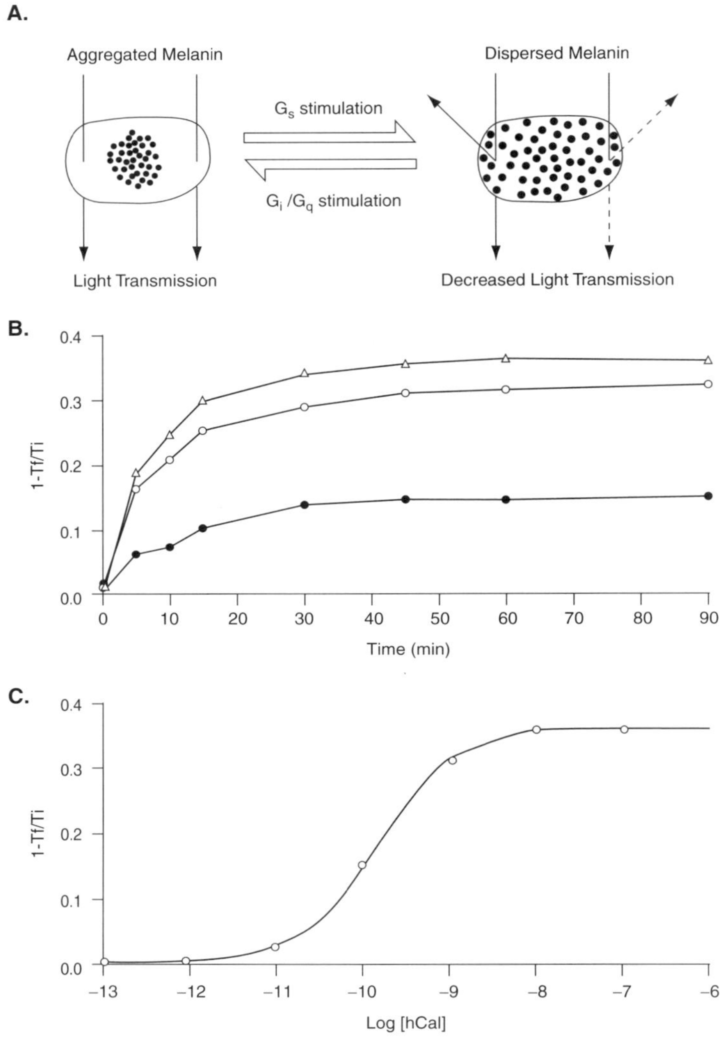

Fig. 4.10. High-throughput screening with *Xenopus laevi* melanophores. (A) Activation of Gi- or Gq-protein produces aggregation of melanin. The melanophore is transparent to visible light. Activation of Gs-protein causes melanin dispersion, making the cell opaque. The relative activity of these G-proteins can be monitored in real-time by measurement of light transmission. (B) Reduction in light transmission produced by human calcitonin activating a transfected calcitonin receptor in melanophores. Responses are shown to 0.1 nM (filled circles), 1 nM (open circles) and 10 nM (open triangles) human calcitonin. (C) Dose–response curve to human calcitonin (hCal) in a recombinant calcitonin receptor system in melanophores

assays make use of the fact that some second messengers, such as cyclic AMP, activate protein kinase A to release catalytic subunits that act in the nucleus to phosphorylate substrate proteins. Some of these proteins, such as cyclic AMP response element binding protein, subsequently bind to an 8-base-pair palindrome sequence called cyclic AMP response element (CRE; see Fig. 4.11A). This element is located in the p-promoter region of cyclic AMP-inducible genes and results in an increase in transcription. For example, in cells transfected with CRE-luciferase promoter, cyclic AMP causes an increase in the expression of luciferase, a protein that can be detected through the production of light with appropriate substrate (Fig. 4.11A). Under these circumstances, all of the cyclic AMP produced by the receptor activation from the beginning of the experiment has been used to produce luciferase; the response is the area under the curve described by the drug–receptor interaction (Fig. 4.11B). These types of assay can be exquisitely sensitive because great amplification of weak signals can be gained by letting the reaction run for sufficient periods of time. On the other hand, this mode also is particularly sensitive to the kinetics of response because temporal inequilibrium produces a lasting interference in the observed response. There are two instances where this can lead to serious artifacts.

The first is the effect of transient agonist responses on the readout of a historical assay. Because the total response measured is the area under the curve, and if the response peaks but wains and is not sustained, then this area will be considerably smaller than the corresponding response to an agonist with a sustained response. Thus, the apparent magnitude of the response to an agonist will also depend on the kinetics of the agonist. This effect is exacerbated by the presence of a basal response in the preparation. Figure 4.12 shows the temporal response to an agonist with sustained kinetics; the bar graph shows the magnitude of the area under the curve. Figure 4.12 also shows the response to an agonist producing a transient response; as can be seen from the bar graph, the resulting response in a historical reporter assay shows a relatively insignificant effect.

Differences in the transient nature of response with concentration also can result in a lack of a dose–response curve for agonists. Figure 4.13A shows the area under the curves (as a response readout in a historical assay) for an agonist that produces sustained responses; a dose–response relationship is observed. On the contrary, an agonist that produces a transient response may demonstrate no visible dose–response relationship in a historical assay. For agonists that produce a transient response, the transitory nature of the response often is related to the magnitude of the response. Thus, the response to low concentrations of agonist (which tend to be more sustained) may produce as much product as much higher concentrations of the same agonist if the responses to the higher doses are transient. Under these conditions, no obvious relationship between dose and response will be obtained. This can be detected in functional studies by obtaining responses at different time points. If the production of the

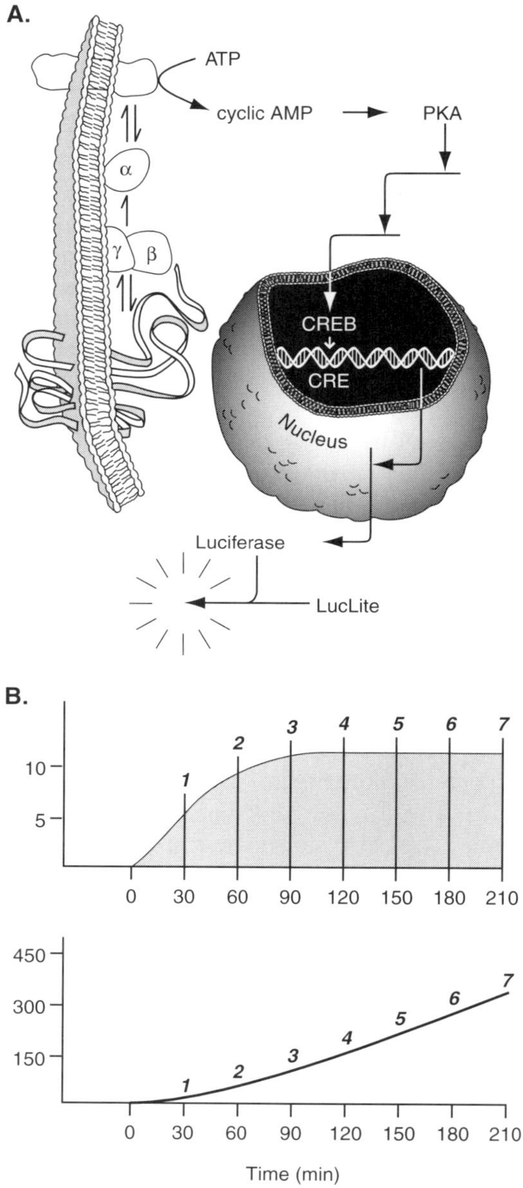
A.
ATP
cyclic AMP
PKA
α
γ
β
CREB
CRE
Nucleus
Luciferase
LucLite
B.
1
2
3
4
5
6
7
10
5
0
30
60
90
120
150
180
210
450
300
150
Time (min)

receptor–product is linear with time, then no change in the location parameter of the dose–response to the agonist should be observed (see Fig. 4.13B for example).

4.4 LIGANDS FOR ORPHAN RECEPTORS

As the study of the human genome progresses, new receptor targets, found from homology to other receptors, will be discovered. These receptors form the bulk of the new therapeutic targets of the future. The endogenous ligand for these new receptors will not be known (hence the term 'orphan' for these receptors) and this absence of knowledge can affect how these new targets are screened for therapeutic ligands. Conventional screens for seven-transmembrane (7TM) receptors are reactive in that they chemically induce signals with ligands and seek changes in the resulting signal with unknown ligands. This approach requires prior knowledge of the probing ligand, i.e. there must be a radioligand to displace or an agonist signal to block. By definition, the probing ligands for orphan receptors are unknown, therefore another strategy must be found to screen these targets. One approach would be to place the orphan receptor in an appropriate responding system and search chemical space for ligands that induce response. This reduces the potential hits to those that both bind to and activate the receptor (i.e. agonists). Theoretically, one method whereby *all* ligands that bind to the receptor will be detected is with constitutive receptor assays.

4.4.1 SCREENING ORPHAN RECEPTORS WITH CONSTITUTIVE RECEPTOR ASSAYS

As discussed in Chapter 3, receptor conformations that activate response systems are part of a library of spontaneously formed states and the ligands can enrich the natural proportion of the active state by selectively binding to it. An important consequence of this idea is that the various conformation states of the receptor

---

Fig. 4.11. The CRE–Luciferase reporter assay. Receptor activation by an agonist results in production of cyclic AMP, which in turn activates protein kinase A (PKA). Active subunits then diffuse to the cell nucleus and bind to cyclic AMP response element binding (CREB) protein, which then binds to a promoter region of cyclic AMP-inducible genes. This results in increased transcription in response to cyclic AMP. Cells can be made to induce transcription of the protein luciferase, which, when brought into contact with an appropriate substrate (LucLite), produces light. The amount of light is directly dependent upon the amount of luciferase produced, which in turn is directly dependent upon the amount of cyclic AMP elevation (receptor-induced stimulus). (B) The amount of luciferase produced is represented by the area under the curve of the upper tracing. These areas, as a function of incubation time, are shown in the lower tracing. After an initial non-linear portion, the function is linear. Responses should be taken only from the linear portion of this curve

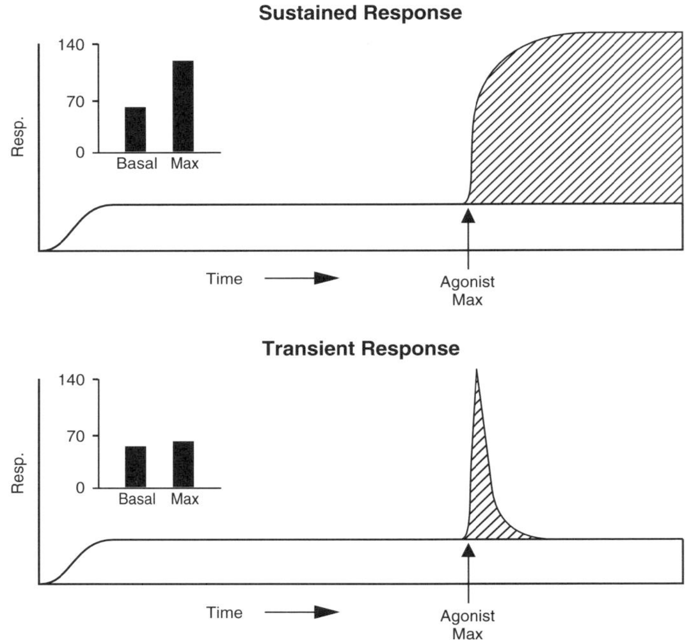

Fig. 4.12. Responses (Resp.) to agonists in stop-time mode. Top tracing shows the elevation of luciferase production in real-time. The area under the curve represents the amount of product formed. Response is taken to be the product of the agonist–receptor reaction (e.g. amount of second messenger, luciferase generated by reporter gene, etc.) and is measured as the area under the curve. A sustained response produces an increased area substantially larger than the baseline response. In contrast, the bottom tracing shows a transient response; the area under the curve is nearly insignificant compared to the baseline response

exist in an interactive thermodynamic equilibrium with each other and with G-proteins. This is shown schematically in a model of 7TM receptor systems (see Fig. 4.14A). Thus the inactive (R_i) and active state receptor (R_a) coexist in an equilibrium and the active state goes on to couple to G-protein spontaneously, with the consequence that cellular response ensues. Systems can be perturbed to the point where a sizable amount of constitutively active complex R_aG is formed and such systems demonstrate constitutive receptor activity. If the relative proportion of *any* of various receptor species is changed, then the species will redistribute according to the equilibrium constants for the system and a change in R_aG will result (see Chapter 3, eq. 3.3). Because R_aG produces a measurable response, the change in receptor distribution will become evident as either an increase or a decrease in constitutive activity of the system.

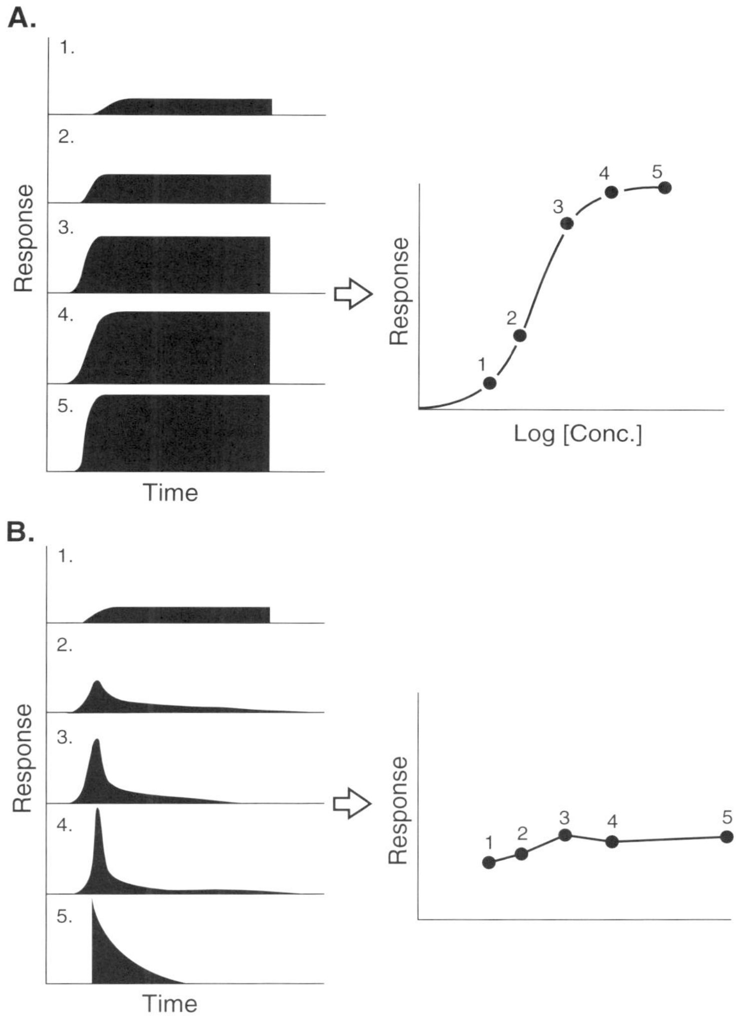

Fig. 4.13. Lack of dose–response relationships for transient agonists in stop-time assays. (A) Agonists that produce sustained responses yield areas that reflect the peak height response. There is a linear increase in area in response to a steady-state agonist response. (B) Poor reflection of transient response as measured by lack of increase in area. There is no linear increase in measured product response with time. If the shape of the curve or the location changes with a change in the time of exposure, this suggests that the production of the response is non-linear. Under these circumstances it is possible that the responses to the agonist are transient (tachyphylaxis occurs). Although this is a problem for characterizing the potency of agonists, it should not hamper the detection of agonists in screening mode

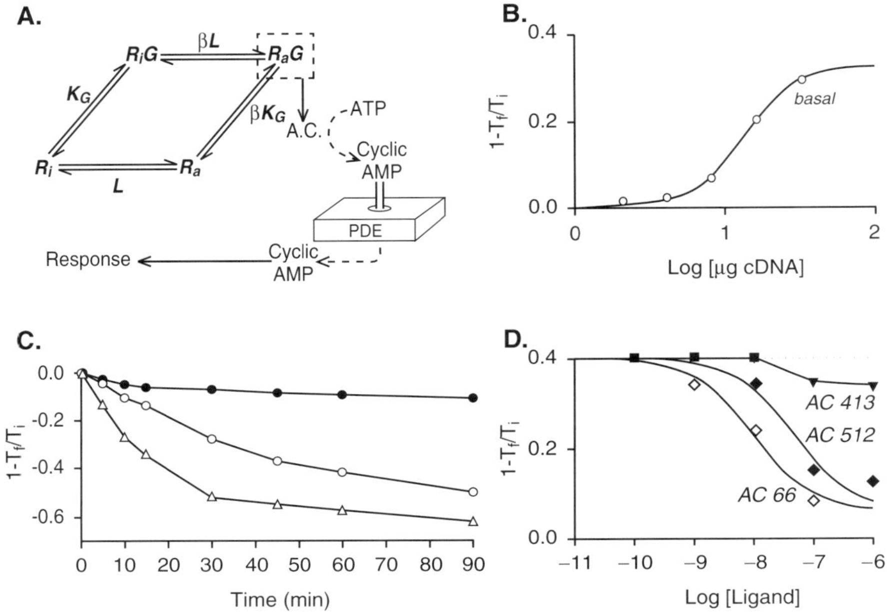

Fig. 4.14. The cubic ternary complex model for seven-transmembrane (7TM) receptors. The inactive receptor R_i can spontaneously form an active state R_a that then goes on to couple to G-proteins to form a complex R_aG. The system constants are the allosteric constant *L*, K_G, and β. For Gs-coupled receptors, formation of R_aG leads to activation of adenylate cyclase and a resulting increase in levels of cyclic AMP. In melanophores this leads to melanin dispersion (observable decreased light transmission) PDE = phosphodiesterase. (B) Increases in basal melanin dispersion due to increased transient transfection of human calcitonin receptor. Response results from increased formation of spontaneously activated receptor with increasing concentrations of receptor. (C) Inhibition of constitutive response (produced by spontaneously activated human calcitonin receptor) by the inverse agonist AC512: 10 nM (filled circles), 100 nM (open circles) and 1 μM (open triangles) in real-time. (D) Dose–response curves to the inverse agonists AC413, AC512 and AC66 in melanophores transfected with human calcitonin receptor under conditions of constitutive receptor activity.

An ideal system for the study of constitutively active G-protein-coupled receptors is the melanophore. All 7TM receptors known to couple to Gs, Gi and Gq that have been transfected into the melanophore system have induced changes in pigment dispersion. Melanophore pigmentation is a sensitive indicator of G-protein activation, which in turn is an indicator of G-Protein-coupled receptor activation state. A major hurdle to developing a constitutive active screening system is the detection of elevated basal activity. Melanophores are ideal from this standpoint because the expression level is controlled by the quantity of DNA transiently exposed to the cells and the resulting constitutive activity (or lack of it) is visualized directly. Thus, the cells can be exposed to a range of orphan receptor DNA concentrations and a sensitive constitutively active system is chosen. For example, Fig. 4.14B shows the constitutive activity produced by transient expression of various levels of human calcitonin receptor. The relevance of such a situation to screening stems from the fact that, unless a ligand recognizes the species R_i, R_a and R_aG as indistinguishable (and for this to occur the equilibrium constant factors α, γ and δ must be unity), then selective binding (the consequence being stabilization) to any of the states will cause redistribution of the states and a change in constitutive activity. In fact, there is a thermodynamic rationale for the assumption that most antagonists will be found to be inverse agonists (Kenakin 1994) in that a differential affinity of a ligand for a range of receptor conformations would be *expected* (Kenakin *et al* 1995). For example, Fig. 4.14C shows the real-time depression of constitutive calcitonin receptor activity in melanophores transfected with human calcitonin receptor by the inverse agonist AC512. Figure 4.14D shows the dose–response curves to three inverse agonists in the same constitutively active system. A partial list of receptors for which inverse agonists have already been found is given in Table 4.3.

It should be noted that some receptors are more prone than others to form the active state spontaneously. For example, constitutive activity in COS cells

Table 4.3. Receptors for which apparent antagonists were found to be inverse agonists in constitutively active systems

Receptor type	Reference
β_1-Adrenergic	Mewes *et al*, 1993
β_2-Adrenergic	Samama *et al*, 1993
	Bond *et al*, 1995
	Chidiac *et al*, 1994
	Gotze and Jakobs, 1994
α_2-Adrenergic	Tian *et al*, 1994
	Jagadeesh *et al*, 1990
	Jagadeesh and Deth, 1992
	Shi and Deth, 1994
α_1-Adrenergic	Cotecchia *et al*, 1995
Muscarinic	Hanf *et al*, 1993
	Spalding *et al*, 1995
Opioid	Costa and Herz, 1989
Dopamine	DeLean *et al*, 1982
Adenosine	Freissmuth *et al*, 1991
Bradykinin	Leeb-Lundberg *et al*, 1994
Serotonergic	Barker *et al*, 1994
	Westphal and Sanders-Bush, 1994
	Lebrecque *et al*, 1995

expressing various levels of thyrotropin receptor is very tightly connected to receptor expression level (Van Sande *et al* 1995). In contrast, other 7TM receptors can be expressed to high membrane density with little visible resulting constitutive activity.

A useful system then would be defined as having elevated basal but submaximal activity (to detect possible positive agonists). Thus, ligands interacting with the orphan receptor would either increase or decrease basal activity and this could be visualized in a high throughput format. If constitutive activity is not observed (with DNA transfection), this would suggest that either the stimulus–response machinery contained in the melanophore is unsuitable for the detection of orphan receptor activation or that the propensity of the orphan receptor to form spontaneously the active state is low. In general, current evidence strongly suggests that it is highly unlikely that a 7TM receptor would not produce responses in this system. In the case of a high thermodynamic barrier to active state formation, a strategy of random saturation mutagenesis could be employed (*vide infra*).

4.4.2 CONSTRUCTION OF CONSTITUTIVE RECEPTOR ASSAYS

There are three basic approaches that have been taken to make 7TM receptor systems constitutive. The simplest is to manipulate the stoichiometry between receptors and G-proteins. As seen in Fig. 4.14A, there is an intrinsic allosteric constant (referred to as L) that controls the spontaneous formation of the active state from R_i (Weiss *et al* 1996a, b). Therefore, most likely there is always a low level of R_a and subsequently R_aG present in the system. This may or may not be detectable. If the amount of receptor is increased or the amount of G-protein increased, this necessarily will increase R_aG according to L and βK_G. This, in turn, usually results in an observable constitutive cellular activity emanating from the receptor. It should be noted that, because L and βK_G control this effect, how constitutive a system is depends upon both the type of receptor (L) and the host cell (what G-protein is present, βK_G).

A second approach that works with some receptors is actually to manipulate biochemically L. Thus, Costa and Herz (1989) showed that removal of sodium ion made opioid receptor very constitutively active. This has been observed also with α_2-adrenergic receptors (Tian *et al* 1994).

A third approach is to produce a receptor mutant. There is evidence to suggest that the inactive form of the receptor is the special conformation and that disruption of total receptor structure with a mutation lowers the energy barrier to the spontaneous formation of R_a. For example, Fig. 4.15A shows the responses of positive and inverse β_2-adrenergic receptor agonists on a constitutively active β_2-adrenergic receptor (Samama *et al* 1993). The important aspect of these data is the fact that the mutant receptor is not a locked version of the natural active state receptor but rather that it is *some relaxed conformation that is more thermodynamically free to form the active state*. Therefore, ligands can still stabilize

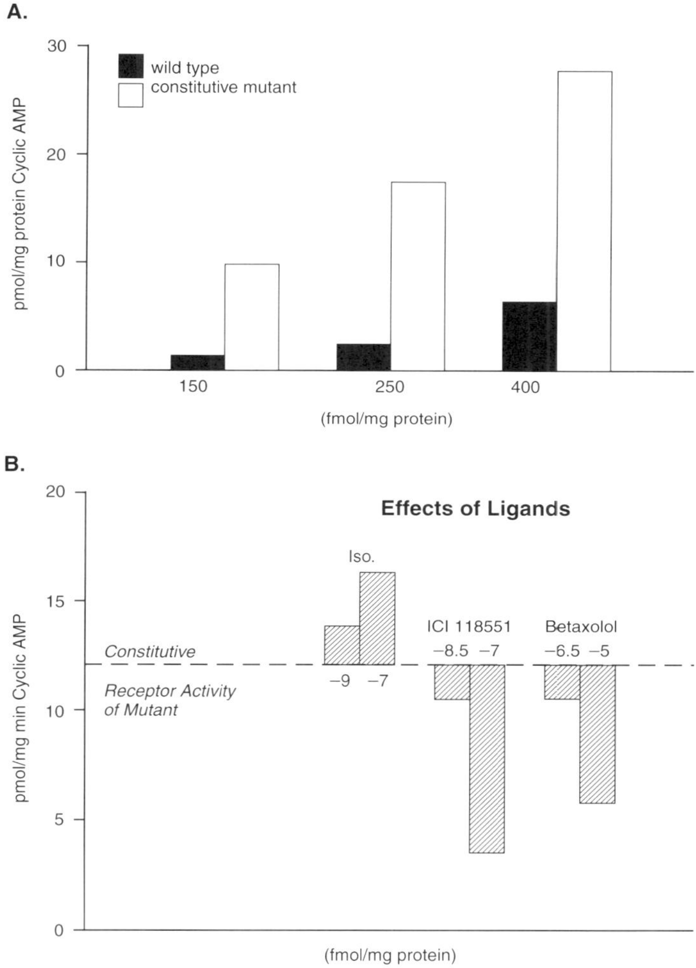

Fig. 4.15. Constitutive receptor activity of mutant β_2-adrenoceptors. (A) Increasing constitutive activity (spontaneous formation of cyclic AMP) in CHO cells transfected with 150, 250 and 400 fmol/mg of protein wild-type β_2-adrenoceptors (filled bars) and mutant β_2-adrenoceptors (open bars). (B) Effects of positive agonist isoproterenol and inverse agonists ICI 118551 and betaxolol on the constitutive activity of CHO cells transfected with mutant β_2-adrenoceptors. (Data from Samama *et al* (1993).)

either the active or inactive form and thereby make known their interaction with the receptor for purposes of screening. As shown in Fig. 4.15B, the mutant β-adrenergic receptor is much more constitutive than the wild type, suggesting a lowered energy barrier to the formation of an active receptor state.

This principle can be exploited by a high-throughput method of creating mutant spontaneously active receptors, called random saturation mutagenesis. This technique has been used to produce high-volume mutation of protein (Bowie and Sauer 1989; Oliphant and Struhl 1989; Yaghmai and Hazelbauer 1992; Burstein *et al* 1995). The major drawback of site-directed mutagenesis for the production of constitutively active 7TM receptors is the need for prior knowledge of mutation sites. Random saturation mutagenesis allows for a generalized region to be explored in an unbiased manner. Because the third intracellular loop of 7TM receptors has been the site of the bulk of known constitutive mutations, it is an ideal starting point for this technique. Thus, a large volume of mutants can be made *en masse*. A shortcoming of this approach is that an equally high-volume method of detection of constitutive activity must be used. This technique has been pioneered by exploitation of the fact that receptor agonist activity in NIH-3T3 cells produces cell transformation that can be detected in a high-volume format (Spalding *et al* 1995). It is noteworthy that the testing of a number of available muscarinic antagonists (i.e. atropine, quinuclidinyl benzylate, *N*-methyl scopolamine, 4-diphenylacetoxy-*N*-methylpiperidine and pirenzepine) demonstrated inverse agonism on the mutant receptor. This is becoming a common finding in constitutively active systems. Accordingly, it is proposed that for intractable receptors (those that do not spontaneously demonstrate elevated basal activity upon overexpression in melanophores) a strategy of random saturation mutagenesis be employed to produce a constitutively active mutant for the purposes of screening.

As shown in Fig. 4.16, the three prerequisites for a constitutive 7TM receptor assay are: an intrinsic ability of the receptor to form the active state; a G-protein that responds to the active state, and a method to observe the product of

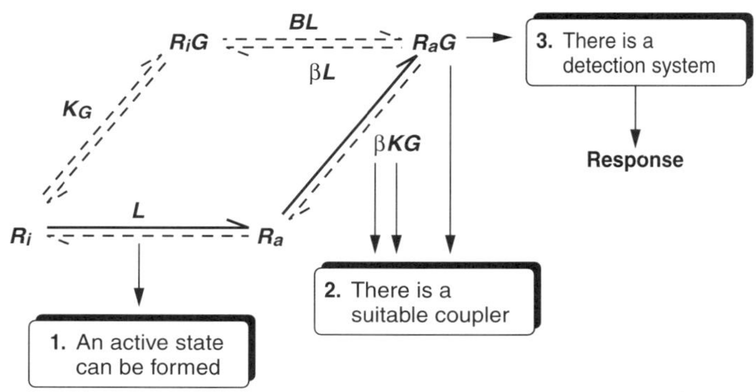

Fig. 4.16. Three prerequisites for the construction of a constitutive 7TM receptor system. The magnitude of L controls the spontaneous rate of formation of active state receptor R_a. There must be a responsive G-protein available in the membrane to couple to the active receptor state, and the activated G-protein must have a response element present in the cell to show response

the G-protein activation. In melanophores, melanin aggregation results from Gi-protein activation and this can be used as a sensitive assay for Gi-coupled receptors. For Gi-protein-coupled receptors, the constitutive activity can be verified by showing that the basal receptor-mediated response is removed by treatment of the cell with pertussis toxin (to inactive Gi protein). Gs-Protein activation causes melanin dispersion in melanophores and this can be used to detect Gs-coupled receptor constitutive activity. In this case, however, it is more difficult to verify true constitutive activity unless an inverse agonist already is available.

4.4.3 CONSTITUTIVE ACTIVITY AND HISTORICAL (REPORTER) ASSAYS

The observation of inverse agonism may not be possible in highly constitutively active stop-time assays such as reporters because the amount of product (response) blocked acutely by the inverse agonist may be inconsequential when compared to the historical constitutive response recorded by the receptor system. Under conditions of constitutive receptor activity where there is a steady-state level of active receptor spontaneously coupled to G-proteins, there is a corresponding basal level of response in the system. The addition of an inverse agonist removes this basal tone and a negative response is observed. In the case of real-time assays, the addition of an inverse agonist brings a decrease in constitutive response and this can be observed directly (i.e. see Fig. 4.15C). In reporter assays or any assay in which the product of receptor response is assayed, it may be difficult to observe inverse agonism. This is because these assays record a history of the constitutive activity and this may be considerable in the period of time before addition of the inverse agonist. Thus, although the inverse agonist may immediately block the constitutive activity upon addition, the decrease in reaction product may be insignificant when compared to the mass of product formed before addition of the inverse agonist. This is shown schematically in Fig. 4.17.

4.5 SECONDARY TESTING IN SPECIALIZED ASSAYS FOR LIGAND TAXONOMY

It is clear that typical binding and functional assays will detect the rudimentary properties of ligand affinity and efficacy. However, once a ligand has been shown to interact with a receptor, the subtle nuances of this interaction with the physiological host can be determined further. The advantage to this process is the classification of the ligand and possibly better targeting of the ligand to a therapeutic area. There are two basic refinements of the initial screening process that can be exploited in secondary screens of novel ligands. The first relates to the possible detection of selective efficacy. Although a functional receptor assay can determine that a molecule produces physiological response, it may not be sensitive to the exact

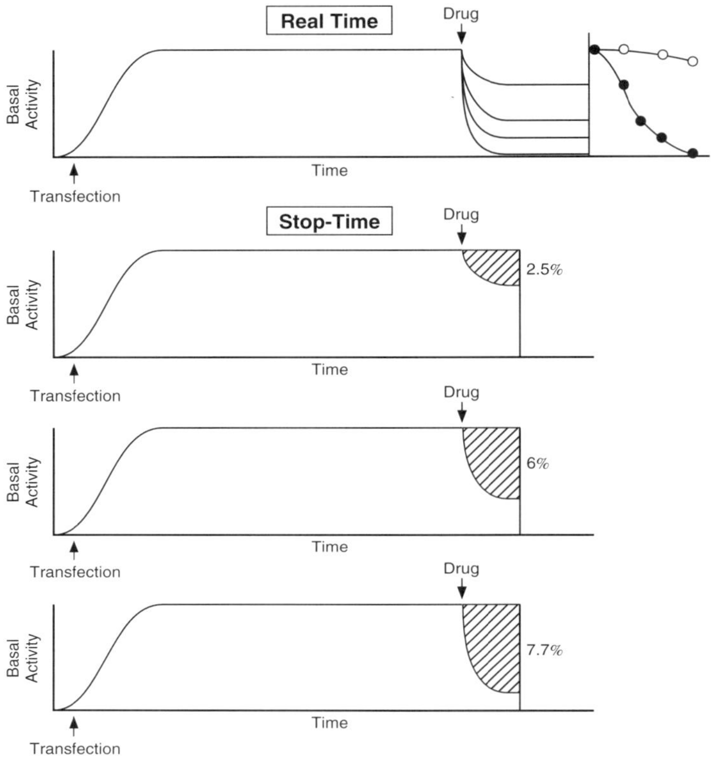

Fig. 4.17. The effects of measuring inverse agonism in stop-time mode in constitutively active receptor systems. In real-time, the decrease in constitutive activity can be observed as a decrease in the steady-state constitutively active receptor response; this is the dose–response curve, shown as filled circles. In a hypothetical constitutive receptor system where the measured product of the receptor activity is the area under the curve, it can be seen that acute blockade of the constitutive receptor activity by an inverse agonist produces a very small diminution in the total area under the curve. Thus, the apparent response to the inverse agonist when measured in this way will appear to be inconsequential; this is the dose–response curve shown as open circles for the same drug

manner in which this response is obtained. As noted in Chapter 3, ligands may produce different receptor active states that may interact differently with effectors such as G-proteins. Cellular functional assays with different G-protein bias (different levels of various G-proteins obtained in recombinant assays, *vide infra*) may allow various ligands to be classified further. If agonists produce G-protein selective activated receptor states, then their potency, relative to standard agonists, will vary as they are tested in different G-protein-enriched cells. These types of assays are referred to as stimulus-biased assays for agonists (see Fig. 4.18A).

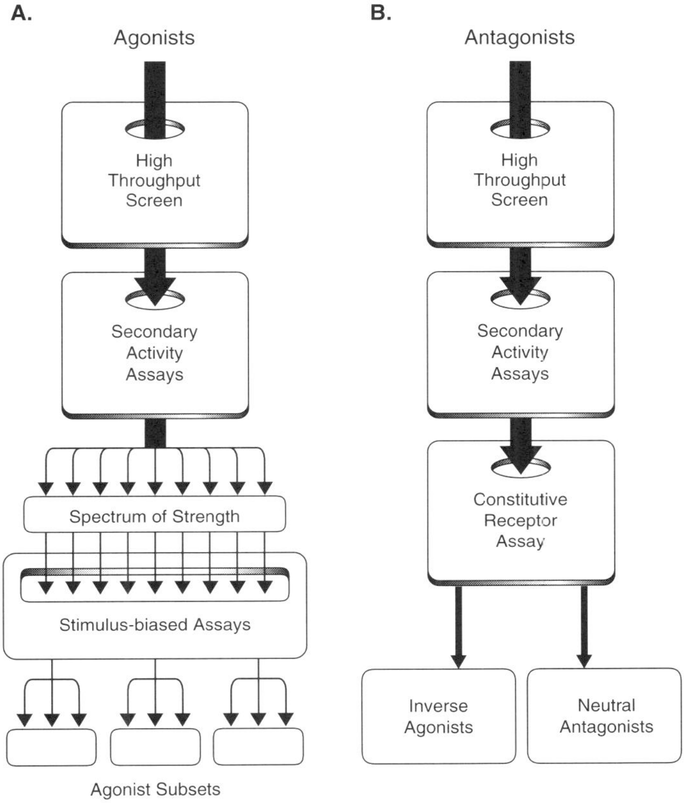

Fig. 4.18. Specialized assays for: (A) agonists and (B) antagonists. (A) Conventional approaches to screening for agonists is to make the assumption that all of the agonist lead compounds produce response by the same mechanism and sort the lead compounds as a function of strength of signal. However, the insertion of specialized secondary assays that can detect complex interactions within response elements may indicate that the lead agonists can be characterized further into groups. This latter characterization process may be useful in safety and efficacy profiles of the agonists therapeutically. (B) The insertion of a constitutive receptor assay to detect possible negative efficacy in antagonists may be useful as a predictor of possible tolerance to the antagonists through receptor upregulation

Another possible area where further determination of the molecular mechanism of a ligand can be beneficial is for antagonists. Specifically, it can be shown that antagonists can simply occupy agonist binding sites and block response, and also that antagonists can themselves induce an inactive form of the receptor and block agonist-induced response and constitutive receptor response. In quiescent receptor systems where there is no appreciable constitutive receptor basal response, these two types of ligand are indistinguishable. However, neutral antagonists and inverse agonists can be distinguished in constitutively active receptor assays (Fig 4.14B). If few natural systems are constitutively active, why might it be relevant to determine if an antagonist in a quiescent assay is really an inverse agonist?

One possible reason is to predict tolerance to antagonism with chronic use of the ligand. Antagonist lead compounds should be tested in constitutive receptor systems to detect possible negative efficacy that could herald the production of receptor upregulation (tolerance) upon chronic use (see discussion in Chapter 3). In general, negative efficacy has been linked to tolerance for antagonists in a number of receptor systems (MacEwan and Milligan 1996*a*,*b*; Lee *et al* 1997), therefore a specialized assay to detect negative efficacy can be a useful secondary assay for antagonists (Fig. 4.18B). The mechanism is thought to involve the natural ability of 7TM receptors to adopt active conformations spontaneously. If these conformations are more susceptible to phosphorylation (and subsequent internalization), as might be expected to occur with chronic agonist exposure, then a dynamic system of receptor synthesis, membrane localization, formation of active state, phosphorylation, internalization, and degradation can be envisaged (see Fig. 3.15). It should be stressed that these mechanisms would be expected to be receptor specific (i.e. depend upon the intrinsic allosteric constant of the particular receptor) and cell host specific (i.e. the membrane environment, G-protein content, receptor synthesis and degradation activities of the cell). However, these data do suggest that active antagonist lead compounds should be tested for negative efficacy for possible future information in terms of predictions of tolerance.

4.6 HUMAN RECOMBINANT RECEPTOR SYSTEMS

Until the last decade, drug development relied upon animal tissues for the testing of biological activity. As shown in Table 4.4, advances in technology have caused

Table 4.4. Pharmacological screening systems for drug discovery

Animal Receptors–Animal tissue
⇓
Animal genetic material–Animal surrogate cells
⇓
Human genetic receptor material–Animal surrogate cells
⇓
Human genetic receptor material–Human surrogate cells
⇓
Human genetic receptor material–Human target cells
⇓
Human genetic receptor material–Human target cells (Under appropriate pathology)

a trend away from animal tissues toward testing in human biological material. The completely proper setting would be to test drugs in human receptors residing in the target tissue under the influence of the pathology they are aimed at treating; this is impractical at present. The compromise is the expression of human target receptor material in an appropriate cell that can allow the receptor to function normally and signal when it interacts with foreign ligands.

Although the use of human receptor material clearly is an advantage over animal tissues, a balancing caveat is the possibility that the recombinant receptor system does not provide a valid mirror of the physiology mediated by that receptor in the body. The possibility for such dissimulation lies in the possible errors in translation of the receptor cDNA, the processing of the receptor material by the host cell, the assembly of the recombinant system and the relative stoichiometry of the components of the receptor system. Adjustment of all of these parameters can lead to an efficient screening system that adequately reflects human physiology.

The vast science of molecular biology that dictates the procedures involved in the expression of human receptor cDNA in surrogate cell systems is beyond the scope of this book. In general, cDNA of the human receptor is transfected into surrogate cells, which then express this receptor in their own cellular matrix upon replication. A transient transfection can be achieved through a single exposure of cells to cDNA (i.e. through electroporation of melanophore cells); this results in a single passage of cells expressing the receptor. The advantage of this technique is that human cDNA can be introduced into any cellular background acutely and the resulting system can be utilized for screening or other study of drug receptor–interaction. However, the level of receptor expression can be variable with this technique and it is ultimately more labor intensive than the production of stable cell lines that express the human receptor routinely with each replication.

The production of stable cell lines constitutes the choosing of clones that intrinsically incorporate the foreign DNA into their own genome and spontaneously express the receptor of interest for screening with each passage. This process occurs spontaneously, but infrequently, and such cells must be isolated through a process of repeated passage and screening with the ligand of interest. Ultimately, if a screening campaign is to continue for any length of time, the production of stable cell lines is preferred because the resulting level of receptor expression is more uniform than what can be achieved through repeated transient receptor transfection.

The choice of surrogate cellular host depends upon a number of factors, including background receptor population (i.e. an animal receptor of the same class could obfuscate the results of the tests on the human receptor), facility of receptor expression, robustness and response characteristics. This latter factor is relevant in specialized cells that inherently have characteristics that target them as host cells, i.e. melanophores.

The essential aspects of expression of human receptor cDNA in surrogate cell lines is the level of expression and the membrane milieu of interacting proteins awaiting the expressed receptors. The former factor can depend upon the natural transcription and processing ability of the surrogate cell line and also upon the intrinsic properties of the human receptor cDNA. The latter factor can be manipulated, to a certain extent, by the experimenter (i.e. through the use of inducible promoters). Specifically, G-protein-coupled receptors require membrane G-proteins to function physiologically. Therefore, if high-affinity binding or physiological response is required of a given G-protein-coupled receptor, then it must be expressed in a surrogate host cell line containing the appropriate G-protein. Therefore, although high levels of receptors can be expressed through the baculovirus system in insect cells, this leads to receptors in a background devoid of G-proteins. However, this can be an advantage because the appropriate G-protein cDNA can be co-expressed with the receptor and the signaling pathways of different receptors can be studied in such a system. Once a surrogate cellular host is chosen and a range of stable cell lines containing the receptor of interest is available, the next criterion for selection of a suitable screening cell line is the level of receptor expression.

4.6.1 WHAT IS THE OPTIMAL RECEPTOR DENSITY FOR DIFFERENT ASSAYS?

4.6.1.1 Function

In general, it is most advantageous to have the most sensitive functional assay possible to detect agonists. Within a limited range, the level of receptor expression in the host cells of recombinant systems can be used to maximize this effect. However, there is a limit to the amount of increased sensitivity to agonists that can be achieved by this method. This is made evident by examination of the equilibrium equations describing 7TM receptors. Thus, the response (as defined by the agonist-induced activation of G-protein and any constitutive basal response) is given by:

$$\text{Response} = \frac{L[\text{R}]/K_{\text{G}}(1 + \alpha\beta[\text{A}]/K_{\text{A}})}{[\text{A}]/K_{\text{A}}(\alpha\beta L[\text{R}]/K_{\text{A}}) + 1 + L[\text{R}]/K_{\text{G}}} \qquad \ldots(4.6)$$

For this calculation, response has been formulated according to the extended ternary complex model (Samama *et al* 1993) as a fraction of the total G-protein concentration. Under these circumstances, the location parameter of the dose–response curve to an agonist (i.e ED_{50}) is given by:

$$\text{ED}_{50} = \frac{K_{\text{A}}(1 + [\text{R}]/K_{\text{G}})}{\alpha\beta L[\text{R}]/K_{\text{G}}} \qquad \ldots(4.7)$$

It can be seen that as the receptor level becomes infinite (as $[\text{R}] \rightarrow \infty$) ED_{50} reaches a limiting value equal to $K_{\text{A}}/\alpha\beta$ (where K_{A} is the equilibrium dissociation constant of the agonist–receptor complex and α and β are modulator constants defining efficacy of the agonist), i.e. the observed potency of the ligand does not itself approach infinity at infinite receptor expression. From eq. 4.7 it can be seen that, within a limited range, increasing the receptor density (by increased receptor expression) can lead to increased sensitivity to agonists. However, this beneficial effect must be balanced with the potential of obscuring spontaneous constitutive receptor activity by receptor overexpression. The model predicts that the basal constitutive response is given by:

$$\text{Basal} = \frac{L[\text{R}]/K_{\text{G}}}{1 + L[\text{R}]/K_{\text{G}}} \qquad \ldots(4.8)$$

where L is the allosteric constant for the receptor and K_{G} is the equilibrium dissociation constant of the receptor–G-protein complex. It can be seen from eq. 4.8 that increased receptor density can lead to an increased basal constitutive receptor effect (for receptors bearing a favorable allosteric constant) and that this increased basal effect may be dissimulating with respect to an assay designed to detect weak agonists. There are increasing studies that have shown this experimentally for β-adrenergic receptors (Whaley *et al* 1994), thyrotropin (Van Sande *et al* 1995) and other receptor types (see Kenakin (1996) for discussion).

There is another potential hazard of high receptor numbers in functional studies when dealing with recombinant receptor systems. This has to do with the increased potential for G-protein cross-talk and the recruitment of complicating response cascades in the cell due to increased ratios of receptors to G-protein. It has been shown that the level of receptor expression can be linked to G-protein pleiotropic activation. For example, a transient expression level of 1 pmol/mg protein of α_2-adrenergic receptors into COS cells leads to selective activation of Gi protein. However, a fivefold increase to 5 pmol/mg leads to dual activation of Gi and Gs protein by the same receptor. This can lead to obfuscating functional effects. For example, a low expression level of hCTR2 (65 fmol/mg) in HEK 293 cells leads to a stable steady-state response to human calcitonin (cytosensor; see Fig. 4.19A). However, a stable HEK 293 cell line containing a much greater number of receptors (30 000 fmol/mg) produces spiked and transient responses to human calcitonin (see Fig. 4.19B). Thus, although cumulative stable dose–response curves can be obtained in the low-receptor expression cell line, a non-dose-dependent and capricious response is observed in the high-receptor expression cell line.

4.6.1.2 Binding

It might be supposed that because radioligand binding experiments directly measure the quantity of ligand–receptor complex, then high levels of receptor expression will lead to a strong signal. This, in turn, should lead to a higher signal-to-noise ratio and a better assay. Although this probably is logical for binding of antagonist radioligands, it is not necessarily the case for agonist

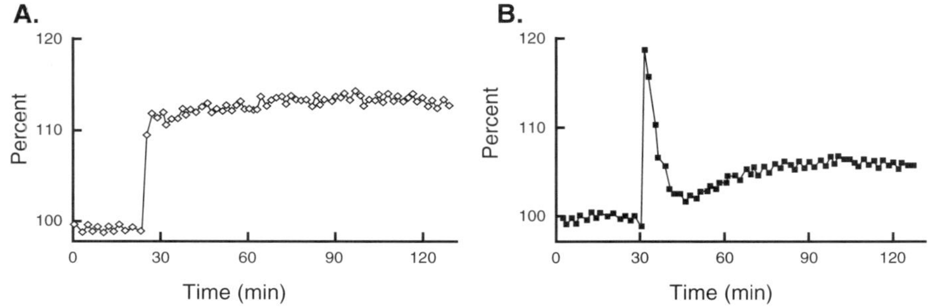

Fig. 4.19. Responses to human calcitonin in HEK 293 cells stably transfected with human calcitonin receptor type 2. (A) Response to 1 nM human calcitonin in a low-receptor expression cell line (65 fmol/mg protein); a sustained response results. (B) Response to the same concentration of human calcitonin in a high-receptor expression cell line (30 000 fmol/mg protein). Responses are transient and capricious with respect to concentration. (From Chen *et al* (1997).)

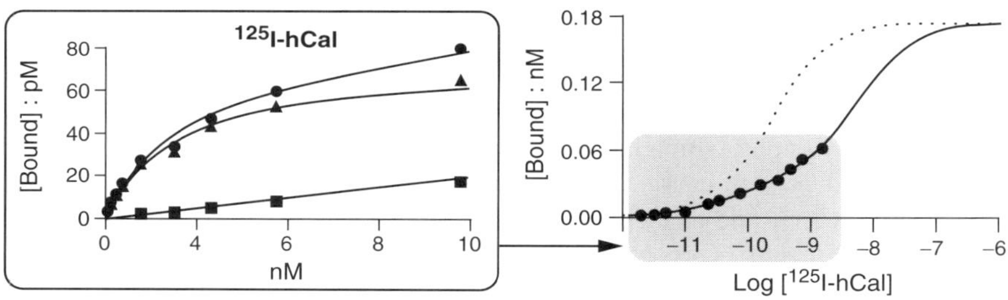

Fig. 4.20. Saturation binding curves for radiolabeled human calcitonin ([^{125}I]hCal) in HEK 293 stably transfected with 30 000 fmol/mg of protein hCTR2. On a linear scale, saturation appears to have taken place but it can be seen that the maximal asymptote observed with [^{125}I]hCAL is considerably lower than that found with an antagonist radioligand [^{125}I]AC512. (From Chen *et al* (1997).)

radioligands. This is because high levels of receptor expression lead to a condition where the G-protein needed for high-affinity ternary complexation with agonists is exhausted and a large degree of low-affinity agonist binding results (to the non-G-protein-complexed receptor). For agonists, this is similar to a high level of non-specific binding. Under these circumstances, a saturation curve to the agonist radioligand often cannot be obtained because the saturation of the high-affinity sites is not evident. For example, the hCTR2 cell line containing 30 000 fmol/mg total receptors (as measured by an antagonist radioligand) produces only 23% high-affinity complexation with radioactive human calcitonin (see Fig. 4.20). Large receptor overexpression also leads to a condition of high-affinity selection, whereby agonists show a higher potency in the displacement of agonist as opposed to antagonist radioligands.

Finally, a possible hazard in creating human recombinant systems concerns the exclusion of critical receptor-active proteins required for physiological function. Small proteins that assist in the coupling of receptors to G-proteins have been discovered for $\alpha_{2A/D}$-adrenoceptors (Sato *et al* 1993) and adenosine receptors (Nanoff *et al* 1995). If these are not included in the recombinant system, then the function of the human receptor may be impaired and/or lead to artifactual receptor data. Similarly, a family of receptor-active mediating proteins (termed RAMP proteins) have been shown to affect the phenotype of CGRP receptors (McLatchie *et al* 1998).

It is clear that recombinant technology may, as yet, fall short of recreating physiology but this situation is constantly changing with the discovery of new players in the physiology of extracellular ligand–receptor interaction.

4.7 REFERENCES

Barker EL, Westphal RS, Schmidt D and Sanders-Bush E (1994) Constitutively active 5-hydroxytryptamine 2C receptors reveal inverse agonist activity of receptor ligands. *J Biol Chem* **269**: 11687–11690.

Bond RA, Leff P and Johnson TD *et al* (1995) Physiological effects of inverse agonists in transgenic mice with myocardial overexpression of the β_2-adrenoceptor. *Nature* **374**: 272–275.

Bowie J and Sauer RS (1989) Identifying determinants of folding and activity for a protein of unknown structure. *Proc Natl Acad Sci USA* **86**: 2152–2156.

Burstein ES, Spalding TA, Hill-Eubanks D and Brann MR (1995) Structure–Function of muscarinic receptor coupling to G proteins. *J Biol Chem* **270**: 3141–3146.

Chen W-J, Armour S and Way J *et al* (1997) Expression cloning and receptor pharmacology of human calcitonin receptors from MCF-7 cells and their relationship to amylin receptors. *Mol Pharmacol* **52**: 1164–1175.

Chidiac P, Herbert TE, Valiquette M, Dennis M and Bouvier M (1994) Inverse agonist activity of β-adrenergic antagonists. *Mol Pharmacol* **45**: 490–499.

Costa T and Herz A (1989) Antagonists with negative intrinsic activity at δ-opioid receptors coupled to GTP-binding proteins. *Proc Natl Acad Sci USA* **86**: 7321–7325.

Cotecchia, Lattion AL, Diviani D and Cavalli A (1995) Signalling and regulation of the α_{1B}-adrenergic receptor. *Biochem Soc Trans* **23**: 121–125.

De Lean A, Kilpatrick BF and Caron M (1982) A ternary complex model explains the agonist-specific binding properties of the adenylate cyclase-coupled β-adrenergic receptor. *J Biol Chem* **255**: 7108–7117.

Freissmuth M, Selzer E and Schutz W (1991) Interactions of purified bovine A1-adenosine receptors with G-proteins. *Biochem J* **275**: 651–656.

Gotze K and Jakobs KH (1994) Unoccupied β-adrenoceptor-induced adenylyl cyclase stimulation in turkey erythrocyte membranes. *Eur J Pharmacol Mol Pharmacol Sect* **268**: 151–158.

Graminski GF, Jayawickreme CK, Potenza MN and Lerner MR (1993) Pigment dispersion in frog melanophores can be induced by a phorbol ester or stimulation of a recombinant receptor that activates phospholipase C. *J Biol Chem* **268**: 5957–5964.

Graminski GF and Lerner MR (1994) A rapid bioassay for platelet-derived growth factor beta-receptor tyrosine kinase function. *Bio/Technology* **12**: 1008–1011.

Hanf R, Li Y, Szabo G and Fischmeister R (1993) Agonist-independent effects of muscarinic antagonists on Ca^{2+} and K^+ currents in frog and rat cardiac cells. *J Physiol* **461**: 743–765.

Jagadeesh G, Cragoe EJ and Deth RC (1990) Modulation of bovine aortic alpha-2 receptors by Na^+, 5'-guanylimido-diphosphate, amiloride, and ethylisopropylamiloride: evidence for receptor–G protein precoupling. *J Pharmacol Exp Ther* **252**: 1184–1196.

Jagadeesh G and Deth RC (1992) Phobol ester and staurosporine modulation of antagonist and agonist binding to alpha-2 adrenergic receptors: differential influence of Na^+ versus guanylylnucleotide regulation. *J Pharmacol Exp Ther* **262**: 775–783.

Jayawickreme CK, Graminski GF, Quillan JM and Lerner MR (1994*a*) Creation and functional screening of a multi-use peptide library. *Proc Natl Acad Sci USA* **91**: 1614–1618.

Jayawickreme CK, Graminski GF, Quillan JM and Lerner MR (1994*b*) Discovery and structure–function analysis of α-melanocyte-stimulating hormone antagonists. *J Biol Chem* **269**: 29846–29854.

Karne S, Jayawickreme CK and Lerner MR (1993) Cloning and characterization of an endothelin-3 specific receptor (ETC receptor) from *Xenopus laevis* dermal melanophores. *J Biol Chem* **26**: 19126–19133.

Kenakin TP (1994) On the definition of efficacy. *Trends Pharmacol Sci* **15**: 408–409.

Kenakin TP (1996) The classification of seven transmembrane receptors in recombinant expression systems. *Pharmacol Rev* **48**: 413–463.

Kenakin TP, Morgan PH and Lutz MW (1995) On the importance of the 'antagonist assumption' to how receptors express themselves. *Biochem Pharmacol* **50**: 17–26.

Lebrecque J, Fargin A, Bouvier M, Chidiac P and Dennis M (1995) Serotonergic antagonists differentially inhibit spontaneous activity and decrease ligand binding capacity of the rat 5-hydroxytryptamine type 2C receptor in Sf9 cells. *Mol Pharmacol* **48**: 150–159.

Lee TW, Cotecchia S and Milligan G (1997) Up-regulation of the levels of expression and function of a constitutively active mutant of the hamster α_{1B}-adrenoceptor by ligands that act as inverse aognists. *Biochem J* **325**: 733–739.

Leeb-Lundberg LMF, Mathis SA and Herzig MCS (1994) Antagonists of bradykinin that stabilize a G-protein-uncoupled state of the B2 receptor act as inverse agonists in rat myometrial cells. *J Biol Chem* **269**: 25970–25973.

Lerner MR (1994) Tools for investigating functional interactions between ligands and G-protein-coupled receptors. *Trends Neurosci* **17**: 142–146.

MacEwan DJ and Milligan G (1996*a*) Up-regulation of a constitutively active form of the β_2-adrenoceptor by sustained treatment with inverse agonists but not antagonists. FEBS Lett. **399**: 108–112.

MacEwan DJ and Milligan G (1996*b*) Inverse agonist-induced up-regulation of the human β_2-adrenoceptor in transfected neuroblastoma X glioma hybrid cells. *Mol Pharmacol* **50**: 1479–1486.

McClintock TS, Graminski GF, Potenza MN, Jayawickreme CK, Roby-Shemkovitz A and Lerner MR (1993) Functional expression of recombinant G-protein-coupled receptors monitored by video imaging of pigment movement in melanophores. *Anal Biochem* **209**: 298–305.

McLatchie LM, Fraser NJ and Main MJ *et al* (1998) RAMPs regulate the transport and ligand-specificity of the calcitonin-receptor-like receptor. *Nature* **393**: 333–339.

Mewes T, Silke D, Ravens U and Jakobs KH (1993) Activation of calcium currents in cardiac myocytes by empty β-adrenoceptors. *Circulation* **88**: 2916–2922.

Motulsky HJ and Mahan LC (1984) The kinetics of competitive radioligand binding predicted by the law of mass action. *Mol Pharmacol* **25**: 1–9.

Nanoff C, Mitteraurer T, Roka F, Hohenegger M and Freismusth M (1995) Species differences in A1-adenosine/G-protein coupling: identification of a membrane protein that stabilizes the association of the receptor/G-protein complex. *Mol Pharmacol* **48**: 806–817.

Oliphant AR and Struhl K (1989) An efficient method for generating proteins with altered enzymatic properties: application to β-lactamase. *Proc Natl Acad Sci USA* **86**: 9094–9098.

Pausch MH (1997) G-protein-coupled receptors expressed in *Saccharomyces cerevisiae*: implementation of high-throughput screening assays for drug discovery. *TiBTech* **15**: 487–494.

Potenza MN, Graminski GF and Lerner MR (1992) A method for evaluating the effects of ligands upon Gs protein-coupled receptors using a recombinant melanophore-based bioassay. *Anal Biochem* **206**: 315–322.

Potenza MN, Graminski GF, Schmauss C and Lerner MR (1994) Functional expression and characterization of human D2 and D3 dopamine receptors. *J Neurosci* **14**: 1463–1476.

Price LA, Kajkowski EM, Hadcock JR, Ozenberger BA and Pausch MH (1995) Funcitonal coupling of a mammalian somatostatin receptor to the yeast pheromone response pathway. *Mol Cell Biol* **15**: 6188–6195.

Price LA, Strnad J, Pausch MH and Hadcock JR (1996) Pharmacological characterization of the rat A2a adenosine receptor functionally coupled to the yeast pheromone response pathway. *Mol Pharmacol* **50**: 829–837.

Samama P, Cotecchia S, Costa T and Lefkowitz RJ (1993) A mutation-induced activated state of the β_2-adrenergic receptor: extending the ternary complex model. *J Biol Chem* **268**: 4625–4636.

Sato M, Kataoka R, Dingus J, Wilcox M, Hildebrandt JD and Lanier SM (1993) Factors determining specificity of signal transduction by G-protein-coupled receptors. *J Biol Chem* **270**: 15269–15276.

Shi A-G and Deth RC (1994) Precoupling of alpha-2B adrenergic receptors and G-proteins in transfected PC-12 cell membranes: influence of pertussis toxin and a lysine-directed cross-linker. *J Pharmacol Exp Ther* **271**: 1520–1527.

Spalding TA, Burstein ES, Brauner-Osborne H, Hill-Eubanks D and Brann MR (1995) Pharmacology of a constitutively active muscarinic receptor generated by random mutagenesis. *J Pharmacol Exp Ther* **275**: 1274–1279.

Tian W-N, Duzic E, Lanier SM and Deth RC (1994) Determinants of α_2-adrenergic receptor activation of G-proteins: evidence for a precoupled receptor/G protein state. *Mol Pharmacol* **45**: 524–531.

Van Sande J, Swillens S, Gerard C, *et al* (1995) In Chinese hamster ovary K1 cells dog and human thyrotropin receptors activate both the cyclic AMP and phosphatidylinositol 4,5-bisphosphate cascades in the presence of thyrotropin and cyclic AMP cascade in its absence. *Eur J Biochem* **229**: 338–343.

Weiss JM, Morgan PH, Lutz MW and Kenakin TP (1996*a*) The cubic ternary complex receptor-occupancy model. I. Model description. *J Theor Biol* **178**: 151–167.

Weiss JM, Morgan PH, Lutz MW and Kenakin TP (1996*b*) The cubic ternary complex receptor-occupancy model. II. Understanding apparent affinity. *J Theor Biol* **178**: 169–182.

Westphal RS and Sanders-Bush E (1994) Reciprocal binding propoerties of 5-hydroxytryptamine type 2C receptor agonists and inverse agonists. *Mol Pharmacol* **46**: 937–942.

Whaley BS, Yuan N, Birnbaumer L, Clark RB and Barber R (1994) Differential expression of the beta-adrenergic receptor modifies agonist stimulation of adenylyl cyclase: a quantitative evaluation. *Mol Pharmacol* **45**: 481–489.

Yaghmai R and Hazelbauer G (1992) Ligand occupancy mimicked by single residue substitutions in a receptor: Transmembrane signalling induced by mutation. *Proc Natl Acad Sci USA* **89**: 7890–7894.

4.8 FURTHER READING

Binding

Hulme EC (1990) Receptor binding studies: a brief outline. In *Receptor–Effector Coupling—A Practical Approach*, ed. by EC Hulme, pp. 203–215. IRL Press, Oxford.

Functional Studies and Amplification of Receptor Signals

Goldberg ND (1975) Cyclic nucleotides and cell function. In *Cell Membranes, Biochemistry, Cell Biology, and Pathology*, ed. by G. Weissman and R. Claiborne, pp. 185–202. HP Publishing, New York.

Hedberg A, Carlsson E, Fellenius E and Lundgren B (1982) Cardiostimulatory effects of prenalterol, a beta-1 adrenoceptor partial agonist, *in vivo* and *in vitro*. *Naunyn Schmiedebergs Arch Pharmacol* **318**: 185–191.

Mueller H, Weingarten R, Ransnas LA, Bokoch GM and Sklar LA (1991) Differential amplification of antagonistic receptor pathways in neutrophils. *J Biol Chem* **266**: 12939–12943.

5 Finding the Optimal Assay Format for the Chemical Target

5.1 ABOUT TARGETS AND TARGET MOLECULES

As stated in Chapter 1, the biological target for therapeutic intervention is the physiological receptor for the drug; the drug is the chemical target. An inspection of the naturally occurring pairs of biological and chemical targets shows a great diversity. Although the size of receptor molecules generally is rather uniform (i.e. between 350 and 490 amino acids, which is a variance of 140%), the range for the endogenous molecules that interact with these targets is vast. The endogenous ligands for receptors can be as small as simple amines like acetylcholine (MW = 226) or as large as the protein SDF (stromolysin-derived factor, MW ~ 11 000) which is a variance of 4800% (see Fig. 5.1). Moreover, many of the therapeutic applications of drugs involve the modification of large protein–protein interactions. For example, mutation studies of the co-receptor required for HIV-1 entry into cells, chemokine receptor type 5 (CCR-5), show that a complex array of domains of the receptor must interact with an equally large range of domains on the viral coat protein gp120 when forming a complex with HIV (Bianiasz *et al* 1997). When such a large area of the receptor is involved, it is clear that a small molecule cannot block the interaction through steric hindrance. In view of the fact that therapeutic drugs (which need to be absorbed by the oral route) must be small molecules (i.e. MW < 500), it is evident that these drugs probably will not block the interaction of receptors with large endogenous ligands (such as SDF) or protein–protein interactions (such as the CCR-5 receptor and the HIV viral coat proteins) by steric hindrance. However, allosteric interactions, whereby small molecules form a specific tertiary conformation of the receptor, may cause the drug effect. As shown in Fig. 5.2, if a small molecule freezes the receptor into a different conformation then the position of numerous parts of the receptor, previously available to bind to a protein, may be altered to the point where the interactions are weak. In this way, small molecules can affect large protein–protein interactions.

Given this idea, it is clear that a drug screening program should accommodate, and in fact encourage, the possibility of finding allosteric ligands. There are strategic differences between screens designed to detect steric interactions and those designed to detect allosteric interactions. As a preface to a discussion of

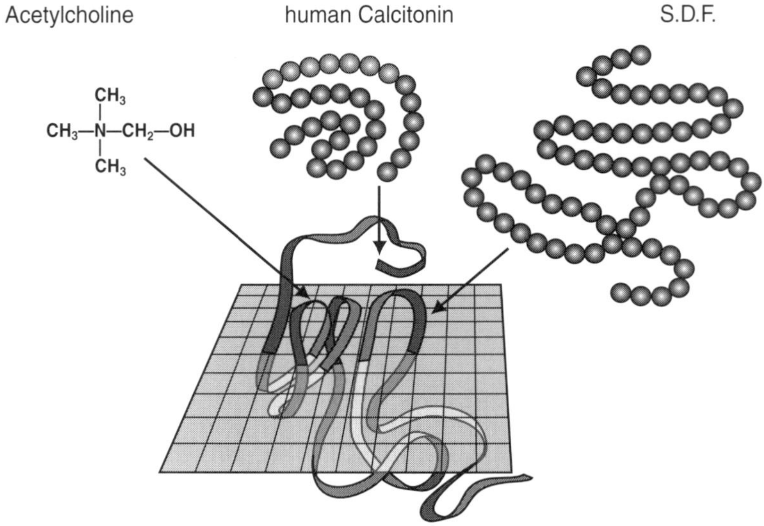

Fig. 5.1. Varying sizes of endogenous ligands for receptors. Ligands can be small amines (i.e. acetylcholine), peptide hormones such as calcitonin (32 amino acids) or larger proteins such as stromolysin-derived factor (S.D.F.)

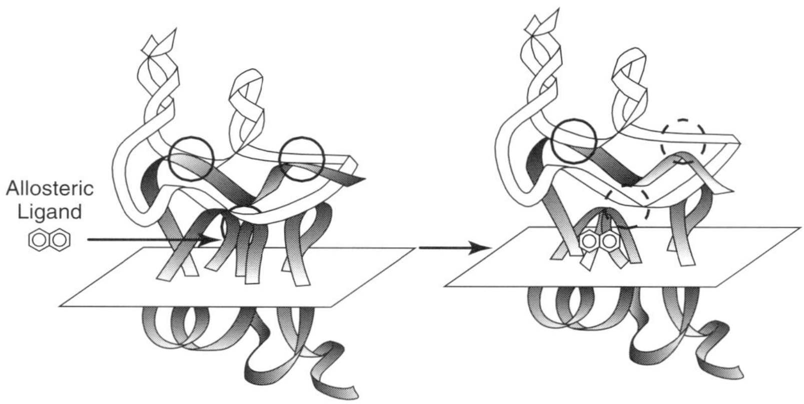

Fig. 5.2. Allosteric mechanisms for receptors. Small ligands can induce a receptor conformation that can disrupt large areas of protein–protein interaction. The ligand, protein and receptor make contact in the three circled regions. The induction of a new conformation by the allosteric ligand disrupts two of these contact domains, thus influencing the interaction between the two large proteins

these differences, it is useful to discuss the nature and usefullness of allosteric ligands in general.

5.2 THE DEFINITION OF FOUR TYPES OF ALLOSTERIC LIGAND

The most relevant basic feature of allosteric–receptor interactions to the drug discovery process is the strength of the interaction between the allosteric and natural ligands on the receptor when they are bound. An allosteric factor termed α describes the effect that the allosteric ligand has on the interaction of the receptor on another ligand (or protein species). An α factor of >1 indicates an enhancement of affinity, whereas $\alpha < 1$ indicates a reduction of affinity. It is critical to note that the magnitude of the effect of an allosteric ligand will be unique to each and every species that interacts with the receptor (especially if the geography of the binding sites is different). As shown in Fig. 5.3, an allosteric

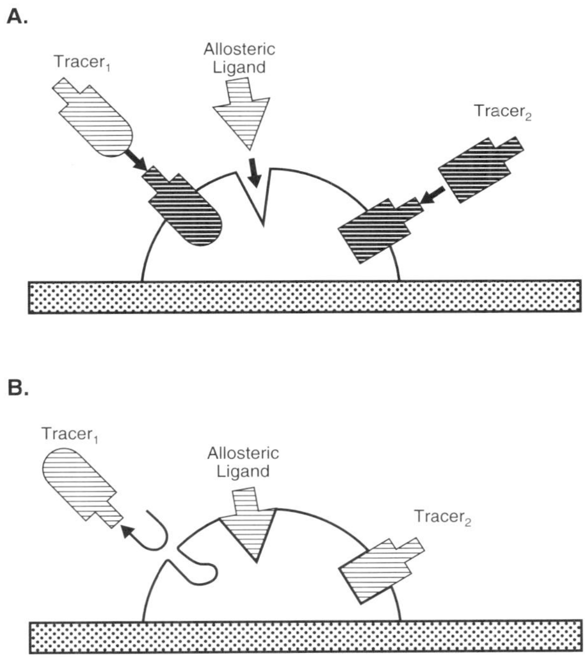

Fig. 5.3. Schematic diagram of a receptor with two distinct sites for tracer ligands (agonist or radioligand) and an allosteric binding site that affects only one of the tracer binding sites. (A) The allosteric ligand and both tracers have distinct binding sites on the receptor. (B) Tracer 1 would detect the allosteric antagonism but Tracer 2 would not

ligand may affect the binding of one tracer but not another. For example, on human muscarinic m2 receptors the allosteric ligand strychnine enhances affinity by a cooperativity factor of $\alpha = 2.43$ for [^{3}H]NMS and *decreases* affinity 20-fold for acetylcholine binding to the same receptor. Thus, a difference in the sensitivity of the screen would result by the choice of tracer ligand used for this particular class of allosteric ligand (Lazareno and Birdsall 1995). A striking array of ligand-dependent cooperativities for various allosteric ligands is shown in Table 5.1. Here it can be seen that whereas the allosteric antagonist alcuronium produces a nearly undetectable change in the affinity of arecoline (1.6), 10-fold changes in the affinity of acetylcholine and bethanechol are observed.

A distinction should be made between allosteric ligands that have α values for external ligands versus those for G-proteins. Four possible types of allosteric ligand can be defined:

(1) *Allosteric antagonist with $\alpha < 1$ for natural agonist*: This ligand reduces the affinity of the receptor for the natural agonist and produces physiological antagonism by doing so. It can be detected by radioligand binding (natural agonist radioligand) and/or functional assays.

(2) *Allosteric antagonist with $\alpha < 1$ for G-proteins*: This ligand reduces the affinity of the receptor for G-proteins and produces physiological antagonism by doing so. It may be detected by radioligand binding (natural agonist radioligand) if it destabilizes the high-affinity ternary complex. However, radioligand binding may not detect this type of antagonist, whereas a functional assay should.

(3) *Allosteric enhancer with $\alpha > 1$ for natural agonist*: This ligand increases the affinity of the receptor for the natural agonist. It produces potentiation of physiological function by doing so. It can be detected by radioligand binding (natural agonist radioligand) and/or functional assays.

(4) *Allosteric agonist with $\alpha > 1$ for G-proteins*: This ligand increases the affinity of the receptor for G-proteins. It produces physiological response by doing

Table 5.1. Cooperativity factors for three allosteric ligands for m1 m2 and m3 muscarinic receptors

	Alcuronium		Strychnine		Vincamine	
Ligand	m2	m3	m1	m3	m1	m2
Arecaidine propargyl ester	1.6	34	6.8	2.9	16.4	5.7
Arecoline	1.7	8.7	5.8	4.4	16.9	10
Acetylcholine	10	5.4	1.2	6.6	4.8	18
Bethanechol	10	23	1.1	49	7.6	13
Carbachol	9.5	9.9	1.4	6.7	1.7	11
Furmethide	8.4	8.4	1.6	6.1	3.2	8.4
Methylfurmethide	7.3	29	1.9	2.5	3.4	8.5

Data from Jakubik *et al* (1997).

so. It may be detected by radioligand binding (natural agonist radioligand) if it stabilizes the high-affinity ternary complex. However, radioligand binding may not detect this type of antagonist, whereas a functional assay should.

5.2.1 WHAT ARE ALLOSTERIC LIGANDS AND WHY ARE THEY USEFUL?

Ligand-induced allosteric effects on receptors are well known and refer to the binding of ligands on the receptor to stabilize a conformation of that receptor such that its interaction with other ligands is altered. The effects can be inhibitory, as with muscarinic allosteric blockers such as alcuronium, gallamine and strychnine (Jakubik *et al* 1996), or stimulatory, such as the allosteric enhancer brucine (Lazareno *et al* 1998) or the adenosine receptor enhancer PD 81 732 (Kolias-Baker *et al* 1994). This latter ligand stabilizes the agonist-promoted binding of receptor to G-protein, thereby potentiating agonism.

There are some advantages to allosteric mechanisms. One is that allosteric antagonism or potentiation is saturable, i.e. there is a ceiling to the effect. This can have the advantage of increasing the half-time without risk of overdose. For example, consider a conventional competitive antagonist with $pK_B = 8.0$; to achieve a therapeutic effect of a 10-fold inhibition of the endogenous agonist, a concentration in the plasma of 100 nM would be required. However, *in vivo*, the concentration increases upon antagonist absorption and decreases with pharmacokinetic elimination. Thus, the kinetics of antagonism are directly related to the pharmacokinetics of absorption and elimination, and a steady state often is not achieved. Figure 5.4A shows the effect, on an agonist dose–response curve, of the absorption of a competitive antagonist *in vivo* with time. A series of dextral displacements are observed with antagonist absorption and then, after the peak plasma concentration it reached, the antagonism begins to decay and the dose–response curve to the agonist returns to the control condition. The blockade (expressed as dose–ratios for a given concentration of agonist) as a function of time is shown in Fig. 5.4C. In contrast, an allosteric antagonist with a cooperativity factor $\alpha = 0.1$ (which produces a 10-fold decrease in the affinity of the agonist) and $pK_B = 8$ for binding to the allosteric site can be given in a much higher concentration (i.e. 10 μM). The effect would not exceed a 10-fold antagonism of the endogenous agonist but the higher concentration would saturate the allosteric binding sites and greatly increase the time of antagonism. Figure 5.4B shows the shift of the agonist curve with the allosteric antagonist. A high dose would not produce more than a 10-fold shift to the right but the curve would be shifted there for a much longer time. Figure 5.4C shows the timecourse of antagonism by the allosteric modulator. A much more even antagonism of longer duration is possible.

This effect is illustrated further by a predicted profile of tachycardia by two types of β-blockers: one a simple competitive antagonist and the other

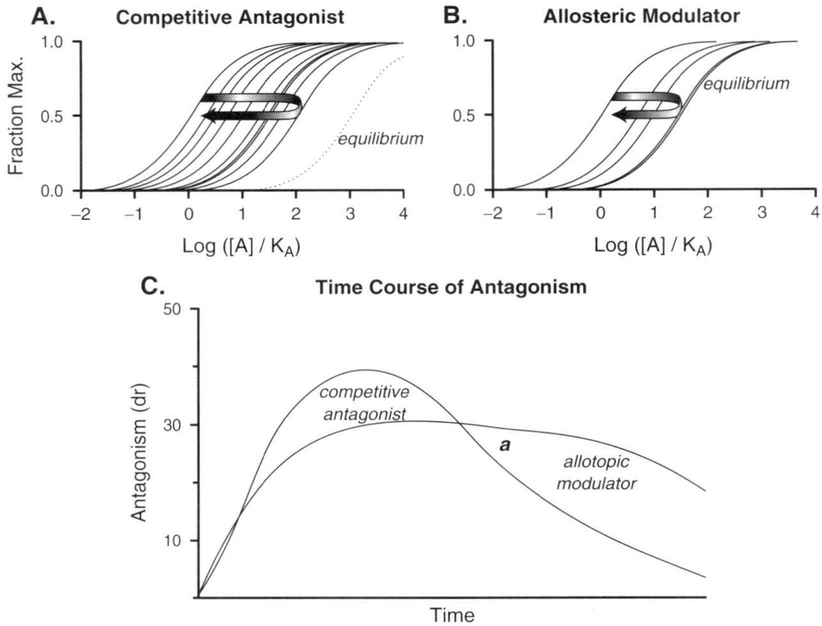

Fig. 5.4. Differences in *in vivo* effects of competitive and allosteric antagonists. (A) Effects of a competitive antagonist on an agonist dose–response curve . As the antagonist is absorbed, the curves shift to the right. As the antagonist is eliminated, the curves again shift to the left. The equilibrium dose ratio may never be attained. (B) Effects of an allosteric modulator. The curve shifts to the right to a maximum allowed by the allosteric constant; it may stay there for a time until the concentration of antagonist diminishes to a point where antagonism is removed. The curves then shift back to the left. (C) The dose ratios (dr) for an approximate ED_{50} agonist concentration for the two types of antagonists as a function of time. Note how the antagonism (dose ratio) for the competitive antagonist peaks with time and then wanes as the drug is eliminated from the receptor compartment, whereas the antagonism to the allosteric antagonist remains relatively constant

an allosteric modulator. Figure 5.5A shows the effect of a moderate dose of standard competitive β-blocker on tachycardia; the tachycardia is reduced for a time and then resumes as the β-blocker washes out of the receptor compartment. The duration of blockade can be increased by increasing the dose of β-blocker (Fig. 5.5B), but excessive β-blockade may lead to heart failure as inotropy is compromised. As seen in Fig. 5.5C, a high dose of allosteric modulator (provided that magnitude of α is suitable) theoretically can control the tachycardia for a long duration of time without producing heart failure.

Another advantage of allosteric ligands is greater selectivity of effect. There is a fundamental difference between the patterns of agonism produced by agonists

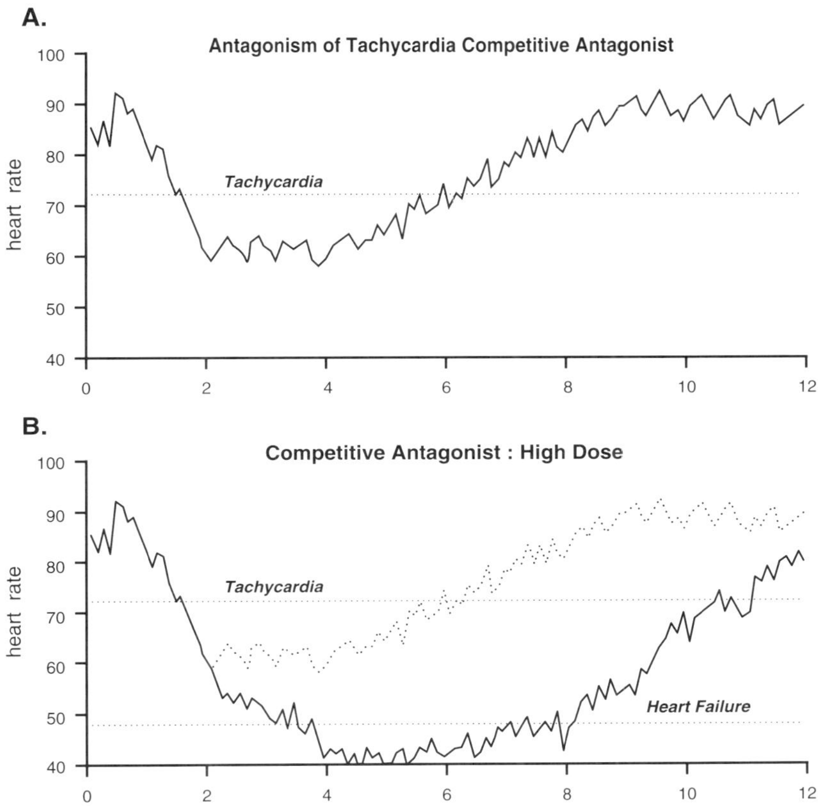

Fig. 5.5. Effects of two types of β-blocker on heart function. (A) A competitive β-blocker decreases tachycardia to acceptable levels but the initial response returns as the antagonist washes out of the receptor compartment. (B) Higher doses of a competitive β-blocker prolong the effect on tachycardia but also could drive the heart into failure due to decreased inotropy. (C) An allosteric antagonist with factor $\alpha = 0.1$ could be given in a high dose (to ensure a long-lasting effect) with no danger of heart failure

and allosteric enhancers. Figure 5.6 shows a hypothetical dose–response curve to a neurotransmitter. An increased response can be achieved by augmentation of the agonist concentration; in this case the response moves along the dose–response curve. In contrast, an enhancer produces a new dose–response curve for the agonist and the response is increased to the same concentration of agonist by virtue of the fact that the effector organ is sensitized. Although the response in this case appears identical, allosteric enhancers allow the existing pattern of selectivity of the tissue to be preserved. Thus, for allosteric enhancers, a more selective site of action would be obtained because the effects would be seen only

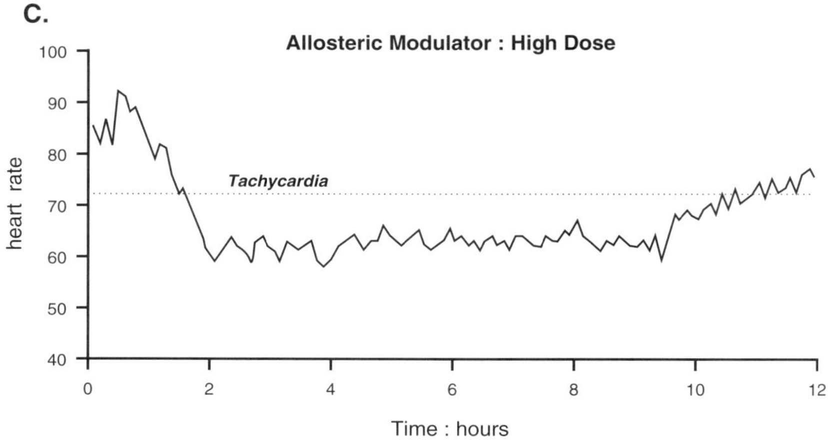

Fig. 5.5. *(continued)*

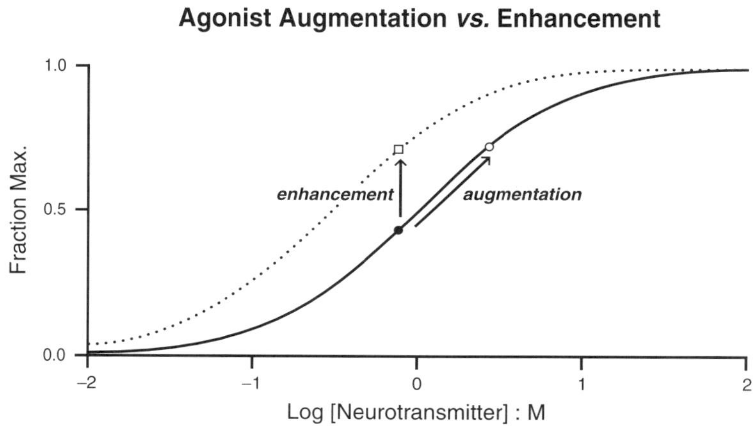

Fig. 5.6. Two ways to increase response. Augmentation of agonist concentration causes response to increase according to the existing dose–response curve for the agonist. Enhancement produces a new dose–response curve for the agonist, whereby the existing dose of agonist can produce increased response

where the endogenous agonist is present. This has been suggested (Lazareno *et al* 1998) as an advantage of muscarinic enhancers that potentiate low concentrations of acetylcholine in Alzheimer's disease — a pathological condition related to a depleted cholinergic transmission in certain areas of the brain. In these brain areas where the neuronal stores of acetylcholine are depleted and stimulation is low, an increase in the efficiency of cholinergic transmission would be beneficial.

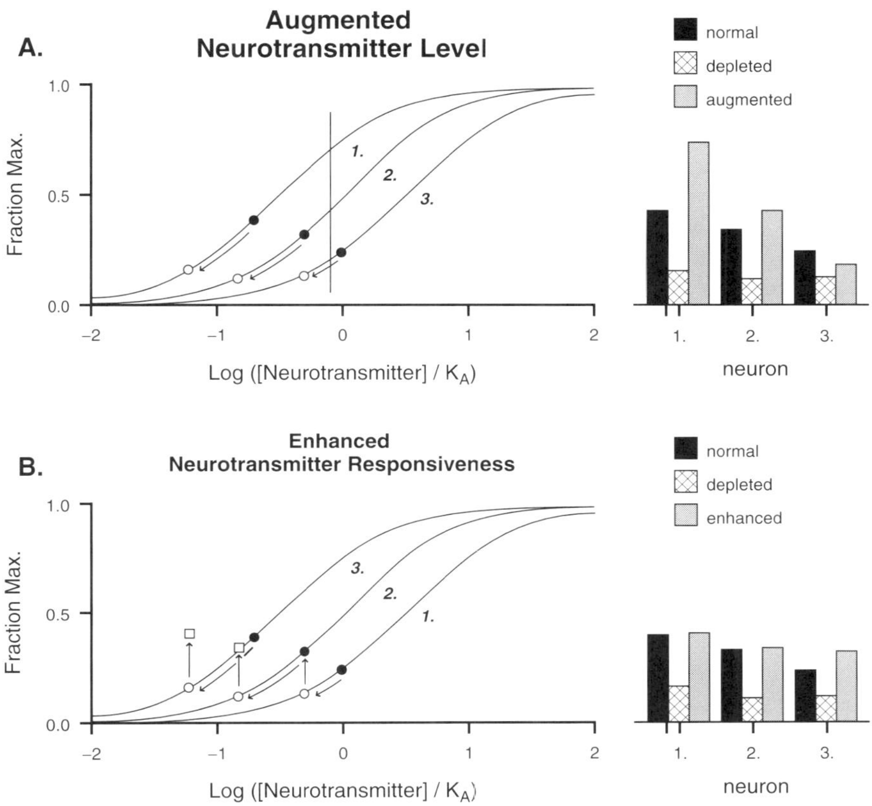

Fig. 5.7. Effects of agonist augmentation (A) and enhancement (B) toward correcting a depletion of existing agonist concentration in three neurons (represented by dose–response curves 1–3). The normal response at each synapse is denoted by the filled circle (see also bars in histogram). Depletion of the transmitter causes response to fall to the level denoted by the open circle (see bar graphs). Augmentation by addition of a given dose of agonist is shown by the vertical line in (A), with the resulting response: neuron 1 is overstimulated, neuron 2 is corrected; neuron 3 is understimulated. In (B), sensitization of the receptors causes near-correction of the depletion

Enhancers, unlike muscarinic agonists, would not produce overstimulation of the system but, rather, would amplify the signaling where it is needed (Jakubik *et al* 1997). This is demonstrated in Fig. 5.7, where neurotransmitter dose–response curves in three neuronal synapses are shown along with a standard response to normal neuron firing (filled circles). Upon depeletion of transmitter, the same firing rate produces a diminished response (open circles). The diminished response can be corrected either by agonist augmentation (adding exogenous agonist) or by enhancing synaptic responsiveness (sensitizing acetylcholine receptors through allosteric enhancement). Figure 5.7A shows the effects of adding a concentration of agonist (vertical line on concentration axis). It can be seen that response is too high in neuron 1, corrected in neuron 2 and too low in neuron 3. In contrast, a threefold potentiation of the sensitivity of the brain produces a much closer correction to control values.

On a broader scale, this can be modeled by randomly constructing 50 dose–response curves for acetylcholine to simulate 50 regions of the brain. For this example a random pattern of neuronal firing was simulated (i.e. random amount of acetylcholine in these regions was released) and response was measured in the 50 regions. The resulting pattern is shown in Fig. 5.8A. Then, the normal

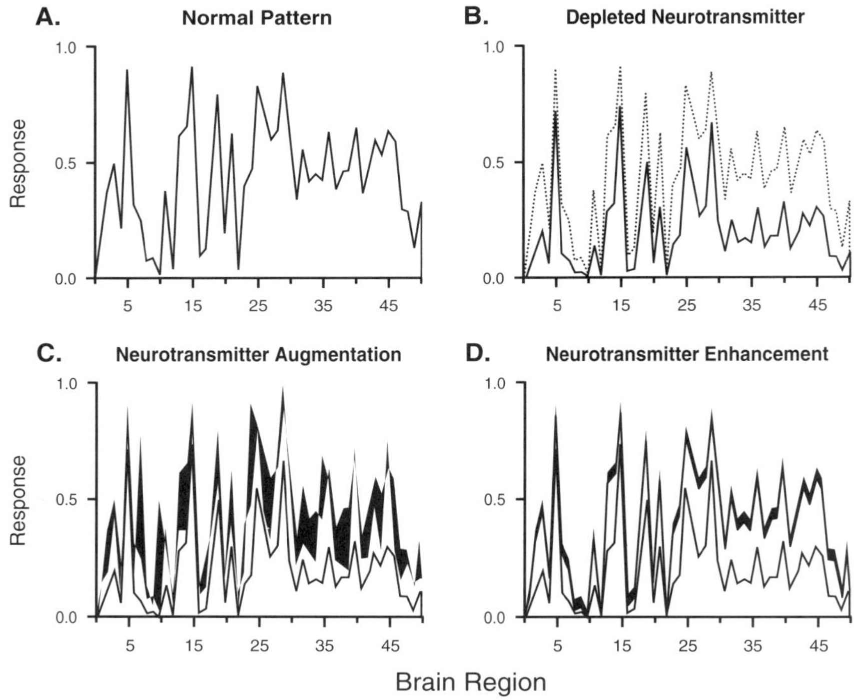

Fig. 5.8. Point responses from 50 randomly constructed dose–response curves given random concentrations of agonist. (A) A hypothetical normal pattern of brain response according to the texture of neurotansmitter release and synaptic sensitivity. (B) The agonist concentration is reduced uniformly for each neuron by a factor of four. (C) The effect of addition of a constant concentration of agonist to the complete system in an attempt to convert the depressed pattern to the normal pattern of (A). The dark regions represent response that either overshoots or undershoots the normal level. (D) The effect of an allosteric enhancer, which increases receptor sensitivity by a factor of three

acetylcholine concentrations were reduced by a factor of 3 and the response was simulated again. As can be seen from Fig. 5.8B, a new and depressed pattern of response was obtained. Fig. 5.8C shows the effect of adding a uniform concentration of agonist to the system. The dark areas represent responses that do not coincide with the control condition (either too much response or too little). In contrast, Fig. 5.8D shows the effects of an allosteric enhancer of with $\alpha = 5$. It can be seen that the allosteric enhancer almost completely corrects the deficit produced by depletion of the neurotransmitter and, importantly, preserves the original pattern of response.

5.2.2 SCREENING FOR ALLOSTERIC LIGANDS

There are special conditions required for screening allosteric ligands. Once an effect is detected, there are quantitative methods available to measure the cooperativity factor of the ligand (see Lazareno and Birdsall 1995). The practical and theoretical problems in designing screens to detect allosteric ligands relate to two general areas.

The first is the previously discussed fact that allosteric effects are ligand specific. Figure 5.3 showed the potential problem of using the wrong tracer probe to detect the allosteric effect. Allosteric ligands must rely on the stable formation of a receptor conformation to induce an effect and, in cases where ligands bind to different sites on the receptor, a given conformational change may affect one site much more than another. (Note the striking array of ligand-dependent cooperativities for various allotopic ligands shown in Table 5.1.)

In the case of some allosteric enhancers, studies with PD 81,732 indicate that the allosteric conformational changes may not directly affect the affinity of the ligand for the receptor but, rather, may stabilize the interaction of the receptor with the G-protein. Because this interaction confers apparent high affinity of the receptor selectively for agonists over antagonists, no changes in antagonist binding are observed. Rather, only agonist binding is increased (see Fig. 5.9). Therefore, if screening were done with an antagonist radioligand, no effects would be observed and the allosteric ligand would not be detected. These type of data clearly indicate that the tracer used for screening should be the physiological ligand, the effects of which are targeted to be changed.

The second feature of allosterism is that the effects may be small. The very advantage that the allosteric effect is saturable (leading to subtle effects, low risk of overdose, concentration increases in $t_{1/2}$) can be a disadvantage too because there could be a disconnection between the concentration range and the effect, i.e. an apparent lack of relationship between dose and response. For example, consider a particularly potent allosteric antagonist (i.e. $pK_B = 1$ nM) that produces a maximal effect of a 10-fold decrease in the affinity of the agonist for the receptor ($\alpha = 0.1$). If this ligand were screened at 1 μM then a 10-fold antagonism would be seen, but the same effect would also be observed at 100 nM, 10 nM, and 10 μM.

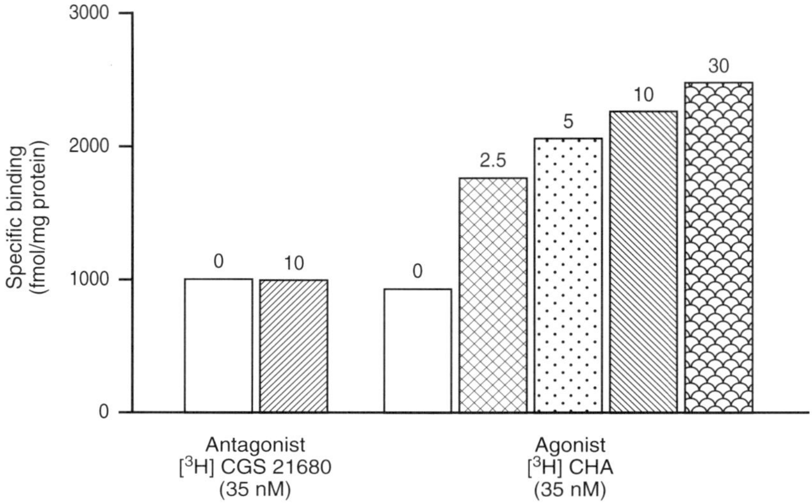

Fig. 5.9. Effects of the allosteric enhancer PD 81,723 on adenosine receptors. No enhancement of the binding of antagonist [^{3}H]cyclopentyl-dipropylxanthine (CGS, present at a concentration of 35 nM) is produced by 10 μM PD 81,723. In contrast, this allosteric enhancer increases the binding of agonist [^{3}H]cyclohexyladenosine (CHA, present at a concentration of 35 nM) at concentrations of 2.5, 5, 10 and 20 μM. (Data from Kolias-Baker *et al* (1994).)

In general, allosteric effects on affinity with both enhancers and antagonists are small, ranging from 2- to 10-fold. The exceptions are allosteric antagonists that completely prevent the receptor from signaling and responding to agonists. In view of this, a low amount of tracer (either radioligand or agonist) should be used in the linear region of the tracer dose–response curve in order to detect allosterism. Allosteric ligands produce changes in the location parameter of the tracer dose–response curve, and in this region of the tracer curve these translate to the largest change in ordinate signal (see Fig. 5.10).

5.2.3 HOW CAN ALLOSTERIC LIGANDS BE DETECTED?

Allosteric effectors produce definitive effects in radioligand binding assays. In the case of allosteric enhancers (producing stabilization of G-protein–receptor interaction), agonist–radioligand binding will be seen to increase instead of diminish in the presence of the non-radioactive ligand (see Fig. 5.11).

An antagonist may or may not demonstrate apparent competitive displacement of a radioligand. Figure 5.12A shows the effects of a range of concentrations of competitive antagonist on a radioligand saturation curve. The increasing concentrations of antagonist eventually produce complete displacement of radioligand and binding diminishes to non-specific binding (NSB) levels. In contrast, an

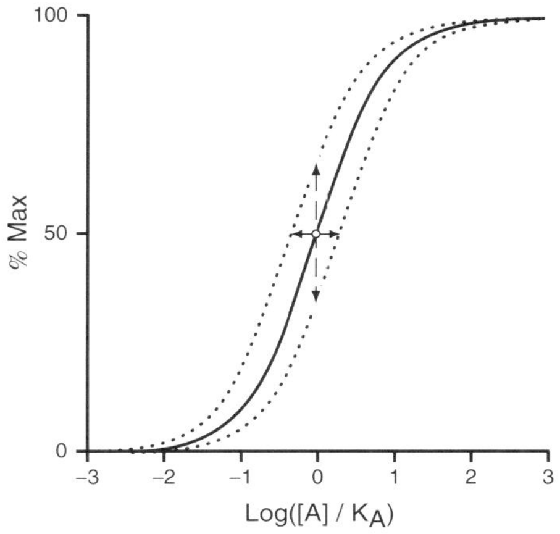

Fig. 5.10. Detection of allosteric effects. The maximal change in the ordinal signal from small shifts to the right or left of dose–response curves produced by allosteric ligands occurs in the linear portion of the dose–response curve to the tracer

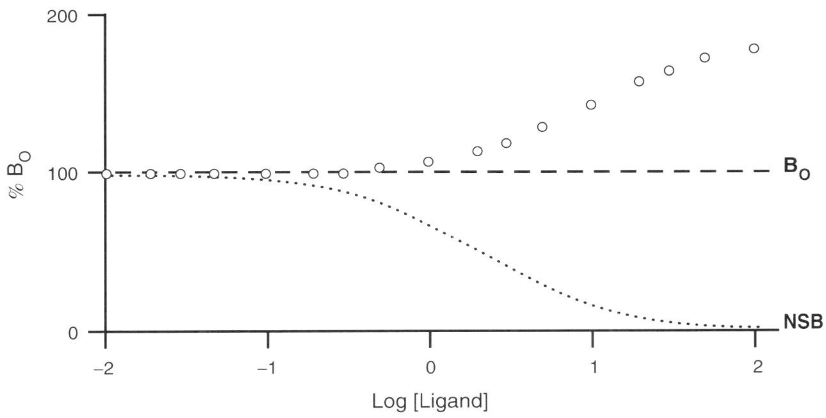

Fig. 5.11. Detection of allosteric enhancers with binding. The binding of an agonist radioligand (present at a level designated B_0, shown by the dotted line) is increased by an allosteric enhancer (open circles). Dotted line shows original level of binding of tracer radioligand.

allosteric antagonist with $\alpha = 0.1$ produces only a 10-fold shift to the right of the saturation curve, irrespective of how much antagonist is placed in the system. As shown in Fig. 5.12B, depending on what concentration of radiolabel is being displaced, the binding may never attain NSB levels. At low doses of radiolabel, a pattern similar to simple competitive antagonism may be observed. However,

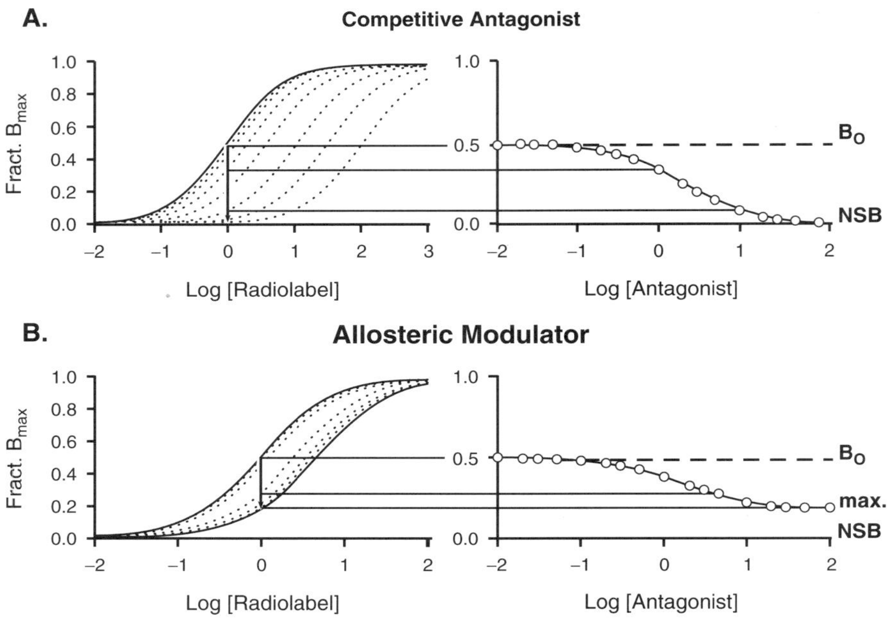

Fig. 5.12. The detection of allosteric antagonists with ligand binding. (A) The effects of a competitive antagonist on radioligand binding. The initial level of binding B_0 is depressed to the level of non-specific binding (NSB). (B) The effects of an allosteric antagonist ($\alpha = 10$) on radioligand binding. In this case, the allosteric antagonist produces a maximal 10-fold shift to the right of the radioligand binding curve. It can be seen that the binding of a half-maximal concentration of radioligand would be depressed by this allosteric antagonist, but not completely to NSB levels

as the concentration of radiolabel is increased, a condition may be attained where NSB is not observed with a maximal concentration of antagonist. Such behavior would be indicative of an allosteric antagonist (see Fig. 5.12B).

5.3 WHEN BINDING AND FUNCTION DO NOT AGREE

As discussed in Chapter 3, if a compound is an agonist, then the EC_{50} for the functional dose–response curve will be either equal to the binding affinity (for a partial agonist) or less than the binding affinity (response curve will be shifted to the left of the binding curve) because of the influence of efficacy. However, there are instances when an apparent paradox can be observed, where the dose–response curve for function is shifted to the *right* of the binding curve. This can occur as a result of the different interactions measured by binding and function, specifically the fact that binding may measure the interaction of the receptor with multiple

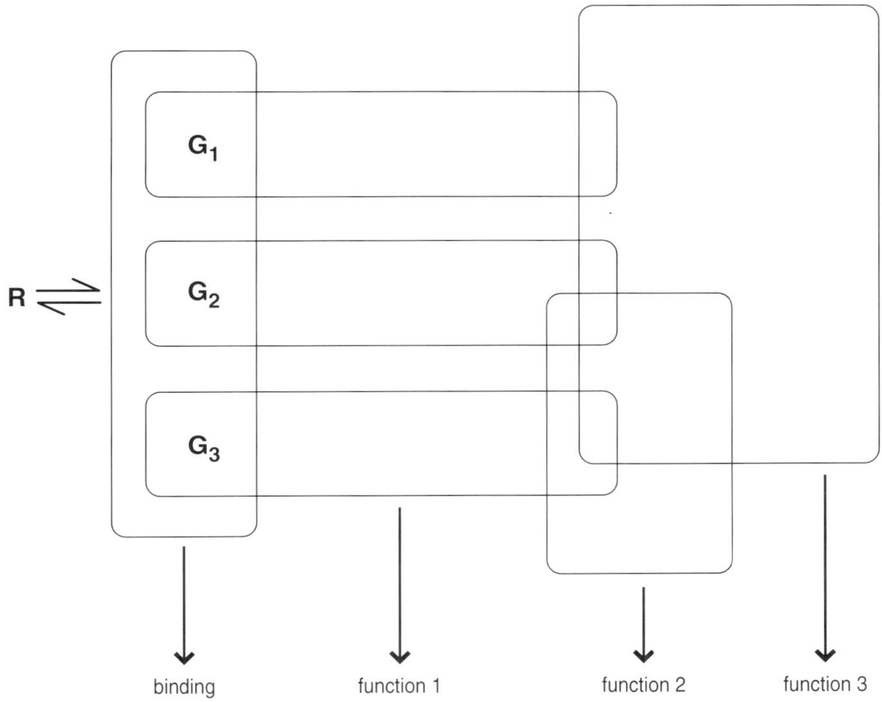

Fig. 5.13. Differences in vantage points can affect the overall sensitivity of the receptor assay. Binding assays may detect interaction with all G-proteins, whereas some functional assays may monitor interaction of the receptor with one G-protein

G-proteins whereas function may select only one (see Fig. 5.13). If the affinities of the active state receptor differ for different G-proteins, then the paradoxical shift to the right of the functional curve can occur.

This can be an important factor in choosing a screen for receptors that promiscuously interact with more than one G-protein. For example, if a receptor interacts with three G-proteins and if the endogenous radioligand (i.e. iodinated peptide) is used for binding, then all three high-affinity states may be reflected in the binding curve for the radioligand (see Fig. 5.14A). The composite binding curve for the radioligand can become extremely broad, spanning nearly seven orders of magnitude. One example of such a binding curve is shown for carbachol on muscarinic receptors (Birdsall *et al* 1987) that interact with more than one G-protein (see Fig. 5.14B).

An EC_{50} value for agonism that is higher in magnitude than the binding affinity describes an apparent situation whereby agonism is produced at concentrations greater than required for full maximal receptor occupancy. For example, Table 5.2

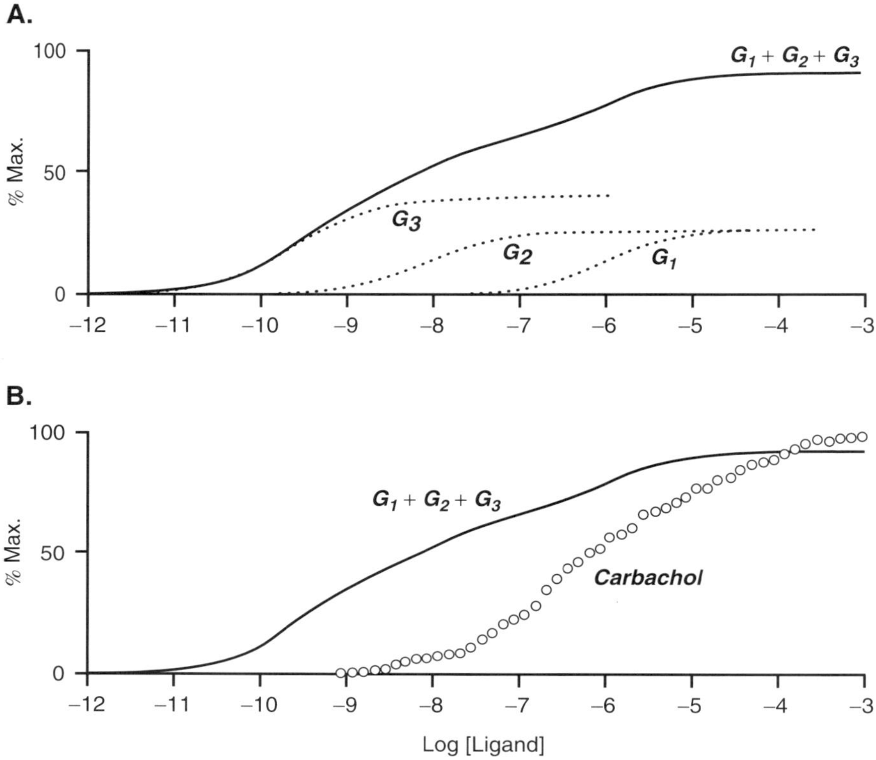

Fig. 5.14. The effects of multiple receptor states (or G-protein coupling) on receptor binding. (A) The receptor binds to three populations of G-proteins in the membrane: the individual binding curves are shown by the dotted lines and the total specific binding of the radioligand is shown by the solid line. It can be seen that the total specific binding spans nearly seven orders of magnitude. (B) Curve shown in (A) with actual data for carbachol binding to muscarinic receptors; redrawn from Birdsall *et al* (1987)

Table 5.2. Differences in agonist potency as observed in binding and function

Agonist	K_i (nM) Binding	EC_{50} (nM) Function[1]	Ratio
5-Hydroxytryptamine	2.6	78 ± 8	30
Tryptamine	57	13 342 ± 2713	234
8-OH-DPAT[2]	66	62 870 ± 7950	952

[1] Elevation of cyclic AMP in transfected LM(tk⁻) cells.
[2] 8-Hydroxy *N*,*N*-dipropyl aminotetralin.
Data from Adham *et al* (1998).

shows the binding K_i and EC_{50} values for functional stimulation of cyclic AMP for some serotonin receptor agonists. As can be seen from this table, for agonists such as 8-hydroxy-*N*,*N*-dipropyl aminotetralin (8-OH-DPAT) the concentrations required for function are 952 times higher than those required for high-affinity binding (Adham *et al* 1998), i.e. agonism apparently begins to be seen after 99.895% receptor occupancy. One possible explanation for such an apparent paradox is that the agonist activates more than one G-protein with differential affinity. This can be modeled with the ternary complex model for receptor activation (Fig. 5.15A), where the affinity factors for two G-proteins have been chosen such that G1 is coupled to the activated receptor at concentrations that are 1/60 of those required to active G2. Under these conditions the binding curve, which reflects receptor interaction with both G1 and G2, is shifted to the left of the agonist dose–response curve for activation of G2 (see Fig. 5.15A). One example of such differential activation of G-proteins can be found for thyroid-stimulating hormone (TSH), a peptide known to activate Gs (to elevate adenylate cyclase) and Gq (to elevate inositol triphosphates, IP_3). Activation of TSH receptors results in an elevation of cyclic AMP and IP_3 but the sensitivity of these systems to TSH is strikingly different (see Fig. 5.15B). Another example can be found in the calcitonin receptor, which is known to activate Gs, Gi and Gq. Figure 5.16C shows that the dose–response curve for rat calcitonin determined by microphysiometry (measuring total cellular respiration and probably an amalgam of total G-protein activation by the calcitonin receptor) lies to the left of the binding curve (as expected for high-efficacy agonists), whereas the dose–response curve for calcium flux (Gq activation) lies to the right of the binding curve. Presumably the activation of Gq (function 1 in Fig. 5.13) is less sensitive to activation by the receptor than the other G-proteins, the activity of which is reflected in changes in cellular metabolism (microphysiometry; function 3 in Fig. 5.13).

This differential and assay-dependent sensitivity to agonists can pose a problem if the screening assay monitors activity different from the therapeutic need. For example, if cyclic AMP elevation were required for a TSH agonist, then a screening assay measuring IP_3 would be far less sensitive, with the resulting possibility of low-efficacy lead molecules not being detected.

Theoretically, the same can apply to antagonism of multiple G-protein receptors if the antagonist produces a receptor conformation that negates the interaction of the receptor with only one of the multiple G-proteins. If some of these receptor–G-protein interactions can be cancelled, then only part of the total radioligand binding curve is affected. For example, Fig. 5.16 shows the binding curve to two different peptide radiolabels for cholecystokinin (CCK) receptors. The CCK receptor is known to bind to multiple G-proteins, some of which are sensitive to inactivation by pertussis toxin. Figure 5.16 shows that when one of the G-proteins, namely Gi-protein, is inactivated by treatment of the system with pertussis toxin, then the binding of only one of the radioligands ([^{125}I]DTGNN) is affected, and then only by approximately 50%. This suggests

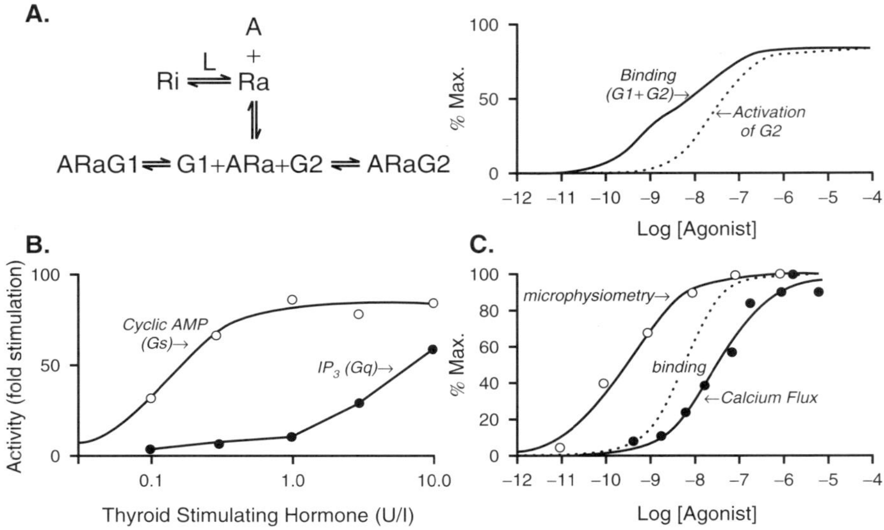

Fig. 5.15. Multiple G-protein interactions causing differences between binding and function. (A) Schematic diagram showing an agonist-stimulated binding of active receptors to two G-proteins. The simulation to the right shows a 60-fold difference in the affinity of the agonist-bound receptor for G1 over G2. The solid line shows the binding curve reflecting the interaction with both G-proteins and the dotted line the activation curve for G2 only. (B) The activation of Gs (increases in cyclic AMP) and Gq (IP_3 turnover) by thyroid-stimulating hormone (TSH) activation of TSH receptors; redrawn from Van Sande *et al* (1995). (C) Differential binding affinity (dotted line) and functional activity for rat calcitonin activation of human calcitonin receptor type 2. Microphysiometry measures total G-protein activation (open circles) and calcium flux reflects only Gq-protein activation (closed circles); unpublished Glaxo Wellcome data, Paul Irving and Chris Watson, Department of Receptor Biochemistry

that [^{125}I]OPE utilizes a G-protein that is not sensitive to pertussis toxin. In contrast, [^{125}I]DTGNN complexes to Gi and to another pertussis-toxin-insensitive G-protein (Molero and Miller 1991).

5.4 MATCHING THE CHEMICAL AND BIOLOGICAL TARGETS WITH A SCREENING ASSAY

There are separate advantages to using binding and function, and the choice of which type of assay is optimum often depends on the type of target ligand and the nature of the biological target. Table 5.3 shows a list of possible chemical target ligands with descriptions.

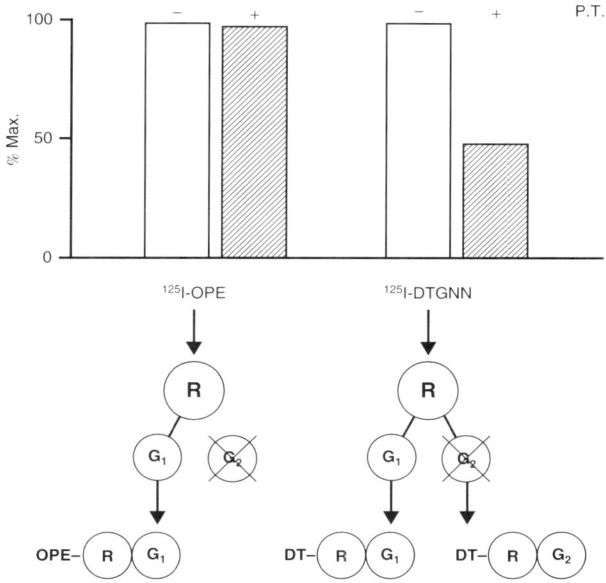

Fig. 5.16. The effects of pertussis toxin (P.T.) on the binding of two ligands for cholecystokinin receptors. When Gi protein is eliminated by pertussis toxin treatment, the binding of the radioligand [^{125}I]OPE (D-Tyr-Gly-[(Nle^{28,31}) CCK-26-32]-phenethyl ester) is unaffected whereas the binding of [^{125}I]DTGNN (D-Tyr-Gly-[(Nle^{28,31})CCK-26-33]) is reduced by 50%. This suggests that the former radioligand promotes coupling to a pertussis toxin insensitive G-protein, whereas the other radioligand promotes coupling to two G-proteins, one of which is pertussis toxin-sensitive. (Data redrawn from Molero and Miller (1991).)

Table 5.3. Types of target ligands in drug discovery

Type of Ligand	Description
Agonist	Produces a positive physiological response
Antagonist	Blocks the physiological response of an endogenous ligand
Inverse agonist	Blocks the constitutive (basal) response of receptors in a tissue that is not mediated by an endogenous ligand.
Allosteric modulator	Blocks the effects of an endogenous agonist through an allosteric mechanism — produces a defined maximal degree of inhibition
Allosteric enhancer	Potentiates the effects of an endogenous ligand

Usually, the type of target molecule is known in the drug discovery process, i.e. in a drug discovery project for treatment of osteoporosis it is known that osteoporosis results from an imbalance in the bone-building vs. bone-resorption pathways in the body. Thus, a parathyroid agonist, which would stimulate the bone-building pathway, would be the desired target. From this point of view, the path to discovery for such a therapeutic agent is dictated by the nature of the target molecule. There is a large body of data describing the advantages and limitations of different pharmacological assays for the detection of different types of pharmacological ligands.

5.4.1 SCREENING FOR AGONISTS

If the desired chemical target is an agonist, there should be some consideration of the molecular mechanism of the agonism. For example, if the receptor is from family A 7 transmembrane receptors (neurotransmitter receptors such as the acetylcholine receptor and α- and β-adrenergic receptors), then the endogenous agonist is a small molecule. Under these circumstances, it is possible that the therapeutic ligand, also a small molecule, can produce agonism by a mechanism similar to the endogenous ligand, and steric occupation of the receptor active site could be a possible mode of action. However, if the receptor is from family B (secretin family of receptors where the endogenous ligands are peptides), then a small molecule may not be able to induce agonism through occupation of the larger active site, and allosteric mechanisms may be operative.

If the agonist produces response through an allosteric mechanism, then there may be problems with validation of the effect. Specifically, conventional agonism is confirmed by using a competitive antagonist of the receptor to block the response and the resulting pK_B determined by Schild analysis for identification of the activated receptor (see Chapter 2). However, the binding domains of an allosteric agonist and the competitive antagonist may not coincide and the presence of the antagonist may not block the effects of the agonist (see Fig. 5.17). Under these circumstances, a dissimulation will occur whereby the receptor activated by the allosteric agonist cannot be identified with common standard competitive antagonists of the receptor. The only means to determine the activated receptor is to show a lack of response in host cells not transfected with the receptor of interest. The proposed scheme for screening agonists (shown in Fig. 5.18), takes into consideration the possibility of agonism through allosteric mechanisms.

5.4.1.1 Functional Assays

As shown in Fig. 5.18, the primary assay for detection of agonists is function. The major advantage of this approach is that the physiological endpoint, namely agonism, is measured directly. Therefore, if a ligand produces response by mimicking the endogenous agonist ligand or allosterically increases the efficiency of interaction between the receptor and G-protein to induce response, all of these effects will be detected in the screen.

In spite of the clear advantage of measuring the desired endpoint, functional assays have the disadvantage of a possibly high endogenous hit rate (i.e. activation

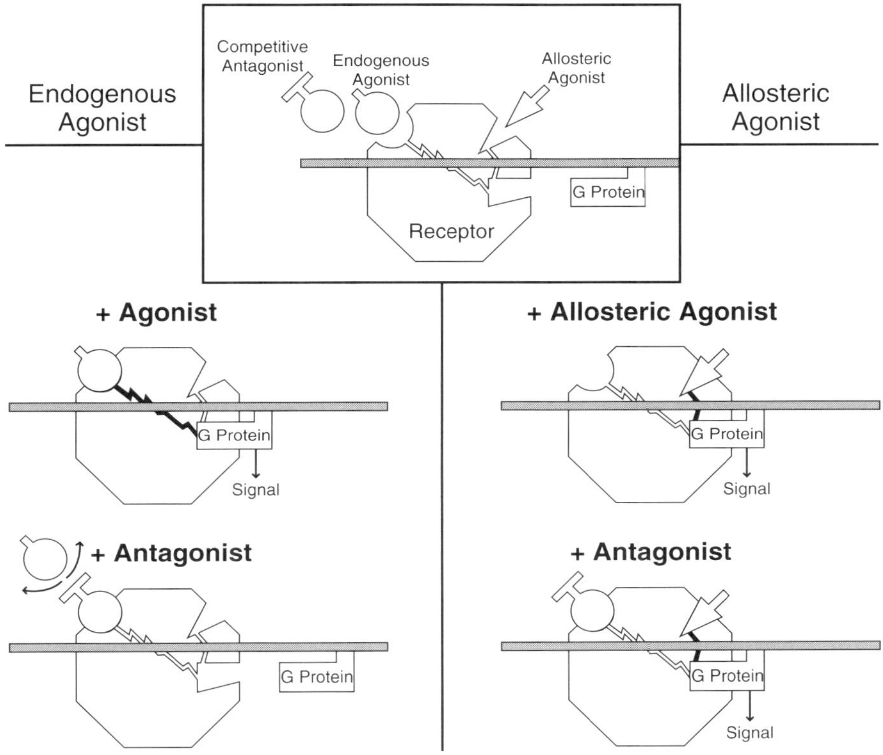

Fig. 5.17. Possible lack of antagonism of an allosteric agonist by a competitive antagonist. In the case of agonism by an endogenous agonist, the antagonist prevents binding of the agonist (and subsequent production of response) sterically (left side panels). However, if the allosteric agonist enhances receptor interaction with the G-protein to produce a response but binds to a domain separate from that of the endogenous agonist, then a competitive antagonist may not affect the response (right side panels)

of other non-target receptors). Host cell systems have many other endogenous receptors, and foreign ligands may activate these to produce responses. Although this would not cause the screen to miss an active lead (follow-up experiments in non-transfected host cells would detect these endogenous hits), it could lead to a considerable added workload. A summary of the positive and negative factors for using a functional assay to detect agonists is given in Table 5.4.

5.4.1.2 Binding Assays (Agonist Radioligand)

Another approach to finding agonists would be to use a binding assay with an agonist radioligand (Fig. 5.18). The importance of using an agonist radioligand

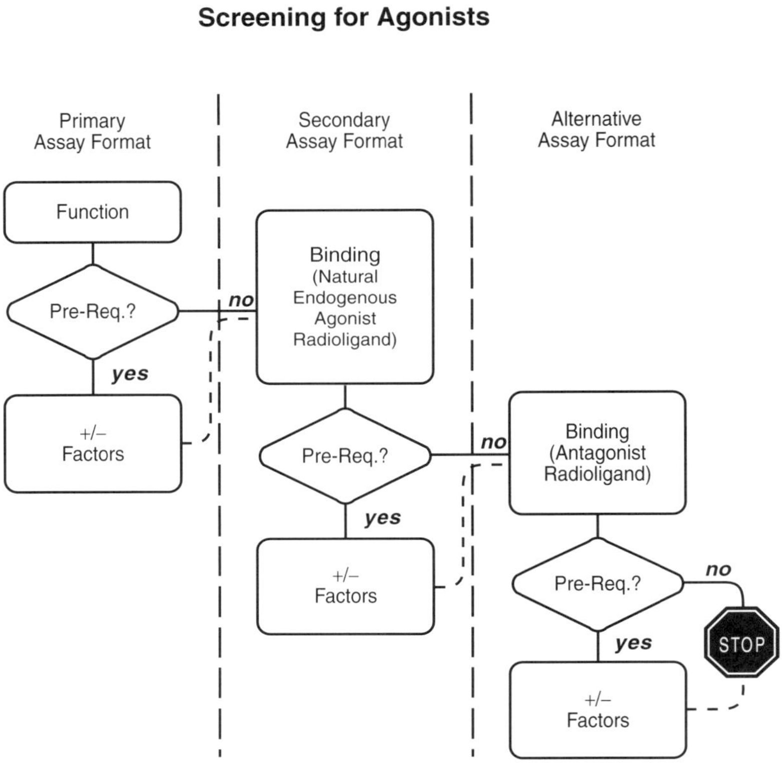

Fig. 5.18. Logical progression map for screening for agonists

Table 5.4. Screening for agonists in a functional assay

Positive factors	Possible negative factors
• Directly measures physiological endpoint (activation of receptor function)	• May have a high endogenous hit rate • Requires a cell-based assay system

over an antagonist radioligand refers to the selection of the high-affinity receptor state by agonists. The use of agonist radioligands may pre-dispose the detection of agonists in displacement experiments. This effect is magnified as the ratio of receptors to G-protein is increased. This phenomenon occurs because agonists form high-affinity ternary complexes preferentially in recombinant systems and have a lower affinity for non-G-protein-complexed receptors. Consequently, if an agonist radioligand is used as a tracer for displacement studies, then an agonist non-radioactive ligand will preferentially seek out the radioactive ternary complex. This signal is observed as a high-affinity displacement in a screen. In contrast, if an antagonist radioligand is used, then a stochastic random sampling of receptors will be bound to radioligand, a small proportion (or, if the receptor is overexpressed, an insignificant amount) of which is G-protein complexed. An agonist radioligand will preferentially form ternary complexes, but this most likely will not detect a significant amount of antagonist radioligand. Therefore, the non-radioactive agonist will not cause observable radioligand displacement until concentrations that interact with the bare receptor are obtained. These typically are 10–100 times greater than the true observed potency of the agonist. Under these circumstances the potency of an agonist in such systems will be considerably lower than what would be observed with receptor function. Usually in natural systems there is sufficient G-protein complexation capability to allow an observable degree of high-affinity complex formation with agonists when displacing antagonist radioligands. However, in recombinant systems where the receptor quantities can outstrip the G-protein content of the host cells (or where the receptors are inserted into the cell membrane in places where G-protein access is limited), very little high-affinity state can be observed. For example, recombinant cell lines containing transfected secretin receptor show only a 1.8% high-affinity state with secretin, This is increased to 15% with co-expression of Gs-protein (Ishihara *et al* 1991).

Another potential advantage of using binding to detect agonists is the possibility of finding very low-efficacy agonists as antagonists. As noted in Chapter 3, the efficiency of the stimulus–response characteristics of a functional assay determines the magnitude of the response to an agonist. If the efficiency of the stimulus–response mechanisms of the host cell is not sufficient to translate the weak stimulus from an agonist into an observable response, then the ligand will bind to the receptor silently. A binding assay will detect this affinity if it is sufficiently high, whereas a functional assay will not. Another positive factor for using a binding assay to detect agonists is that the endogenous hit rate (activation of host cell non-target receptors; see section 5.4.1.1) would be less of a problem.

There are two potential negative factors involved in a binding screen for agonists. The major one is the possibility of not detecting an allosteric agonist. For some biological targets, this is substantially important in order to preclude the use of this assay format for screening. Also, for receptors that bind to more than one G-protein, a binding assay with the natural endogenous agonist radioligand may be less sensitive than the functional assay that reflects the therapeutic need. This is because some allosteric ligands may support activation of only one of the G-proteins; this would make the binding assay only partly blocked. This idea is discussed more fully in section 5.3 (see Fig. 5.15). Another potential problem comes in the detection of numerous false positives in antagonists with no efficacy. Because agonists require the dual properties of affinity and efficacy, whereas antagonists require only affinity, it is probable that more antagonists will be detected in a binding assay than agonists. These factors are summarized in Table 5.5.

Table 5.5. Screening for agonists in a binding assay (with an agonist radioligand)

Positive factors	Possible negative factors
• Should bias detection toward efficacy-possessing ligands by virtue of high-affinity selection of receptor species	• May not detect allosteric agonist
• May detect very low-efficacy ligands not otherwise detectable by a functional assay	• May have many false positives by detecting antagonists
• Endogenous hit rate (activation of non-target receptors) is not a factor	

5.4.1.3 Binding Assays (Antagonist Radioligand)

If an agonist radioligand is not available, agonists can be screened in binding assays with antagonist ligands (see scheme in Fig. 5.18). The positive and negative factors are the same as those for agonist radioligands (see Table 5.5), with one added large possible negative factor. Specifically, high-efficacy/low-affinity ligands may not be detected with this type of screening assay.

It is well known that the dose–response curves for efficacious agonists often lie far to the left, along the concentration axis, of the curve for receptor occupancy. There are two reasons why this occurs: the original receptor stimulus imparted by the agonist to the tissue is processed through a series of saturable cytosolic biochemical reactions, which considerably amplify the original signal; and the agonists themselves possess a chemical property that imparts the stimulus to the system (intrinsic efficacy).

The severity of the differences between concentrations producing functional effects and binding affinity can be dependent upon the type of functional assay. For example, it has been estimated that activation of 0.00001% of the β-adrenergic receptor population leads to 50% maximal production of glucose (Goldberg 1975). In neutrophils, it has been estimated that one β-adrenergic receptor activated by isoproterenol leads to the production of 10 000 molecules of cyclic AMP (Mueller *et al* 1991). Because the amplification is progressive, then where the function is viewed along this cascade can control the sensitivity that one observes for agonists. Thus, whereas no increases in cyclic AMP can be observed with the weak β-adrenergic agonist prenalterol, a powerful inotropy is seen in isolated tissues. This probably occurs because cardiac cells can produce measurable inotropy with increased levels of cyclic AMP that are too small to be detected biochemically (Hedberg *et al* 1982). This principle can be illustrated with HEK 293 cells transfected with hCTR2. When a primary event such as calcium entry is measured, the potency of human calcitonin function is 1.3 times the binding affinity ($ED_{50} = 35$ nM; $K_i = 44$ nM). However, when a more amplified response such as metabolic activity of the cell (cytosensor) is used, there is a 223-fold amplification of the signal and the functional response is more sensitive than affinity ($ED_{50} = 0.2$ nM; $K_i = 44$ nM).

The more efficiently coupled are the receptors in the tissue and the higher the intrinsic efficacy of the agonist, the further to the left the response curve will be with respect to the receptor binding curve. This can cause dissimulation between detection sensitivities of functional vs. binding assays for agonists. For example, the receptor affinity of acetylcholine for muscarinic receptors is of the order of 8 μM, but responses in guinea pig ileum are observed at 6 nM (an amplification factor of 1300). Thus, a binding assay run at 1 μM would not detect acetylcholine, whereas a functional assay would. The difference between physiological function and biochemical binding sensitivity depends very much on the estimate of affinity and how it is obtained. For example, if the affinity is measured in a G-protein-rich system, then there is a likelihood of the production of a high-affinity ternary complex between the agonist, receptor and G-protein. Under these circumstances, the measured potency of binding is actually increased by the efficacy factor because it is efficacy that promotes ternary complexation. This effect can be enhanced greatly if binding is measured with an agonist radioligand (see section 5.4.1.2).

The endogenous agonists for many receptors are known to have low affinity but high efficacy. For example, Table 5.6 shows a number of agonists for α- and β-adrenoceptors and muscarinic receptors. It can be seen from these data that there can be a large discrepancy between the concentrations required for significant binding and those required to produce functional response. In general, if a binding assay is used for the detection of agonists, it will err toward finding agonists, of high affinity and possibly not detect efficacy-driven (low-affinity/high-efficacy; see Chapter 3) agonists.

5.4.2 SCREENING FOR AN ANTAGONIST

There are different considerations when the chemical target is an antagonist. However, as for agonists, the concept of allosterism also can be an important determinant of how the screening assay is designed. There are two types of allosteric mechanisms that could lead to antagonism: inhibition of endogenous agonist affinity for the receptor and inhibition of the receptor affinity for the G-protein:

(1) *Allosteric antagonist with $\alpha < 1$ for natural agonist*: This ligand reduces the affinity of the receptor for the natural agonist and produces physiological antagonism by doing so. It can be detected by radioligand binding (natural agonist radioligand) and/or functional assays.
(2) *Allosteric antagonist with $\alpha < 1$ for G-proteins*: This ligand reduces the affinity of the receptor for G-proteins and produces physiological antagonism by doing so. It may be detected by radioligand binding (natural agonist radioligand) if it destabilizes the high-affinity ternary complex. However, whereas

Table 5.6. Relative values of ED_{50} for agonism and affinity family a receptors

Agonist	Affinity (nM)	ED_{50} (nM)	Ratio
α-Adrenoceptors			
Phenylephrine	3160	1000	3.16[1]
Norepinephrine	340	13	21[2]
Muscarinic receptors			
Carbachol	20 000	316	63[3]
Acetylcholine	7940	59	398[4]
β₃-Adrenoceptors			
Isoproterenol	10 000	2.5	3981[5]
BRL 3744	400	28	14[5]
CL 31643	25 118	251	100[5]
β₂-Adrenoceptors			
Isoproterenol	32	0.16	200[5]
BRL 37344	63	20	3.2[5]
CL 31643	5011	794	6.3[5]
β₁-Adrenoceptors			
Isoprotereol	126	0.85	148[5]
BRL 37344	1584	708	2.2[5]
CL 31643	1 00 000	7080	14[55]

[1] Siegl PKS, McNeill JH (1982) *Can J Physiol Pharmacol* **60**: 1131–1137.
[2] Besse JC, Furchgott RF (1976) *J Pharmacol Exp Ther* **197**: 66–78.
[3] Furchgott RF (1967) *Ann NY Acad Sci* **139**: 553–570.
[4] Sastry R and Cheng HC (1972) *J Pharmacol Exp Ther* **180**: 326–339.
[5] Unpublished data: Conrad Cowan, Glaxo Wellcome, Department of Receptor Biochemistry.

a functional ligand will detect such an antagonist, radioligand binding may not. This type of ligand also may decrease constitutive receptor activity.

5.4.2.1 Binding Assays

A logical scheme for the detection of antagonists, with consideration of allosteric mechanisms, is given in Fig. 5.19. As can be seen from this figure, the primary assay for the detection of antagonists is binding. A consideration in using a binding assay for the detection of an antagonist is the choice of radioligand, either the endogenous agonist or an antagonist radioligand. Endogenous agonist radioligands often are available because an iodinated derivative with high affinity can be made. This may be particularly important if the antagonist works through an allosteric effect, because the degree of allosteric antagonism of one tracer ligand may completely differ from the degree of allosteric antagonism that would be seen for another tracer ligand. For example, although the allosteric ligand alcuronium produces a 10-fold change in the affinity of the muscarinic m2 receptor for acetylcholine, it produces only a 1.7-fold change in the affinity for arecoline

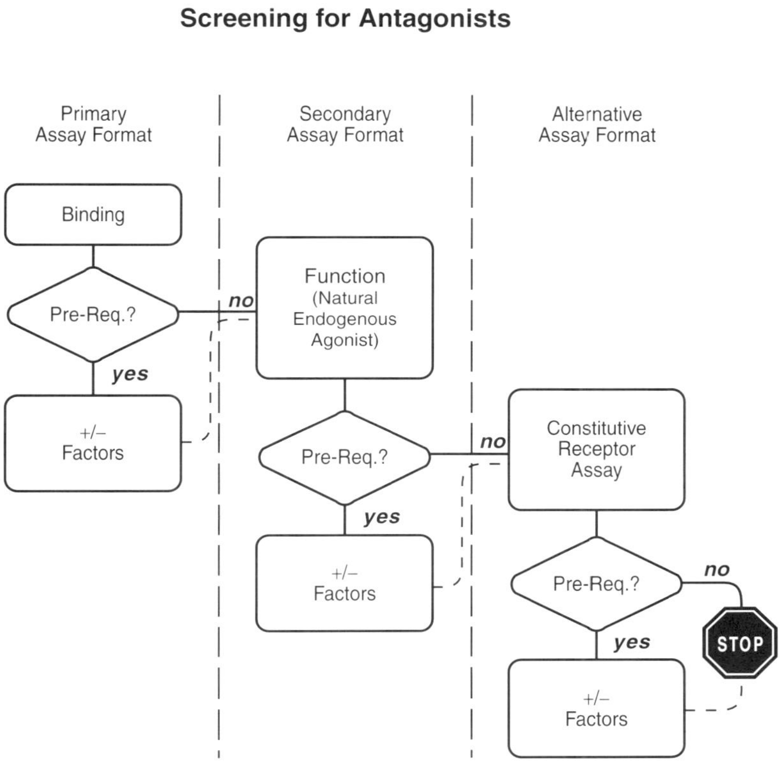

Fig. 5.19. Logical progression map for screening for antagonists

(Jakubik *et al* 1997). Therefore, if arecoline were used for screening, the effect of the endogenous agonist alcuronium may never have been detected.

There are possible disadvantages to the use of endogenous agonist radioligands in binding assays. As pointed out previously, using an agonist radioligand may predispose the assay to detect only agonists (see section 5.4.1.2). Also, the kinetics of the assay required for the detection of antagonists may be relevant. If an agonist radioligand A^* is used for binding, then the signaling species in the assay most likely is the ternary complex A^*R_aG. Normal mass action kinetics would allow for another agonist (which would have a higher affinity for the R_aG species than the uncomplexed receptor) to displace the agonist radioligand:

$$A + A^*R_aG \rightleftharpoons AR_aG + A^* \qquad \ldots(5.1)$$

However, antagonists, especially inverse agonists with negative efficacy, may require the presence of the R_i species for binding. Under normal circumstances the reversibility of receptor systems should allow for receptor species to redistribute rapidly in the presence of ligands, but the formation of antagonist binding sites may require protein–protein transformation, the kinetics of which may be slower than mass action kinetics:

$$\begin{array}{ccccc} B + A^*R_aG & \rightleftharpoons & B + R_aG + A^* & \rightleftharpoons & B + R_a + G + A^* \\ & & \text{Receptor State Interconversion} \Rightarrow & & \upharpoonleft\downharpoonright \\ & & & & \upharpoonleft\downharpoonright \\ & & BR_i + A^* + G & \rightleftharpoons & B + R_i + G + A^* \end{array} \qquad \ldots(5.2)$$

Thus, the receptor might be required to convert into the G-protein-free inactive form (in the presence of A^*, which would drive the reaction in the opposite direction) before antagonist–receptor binding could occur. For some receptor systems the kinetics of these transformations may be slow enough to bias the detection of only agonists; this would depend upon the magnitudes of L and βK_G (see Chapter 4, Fig. 4.16).

For receptors that activate more than one G-protein, an allosteric antagonist may block the interaction of the receptor with only one of the G-proteins. This would make a binding assay that measured interaction with all of the G-proteins (presumably what would occur with the natural agonist radioligand) less sensitive than if it used an antagonist radioligand (see section 5.3 and Fig. 5.16 for further discussion). Finally, although iodinated analogs of natural agonist peptides often are readily available, they may lack the stability needed for robust binding assays.

The alternative is to use an antagonist radioligand for binding. This has the possible advantage of presetting the assay to detect antagonists (i.e. the kinetics of protein–protein interaction are not relevant and only mass action kinetics apply). The possible disadvantage of this approach is that it may not detect an allosteric ligand that blocks receptor function but does not interfere with the binding of the antagonist tracer ligand (see Figs. 5.3 and 5.17). This latter possibility is quite important because potentially useful ligands would never be detected under these circumstances. The advantages and disadvantages of various radioligands for the detection of antagonists are summarized in Table 5.7.

Once the choice of radioligand has been made, the relative positive and negative factors of using the binding assay can be considered. On the positive side, binding assays have a better sensitivity of detection for low-affinity ligands.

Functional assays inherently have a slightly less sensitive detection window for antagonists. This is due to the capability of radioligand assays to expand extremely small windows of detection (assuming favorable signal-to-noise ratios). To observe antagonism in a functional assay, a measurable level of agonism must be obtained; a decrease in signal is then observed according to the Gaddum equation (Gaddum, 1937, 1957):

$$\text{Response} = \frac{[\text{Agonist}]/K_A}{[\text{Agonist}]/K_A + [\text{Antagonist}]/K_B + 1} \qquad \ldots(5.3)$$

Table 5.7. Choices of radioligand for detection of antagonists

Advantage	Possible disadvantage
Agonist radioligand (natural endogenous agonist)	
• May detect allosteric antagonist that does not necessarily displace ligand but does interfere with receptor interaction with G-protein	• May predispose assay to detect only agonists if the kinetics of protein–protein interconversion are slower than mass action kinetics • May predispose assay to detect agonists through high-affinity selection • Natural agonist may not be sufficiently chemically stable for assay • If receptor interacts with more than one G-protein, an allosteric antagonist that blocks interaction with only one G-protein may show little effect
Antagonist radioligand	
• Screen may be preset to antagonist mode by radioligand binding to inactive receptor state: does not require protein–protein interconversions	• May not detect allosteric ligand that blocks receptor function (and/or interaction with G-protein) but does not alter affinity of another ligand

To affect appreciably a measurable response, the antagonist must be present at the receptor in concentrations exceeding K_B. In contrast, binding windows can be chosen to increase the sensitivity of the measurement window if the specific binding is high. Under these circumstances, the observed IC_{50} of the potential antagonist is related to the K_B of the antagonist by a metameter of the Gaddum equation, also called the Cheng–Prusoff (1973) equation:

$$IC_{50} = (([A]/K_A) + 1) \times K_I \qquad \ldots(5.4)$$

Thus, it can be seen that if an antagonist candidate has a K_I value of 1 μM, then it will demonstrate a comparable IC_{50} in a binding assay where a trace amount of radioligand can be used (i.e. where $[A] \ll K_A$). Under these circumstances, when $[A]/K_A < 1$ then $IC_{50} \rightarrow K_I$. In contrast, in a functional assay, the concentration of agonist usually must be equal to or greater than the ED_{50} (i.e. $[A]/K_A > 1$; note that K_A here does not refer to the equilibrium dissociation constant of the agonist–receptor complex but, rather, is an operational location parameter of the dose–response curve).

This effect is illustrated in Fig. 5.20. The effects of a concentration of antagonist equal to K_I in a functional assay are shown in Fig. 5.20A. It is assumed that the screen utilizes a signal generated by an agonist concentration that is twice the ED_{50}. Under these conditions, the antagonist produces a 23% inhibition of the control signal. In contrast, the same antagonist is shown in a scintillation proximity assay (inset to Fig. 5.20A), where the concentration of radioligand

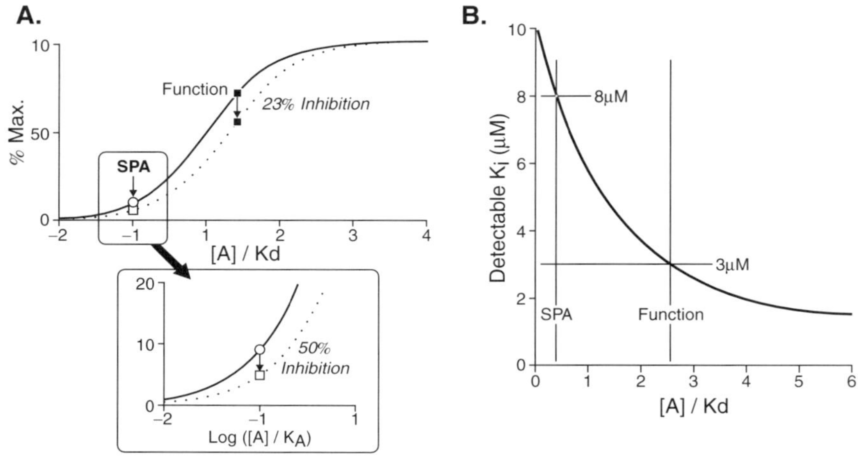

Fig. 5.20. Differences in detection windows for functional and binding assays. (A) The effects of a competitive antagonist on the dose–response curve to an agonist or a radioligand in a scintillation proximity assay (SPA) (enlarged inset). To detect antagonism of response, a concentration of agonist at or greater than the ED_{50} usually is required for an adequate signal-to-noise ratio. The effect of the antagonist shown is to produce a 23% inhibition of functional signal. If a lower level of stimulation of the system could be accessed, as in an SPA for example, then that same concentration of antagonist would produce a larger effect (50% inhibition). (B) The relationship between detectable K_i of antagonists and initial strength of system stimulation. As can be seen from this relationship, an SPA is capable of detecting compounds with K_i as low as 8 μM, whereas a functional assay could detect compounds with K_i lower than 3 μM. Antagonists were tested at 10 μM

used is $0.1 \times K_d$ for radioligand binding. Assuming that the signal-to-noise ratio is adequate, the same concentration of antagonist produces a 50% inhibition of the control signal. In general, this effect translates to a practical difference in the detection windows of functional vs. radioligand assays for antagonists by a factor of 2.5–5. Figure 5.20B shows the relationship between the strength of the initial control signal used for detection and the lower limit of antagonist K_I detectable for compounds tested at a concentration of 10 μM. It can be seen that if a trace amount of radioligand is used, compounds as weak as 8 μM should be detected. In contrast, a functional assay would detect only compounds with $K_I < 3$ μM.

Another potential positive factor is that the assay is not cell-based and a high endogenous hit rate probably would not be a factor. A possible negative factor (although not a very serious one) is the possibility of finding an allosteric ligand that affects radioligand binding but does not block receptor function. This would be detected in secondary testing in functional systems. The general advantages

Table 5.8. Screening for antagonists in binding assays

Advantages	Possible Disadvantages
• Better detection window for low-affinity ligands	• May find false positive in an allosteric ligand that affects receptor occupancy by the radioligand but not agonist function
• Endogenous hit rate is less of a factor	
• Not cell-based	

and possible disadvantages of binding for the detection of antagonism are summarized in Table 5.8.

5.4.2.2 Functional Assays

Clearly an advantage of using a functional assay for the discovery of antagonists is the possibility of detecting allosteric ligands that block receptor function but do not displace radioligands (i.e. see Fig. 5.17). Possible disadvantages include the lower detection window for functional (vs. binding) assays (see section 5.4.2.1) and the possibility of a high endogenous hit rate whereby toxic compounds or compounds that depress agonist response through non-specific and physiological mechanisms will be detected as antagonists. However, in view of the fact that these false positives can be identified in secondary testing, this would only be a problem if the volume were high. The factors summarizing the use of functional assays to detect antagonists are summarized in Table 5.9.

5.4.2.3 Constitutive Receptor Assays

Theoretically, constitutive receptor assays can be used to detect antagonists but the premise for this, as yet, is unproven. The basis for assuming that antagonists will depress constitutive receptor function lies in the assumption that ligands will have differing affinity for the two (possibly three if G-protein binding confers a different conformation to the active state receptor) receptor conformations present in a constitutive assay. These assays contain the inactive receptor species (R_i), the active receptor conformation (R_a) and the G-protein-bound signaling species

Table 5.9. Screening for antagonists in functional assays

Positive factors	Possible negative factors
• Directly measures physiological endpoint (blockade of receptor function), therefore will detect allosteric ligands that do not displace radioligands	• Lower detection window than binding assays
	• Endogenous hit rate may be high (non-specific, toxic compounds)

(R_aG). If the ligand has identical affinities for the different conformations, then it will bind to all receptor species and not affect the relative distribution. However, if the ligand has a selective affinity for any of the conformations, then it will enrich that particular conformation (see Chapter 3, eq. 3.3 for further details). This should be detected as a change in the constitutive activity (see Chapter 4 for further discussion). Thus, if is assumed that most antagonists are truly inverse agonists, then this approach can be used to screen for antagonists. Insufficient data are available at this time to make this assumption (although, in thermodynamic terms, theoretically it is preferred over the idea that a ligand will have identical affinities for different tertiary conformations of a protein). However, it should also be considered that there have been neutral antagonists, found in functional assays with agonists, that do nothing in constitutive assays.

5.4.3 SCREENING FOR INVERSE AGONISTS

As discussed earlier in this chapter (and in Chapter 3), there are ligands that possess negative efficacy and thus can reverse constitutive receptor activity. It is not yet clear whether such a property is therapeutically useful. Theoretically, if there were constitutively active regions of agonism associated with pathology, then an inverse agonist would be the only type of drug that could reverse the effect. This might occur under conditions of receptor overexpression, such as that observed in some cancer cells. Under these circumstances, cellular activation due to constitutive receptor activation could lead to increased cancer cell growth; an inverse agonist could be beneficial in these circumstances. A constitutively active chemokine receptor has been associated with human herpes virus infection (Arvanitakis *et al* 1997). Infection with KSHV leads to expression of the constitutive receptor, which in turn elevates IP_3, leading to cell proliferation and continued viral replication. Again, an inverse agonist for the chemokine receptor would be the only useful treatment.

As well as unique therapeutic applications, it should be remembered that an inverse agonist functions as a simple antagonist in quiescent (non-constitutively activated) systems. Thus, an assay for inverse agonism could simply yield antagonists. Lastly, the use of constitutively active assay systems can increase the spectrum of detection of an assay in cases where no endogenous ligand can be used as a tracer probe; this has been discussed earlier in the context of screening orphan receptors (see Chapter 3 for further discussion).

5.4.3.1 Constitutive Receptor Assays

The scheme for screening for inverse agonists is shown in Fig. 5.21. Clearly the optimal assay is a constitutively active receptor assay (as discussed above) because these are the only assays capable of detecting which ligands possess negative efficacy. The possible negative aspects of these types are assay are the increased cost and effort and possible increased inherent variability. The

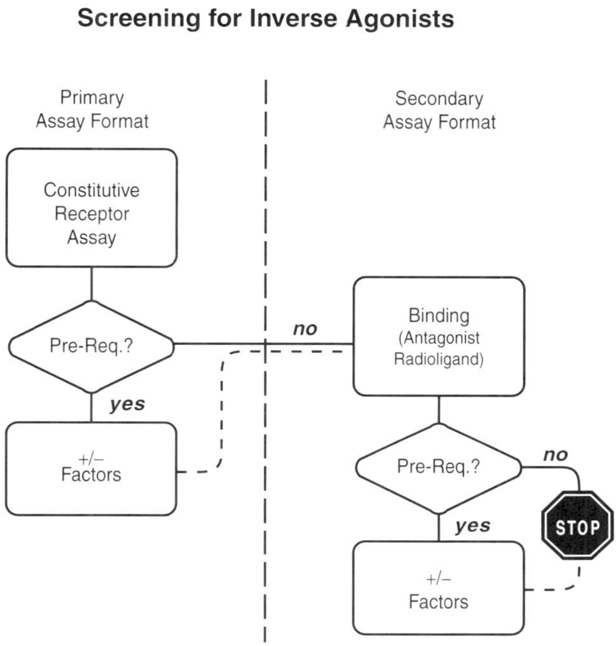

Fig. 5.21. Logical progression map for screening for inverse agonists

cost/effort aspect comes from the fact that constitutive assays usually are transient transfection assays. This is because stable cell lines adjust to constitutive elevated basal activity and, with time, the elevated activity returns to normal. Thus, a new transient cell line must be made for each assay. With this transfection can come variability, because the receptor level varies with efficiency of expression and the magnitude of the constitutive response can vary with the metabolic state of the cell. Finally, a neutral antagonist, of course, will not be detected in a constitutive receptor assay. These factors are summarized in Table 5.10.

5.4.3.2 Binding Assays

Theoretically, inverse agonists can be found in conventional binding assays, the rationale for this being the thermodynamic argument stating that the probability

Table 5.10. screening for antagonists in a constitutive receptor assay

Advantages	Possible Disadvantages
• Very sensitive to negative efficacy	• Greater cost and effort • Possible greater variability • Will not detect a neutral antagonist

that ligands will have differing affinities for different receptor conformations (see section 5.4.2.3). Thus, there need not be a specific design to find inverse agonists in a binding assay, except to use an antagonist radioligand to maximize the probability of finding an antagonist (see section 5.4.2.2) and not an agonist (see section 5.4.1.2). After a ligand with affinity has been found, a secondary test for negative efficacy in a constitutive receptor assay would be done (see Chapter 4, section 4.5 and this chapter, section 5.4.3.1). A possible negative factor would be the detection of false positives in neutral antagonists, but this would be detected in the secondary testing in a constitutively active system.

5.4.4 SCREENING FOR ALLOSTERIC ENHANCERS

As described previously, another class of potentially useful drug are ligands that potentiate the activity of the endogenous agonist. This has the advantage of preserving the response pattern of the system (see section 5.2.1). A scheme for the detection of allosteric enhancers is shown in Fig. 5.22.

The primary assay for the detection of an allosteric enhancer is a functional assay utilizing the endogenous natural agonist. The major advantage of this

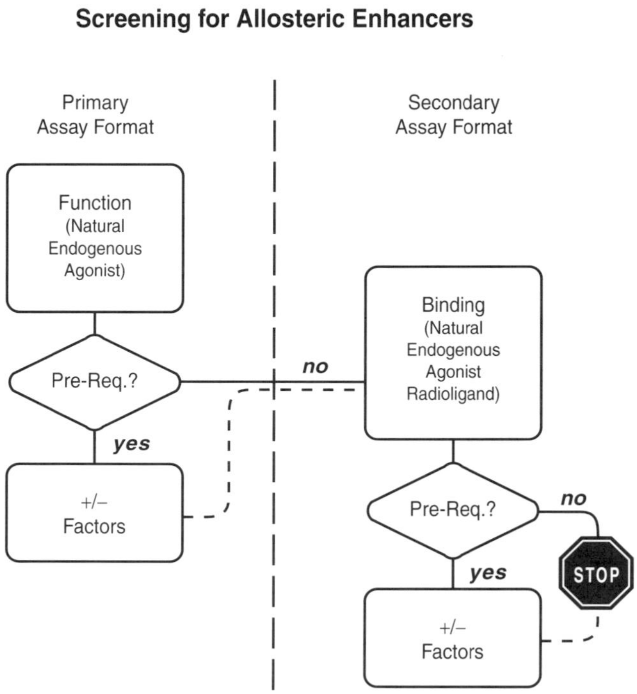

Fig. 5.22. Logical progression map for screening for allosteric enhancers

approach is that the physiological endpoint, namely an enhanced response, is measured directly. In this way, ligands that either increase the affinity of the receptor for the endogenous ligand or increase the efficacy of the agonist (or efficiency of agonist-induced receptor–G-protein interaction) will be detected. It is conceivable that this could occur through a ligand-induced production of a receptor conformation that is more conducive to activation by the agonist than the natural conformation. Another approach would be a binding assay with the endogenous natural agonist radioligand to detect possible increases in affinity with the allosteric enhancer (see section 5.2.3).

5.5 CONCLUSIONS

In general it can be seen that the desired chemical target largely determines the approach taken to screening. There are distinct advantages to using either binding or functional assays in screening and in some cases a particular receptor target may contravene standard choices of assay. In particular, the detection of ligands that have an allosteric mechanism of action on receptors may require direct functional assays for detection.

5.6 REFERENCES

Adham N, Zgombick JM, Bard J and Branchek TA (1998) Functional characterization of the recombinant human 5-hystroxytryptamine$_{7(a)}$ reeceptor isoform coupled to adenylate cyclase stimulation. *J Pharmacol Exp Ther* **287**: 508–514.

Avanitakis L, Geras-Raaka E, Varma A, Gershenforn C and Cesarman E (1997) Human herpesvirus KSHV encodes a constitutively active G-protein coupled receptor linked to cell proliferation. *Nature* **385**: 347–350.

Bianiasz PD, Fridell RA, Aramori I, Ferguson SSG, Caron MG and Culen BR (1997) HIV-1-induced cell fusion is mediated by multiple regions within both the viral envelope and CCR-5 co-receptor. *EMBO J* **16**: 2599–2609.

Birdsall NJ, Hulme MEC, Keen M, Stockton JM, Pedder EK and Wheatley M (1987) Can complex binding phenomena be resolved to provide a safe basis for receptor classification? In *Perspectives on Receptor Classification, Receptor Biochemistry and Methodology*, Vol. 6, ed. by JW Black, DH Jenkinson and VP Gerskowitch, pp. 61–71. Alan R Liss, New York.

Cheng Y and Prusoff WH (1973) Relationship between the inhibition constant (Ki) and the concentration of inhibitor which causes 50 percent inhibition (I_{50}) of an enzymatic reaction. *Biochem Pharmacol* **22**: 3099–3108.

Gaddum JH (1937) The quantitative effects of antagonistic drugs. *J Physiol London* **89**: 7P–9P.

Gaddum JH (1957) Theories of drug antagonism. *Pharmacol Rev* **9**: 211–218.

Goldberg ND (1975) Cyclic nucleotides and cell function. In *Cell Membranes, Biochemistry, Cell Biology, and Pathology*, ed. by G Weissman and R Claiborne, pp. 185–202. HP Publishing, New York.

Hedberg A, Carlsson E, Fellenius E and Lundgren B (1982) Cardiostimulatory effects of prenalterol, a beta-1 adrenoceptor partial agonist, *in vivo* and *in vitro*. *Naunyn Schmiedebergs Arch Pharmacol* **318**: 185–191.

Ishihara T, Nakamura S, Kaziro Y, Takahashi T, Takahashi K and Nagata S (1991) Molecular cloning and expression of a cDNA encoding the secretin receptor. *EMBO J* **10**: 1635–1641.

Jakubik J, Bacakova L, Vera L, El-Fakahany EE and Tucek S (1996) Activation of muscarinic acetylcholine receptors via their allosteric binding sites. *Proc Natl Acad Sci USA* **93**: 8705–8709.

Jakubik J, Bacakova L, El-Fakahany EE and Tucek S (1997) Positive cooperativity of acetylcholine and other agonists with allosteric ligands on muscarinic acetylcholine receptors. *Mol Pharmacol* **52**: 172–179.

Kolias-Baker C, Ruble J, Dennis D, Bruns RF, Linden J and Belardinelli L (1994) Allosteric enhancer PD 81,723 acts by novel mechanism to potentiate cardiac actions of adenosine. *Circ Res* **75**: 961–971.

Lazareno S and Birdsall NJM (1995) Detection, quantitation, and verification of allosteric interactions of agents with labeled and unlabeled ligands at G-protein-coupled receptors: interactions of strychnine and acetylcholine at mucarinic receptors. *Mol Pharmacol* **48**: 362–378.

Lazareno S, Gharagozloo P, Kuonen D, Popham A and Birdsall NJM (1998) Subtype-selective positive cooperative interactions between brucine analogues and acetylcholine at muscarinic receptors: radioligand binding studies. *Mol Pharmacol* **53**: 573–589.

Molero X and Miller LJ (1991) The gall bladder cholecystokinin receptor exists in two guanine nucleotide-binding protein-regulated affinity states. *Mol Pharmacol* **39**: 150–156.

Mueller H, Weingarten R, Ransnas LA, Bokoch GM and Sklar LA (1991) Differential amplification of antagonistic receptor pathways in neutrophils. *J Biol Chem* **266**: 12939–12943.

Van Sande J, Swillens S, Gerard C, *et al* (1995) In Chinese hamster ovary K1 cells dog and human thyrotropin receptors activate both the cyclic AMP and phosphatidylinositol 4,5-bisphosphate cascades in the presence of thyrotropin and cyclic AMP cascade in its absence. *Eur J Biochem* **229**: 338–343.

6 Mathematical and Statistical Framework for Problems in Drug Discovery

6.1 INTRODUCTION

Collection and interpretation of large quantities of data are central to many strategies for drug discovery, e.g. high-throughput screening. There are numerous types of data: results from primary and secondary screening, calculated and measured properties of chemical structures; databases of protein sequences; results of clinical genetics studies; and data from genomics. Although the high-throughput screening data are remarkable for their volume, it is more important to consider such data as one of many different types of information available for drug discovery. Mathematics and statistics have multifaceted roles for the analysis of data in pharmaceutical research. For assay development and optimization, statistical experimental design is a powerful method for rapidly identifying factors that contribute to assay variability and cost. As data become available for structure–activity relationship analysis, the statistics support data-based rather than opinion-based interpretations of the data. These interpretations are based on samples of data and statistical methods, including analysis of variance. Analysis of biological data often involves the use of mathematical models for simulation or fitting to experimental data. These models are used to represent and test mechanistic hypotheses for biological systems such as signal transduction. This chapter and the following three chapters provide a framework for analysis and interpretation of data in terms of the sampling and statistical methods that are relevant for specific applications in the drug discovery process.

6.2 FUNDAMENTAL PRINCIPLES OF DATA ANALYSIS

Statistics are concerned with the study of variability. A specific application of statistics will involve design or optimization of an experiment, providing descriptive or summary measures of observations from an experiment or fitting mathematical models to data to identify relationships in data. In most cases, samples are used to make inferences about a biochemical system or experiment. The first step in any of these analyses is identification of the specific research questions to answer and the appropriate experimental design. The data are interpreted in order to understand more about the source of the data and to draw conclusions about the research questions. The process is often iterative (Fig. 6.1). Data collection, data reduction and, finally, interpretation or association analysis follow experimental design in a cyclic fashion.

Experimental data analyzed in the drug discovery process are diverse in terms of sample size, relationships between the data and sample distribution shape. In a typical preclinical hemodynamic in vivo experiment, data are collected from 10–20 subjects. The data structure consists of over eight variables, sampled at 200 different time points. The variables are highly correlated: as one hemodynamic measure changes, then other measures also change. Multiple hypotheses are tested: are control and treatment groups different at different doses and time points? Is there a trend with respect to time or dose? At what time is there a significant difference between control and treatment groups? At the other end of the spectrum, the results from a high-throughput screening biochemical assay can consist of over a million observations of a single measurement, such as percentage inhibition in an enzyme assay. The observations are associated with the test chemical compounds. Statistical analyses include: identifying compounds

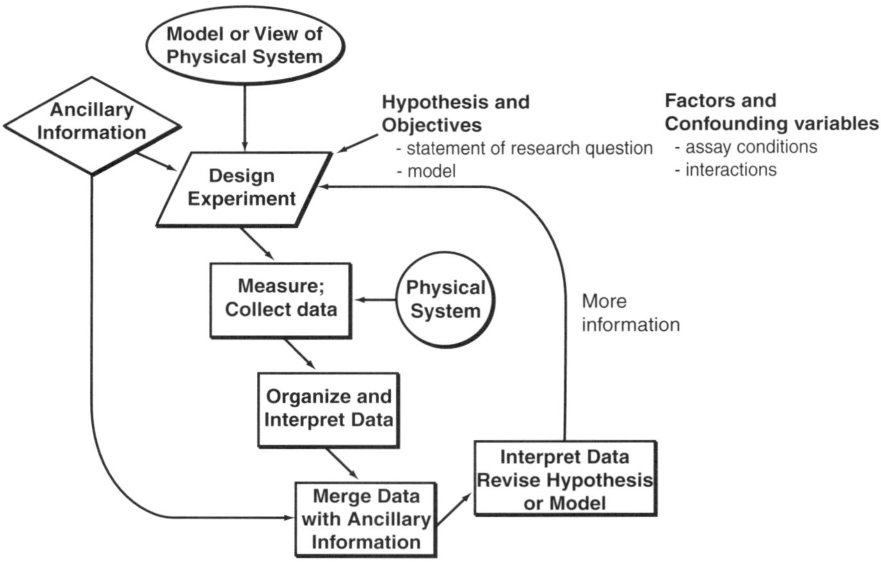

Fig. 6.1. The iterative process of data analysis. Interpretation of data from experiments involves the process of designing experiments, collecting data and making inferences from the data. The process iterates as results from one experiment are used to design subsequent experiments. Ancillary information from previous experiments and knowledge about the physical system are merged with data from the current experiment

to retest in concentration–response format, characterization of compound sets and specific chemistries with respect to biological results and identifying if there are trends in the data with respect to properties of the chemicals, e.g. molecular weight or hydrophobicity.

The different types of data and research questions preclude a cookbook approach to statistical analysis. Instead of associating an analysis (e.g. t-test) with a specific problem in discovery, it is more effective to define a framework for the analysis of data. The framework involves stating the research questions in a clear, quantifiable sense, understanding the data (scales, variability, precision), designing the experiments to obtain maximal information, identifying appropriate statistical tests and methods for analysis of the data and finally reporting the data in a concise format. The components of the framework are broadly stated as:

(1) Clear statement of research problem and questions to address by measurement and interpretation of data.
(2) Use of statistical experimental design to maximize information gained from experiment.
(3) Identification of sources of variability in the experiment.
(4) Characterization of the sample and underlying sampling assumptions. Sample size determination for hypothesis tests.
(5) Appropriate choice of statistical methods for visualization, analysis and interpretation of results.
(6) Concise and clear presentation of results.
(7) Linkage between experimental design and data analysis. Each experiment provides more data for a subsequent experimental design.

This chapter defines the framework and describes the application of statistical methods to problems in drug discovery. It does not attempt to cover all of the theoretical and application-specific issues associated with many different statistical methods. Two methods — resampling and outlier identification — are covered in depth because of their application to problems in analysis of biochemical assay data.

6.2.1 STATING THE RESEARCH QUESTIONS

Stating the research questions may seem like an obvious part of the research process, however it is an essential step both for design of the experiments and for interpreting the data. There are two general classes of questions: experiments designed with a specific objective or set of hypotheses to test; and pattern recognition or generation of rules from historical data. Experimental objectives include determining the characteristics of a biochemical system (e.g. measurement of a K_i, IC_{50} or other variable), modeling the relationship between variables (correlation analysis, regression) and testing the effect of an intervention (comparing treatments to a control or each other). Statistics also plays a role in database mining and pattern recognition, where data acquired over the course of time are to be examined for potential application to new projects. Although there are some specialized methods for data mining, the statistical methods and decision processes that form the basis of the framework are applicable to both problems.

6.2.2 MEASUREMENT SCALES

The scale of a measurement partially determines the types of statistical analysis that are appropriate. Data collected in pharmaceutical research laboratories have scales of ratio, interval, ordinal or nominal (Fig. 6.2). Ratio and interval data are the most common measurement scales. Variables collected on these scales are continuous in that they can assume any value within a range. Equal differences between values on ratio and interval scales correspond to equal amounts, regardless of the position on the scale: for example, the difference between 1 g and 3 g is the same as the difference between 101 g and 103 g. Ratio scales have one additional property not applicable to interval scales: equal ratios of values on these scales are equivalent. The easiest way to differentiate interval and ratio scales is that ratio scales have a natural zero point. Weight and height are examples of ratio scales where the zero point is clearly defined. Centigrade temperature is an interval scale. The zero point is arbitrary and different from the zero point of the Fahrenheit temperature scale; neither corresponds to a true temperature of zero.

Ordinal scales differ from continuous scales in that the variables can take on only a discrete set of values but have a definite order to them. Ranked data, such as ranked potencies, are common forms of ordinal data. Nominal scales are similar to ordinal scales but have no explicit order to them. Variables such as gender and presence or absence of chemical features are expressed on nominal scales. Often, analysis may involve using a variable on different scales, depending on the objective of the analysis. Dose, for example, is treated as a continuous variable for dose–response curve fitting or modeling. Dose is considered on a nominal or ordinal scale in an analysis of variance (ANOVA) experiment to identify, for example, the lowest dose that gives a response significantly different from the control response. Because a different analysis is obtained for different variable scales, it is very important to set scales appropriately when using statistical software.

6.2.3 POPULATIONS AND SAMPLES

Statistics are used to summarize results and to make comparisons and contrasts associated with an experiment. In some cases, the experiment and analysis are set up to take measurements on an entire group of observational units, a general term that includes compounds or subjects. In other cases, the study design is to take a sample from a portion of a larger group. In the first case, the group is termed a population in statistical terms; in the second case, the group is considered a sample. It is important to understand the distinction between a

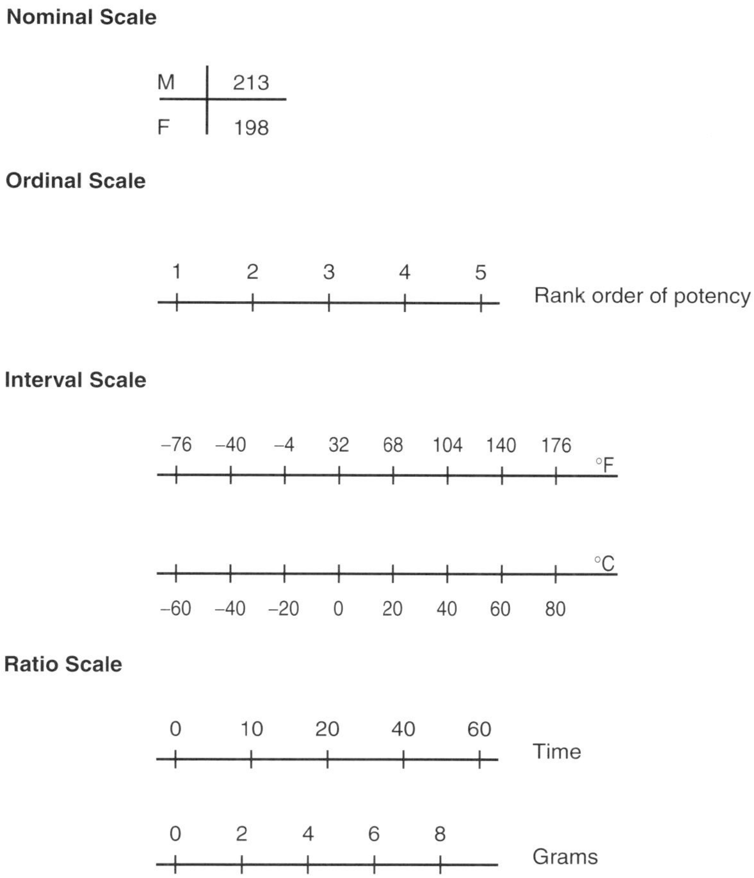

Fig. 6.2. Measurement scales. The scale that is used to obtain measurements is important in the choice of analysis methods. The scale is one determinant in the types of analyses that can be performed. Nominal scales put observations into different bins without any order; they are simply names for observations from a sample. One example of a nominal scale is gender. Ordinal scales put observations into different bins but there is an explicit ordering of the bins. The variables scored on an ordinal scale are discrete instead of continuous. Rank order of potency of compounds is an example of an ordinal scale. Interval scales are continuous. Variables can assume any value in a defined range. Equal differences between values on interval scales correspond to equal amounts regardless of the position on the scale. There is not a natural zero point for an interval scale. Centigrade temperature is an example of an interval scale. Ratio scales are similar to interval scales. Variables measured on ratio scales are continuous, and equal differences between values are equivalent regardless of the position on the scale. Ratio scales have the additional properties that there is a natural zero point and equal ratios of values on the scale are equivalent

population and a sample so that the appropriate statistics are used. Unfortunately, the terms sample, population, parameter and variable are used interchangeably for numerous applications. By the strictest definition, descriptive statistics are calculated for populations. In practice, descriptive statistics are calculated for samples in addition to populations. The most important part of the statistical framework is to understand the collection of observational units and their relationships. e.g. are the observational units (subjects) independent? Are measurements taken on the same unit at different times or conditions? Does variability change as a function of some factor?

Populations are well-defined, enumerated collections of things. Use of the term does not imply size: a population can be large or small. A combinatorial library, all compounds used in a drug discovery project or all people involved in a clinical trial can be defined as populations. Because a population is the fully enumerated collection, descriptive statistics such as the mean, variance and histograms are very useful for characterization. The descriptive statistics are termed the parameters of the population.

Samples are subsets of populations. When samples are chosen carefully, they are used to make inferences about the population. In a clinical trial, a subset of the human population is sampled in order to make inferences about the effect of a therapeutic intervention. When a compound is tested in an isolated tissue experiment, a sample is used to make inferences about the effect of the compound in a larger population. Methods for choosing the sample are important. Many statistical methods assume that the samples are drawn at random. Investigators must be careful to ensure that the subjects in the sample are in fact drawn randomly and are not from a stratified part of the population. Generally, calculations made on samples are designated statistics whereas calculations made on the population are called parameters.

Among the many terms that are used in statistical research that cause confusion are bias and robustness. Technically, the bias of a statistic is calculated by taking the average of the errors of the estimate and including their sign (positive or negative). By definition, this is a comparison between a sample statistic and a true or population-based parameter. A positive bias indicates that the sampling method gives estimates that are, on average, too high; a negative bias indicates that estimates are too low. The mean error of the estimate is zero for an unbiased estimate.

Robustness is a term that implies a resistance or decreased sensitivity to assumptions. A robust statistic, for example, is one that gives consistent and accurate estimates for a variety of population distributions (different shapes and forms). A statistic may be termed robust for non-normality, indicating that it works well for non-normal distributions. Finally, a statistic may be described as being robust to outliers, indicating that the statistic is resistant to the effects of outliers. For example, the median is a more robust statistic to outliers than the mean. When the term robust is used, it is advisable to describe the context.

6.2.4 VARIABILITY

All biological processes have inherent variability. When a compound is tested in a radioligand binding assay or in an animal, the results will vary for different samples and upon retest on different days. When a combinatorial library is synthesized, the activities of the compounds will follow a probability distribution that is scaled to the level of chemical interaction between the library compounds and the biochemical target. Part of the research design, assay development, compound optimization and compound progression involves analysis of the variability of the data. Variation has both a systematic and random component (Fig. 6.3).

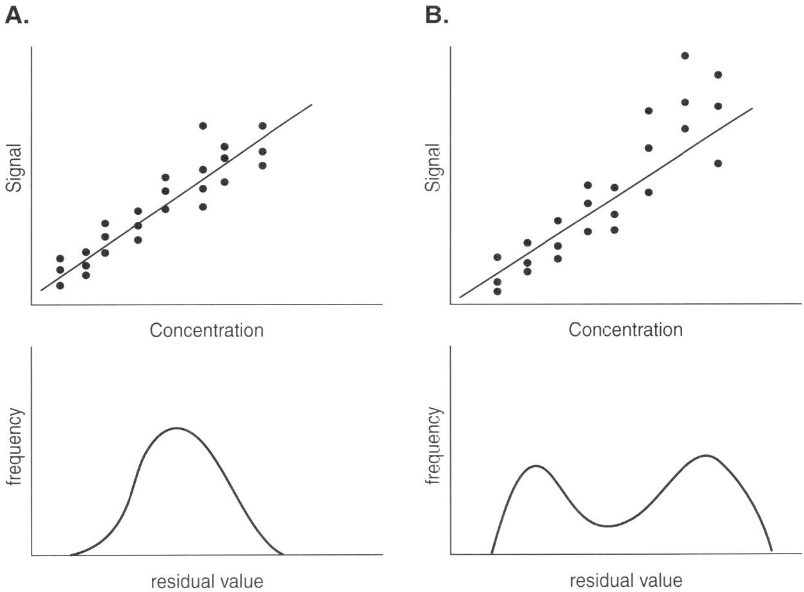

Fig. 6.3. Random and systematic variation. Variation in a process or system can be characterized as having random components, systematic components or both. (A) The relationship between concentration and signal is systematic whereas the error in the relationship between these variables is random. The signal increases in magnitude as a function of increasing concentration. Observations of the signal for different values of concentration are randomly scattered around the line: some below and some above the line. If the residuals (differences between the observations and the prediction from the line) are plotted in a histogram, they are normally distributed. (B) In this case, the relationship between concentration and signal is systematic. The error in the relationship between these variables also has a systematic component. The points do not randomly fall around the line. For low values of concentration the points generally fall below the line, whereas for high values of concentration the points are above the line. A histogram of the residuals shows two peaks for the two characteristic residuals: positive and negative

Systematic variation has a predictable pattern. A liquid-handling system that repeatedly underdispenses a stated volume (e.g. 10 μl) is an example of systematic variation. If plate controls are used to make a calculation and the controls vary in a systematic fashion, then so will all of the calculations based on the controls. If a compound degrades with time, the measured activity will decrease in a systematic fashion.

Random variation does not have a predictable pattern. If a compound is given to animals then, depending on the dose, some animals may not respond to the dose but others may respond with maximal effect. Monomers chosen at random and used to construct a combinatorial library will provide a range of activities that correspond to a sample distribution. For many biological processes, the random component of variation is modeled as a probability distribution, either the normal distribution or a distribution that more closely approximates to the shape of the variability of the underlying process. The probability distribution is characterized by a measure of central tendency, where the majority of the values are recorded, and a measure of variability or dispersion. The mean and variance characterize the shape of the normal distribution (Fig. 6.4).

For an investigator, it is important to recognize sources of variability, test for systematic variation that can be eliminated, use experimental design to minimize random variability and, most importantly, have an estimate of the variability of their results. The estimate of variability is needed to understand whether differences are most likely due to real differences between compounds or due to chance alone.

6.2.5 DETERMINING ASSAY VARIABILITY OR PRECISION

An estimate of assay precision or variability is essential for statistical analysis. Estimates of variability are obtained by repeating the experiment under similar experimental conditions in order to obtain replicate measurements. It is important to understand the conditions under which the experiment was repeated. Repeated tests may be in the same run, between runs on different days or even repeated measurements in the same sample (animal, tissue, biological preparation). All can be reported as an estimate of precision under the given condition, but for an estimate of variability it is essential that the repeated measurements be independent. The essential issue is that the replications must involve independent experimental units or subjects, not a subsample. For example, if an incubation of cell membranes, radioligand and inhibitor is pipetted into multiple wells, the estimate of variability is representative of the error in pipetting more than the biological variability of the assay. Measurements of the heart rate of a subject at different time points are not independent. The variance between sub-samples is generally much lower than the variance between independent samples.

Some experimental designs, particularly for *in vivo* studies involve repeated measurements in the same subject. These designs are often very efficient, because the variance within a subject is usually much lower than the variance between

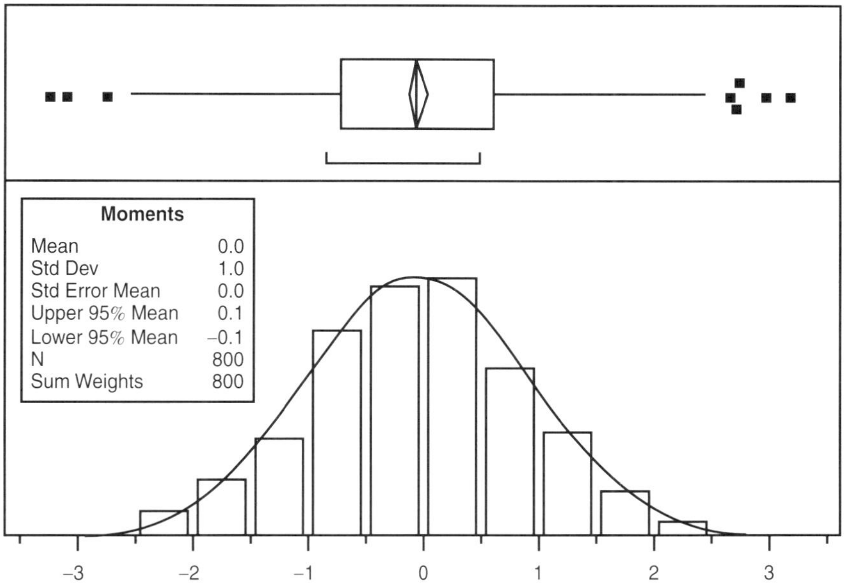

Fig. 6.4. The normal distribution. The normal distribution is characterized by two parameters: a measure of central tendency (mean) and a measure of dispersion (variance). In this case, the mean is zero and the standard deviation is 1. The box plot above the bell-shaped curve shows two different graphical measures of central tendency and dispersion for the distribution. The median (50th quantile) is the line in the center of the box and the quartiles (25th and 75th quantiles) define the edges of the box. The mean and the 95% confidence limits on the mean are defined by the diamond in the middle of the box, the mean being the vertices of the diamond. If the distribution is skewed, there will be differences between the mean and the median that are easily identified from the plots. The lines that extend from the box are set at 1.5 times the difference between the 25th and 75th quartiles. The individual points fall outside this range and can be considered roughly as outliers based on a robust estimate of variation (the difference between the quartiles).

subjects. However, special care must be taken to analyze repeated measures designs. Measurements in the same subject, e.g. at different time points or under different conditions, tend to be correlated. These experiments require analysis that incorporates the correlation structure of these measurements. An experimental design that involves measurements in the same subject is usually analyzed by using repeated measures analysis of variance.

During the course of many drug discovery projects, compounds are tested different number of times. A standard reference compound is often tested on every screening plate. Other compounds are tested once or repeated two to four times. The compounds tested once cannot contribute to an estimate of assay variability, but any compound tested multiple times contributes to the overall estimate of assay variability. Figure 6.5 shows a simple model to estimate overall assay variability. The linear model has a single factor, the compound identifier. In this model, a mean is estimated for each compound. The root-mean-square (RMS) error estimate is the overall or global estimate of variability. Checking for systematic patterns in variation is important with this type of dataset. Variability is plotted as a function of signal strength (Fig. 6.6) to check for heteroscedastic (non-uniform) variance. Often, variance of a biological response increases as concentration decreases. Biologically, the effect is seen if there are threshold effects present: fewer of the experimental units respond at a concentration lower than the threshold compared to a higher concentration. Other possible systematic effects include factors related to compound presentation (row and column position on a plate), robotic effects (systematic changes in volume dispensed), time effects (equilibrium, reagent batches) and readout (edge effects or non-uniform heating of the plate in the reader).

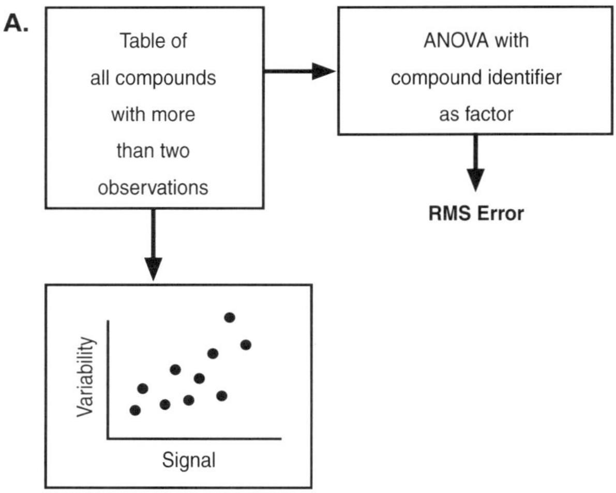

Fig. 6.5. Simple model to estimate overall assay variablity. (A) Schematic of the model. Construct a table of all compounds with more than two observations. Calculate an analysis of variance model using compound identifier as the only factor. The root mean square (RMS) error is an overall estimate of variability across all of the compounds. A plot of variability versus signal (affinity, function or other measure of response) indicates whether this estimate varies as a function of signal. (B) An example of the model using data on α_{1B} affinity (pK_i) for twelve compounds. The RMS error is 0.32, which is approximately 1/3 log unit. The histogram of the standard error of α_{1B} pK_i estimates shows that this number has considerable variability, ranging from 0 to 0.7. Most values are below 0.1. (Data provided by David Saussy, Glaxo Wellcome.)

B.

Compound	N	mean	sem
CMPD 1	3	6.09	0.16
CMPD 2	3	7.92	0.10
CMPD 3	3	7.07	0.09
CMPD 4	3	7.05	0.10
CMPD 5	3	6.48	0.03
CMPD 6	3	8.60	0.12
CMPD 7	3	7.35	0.07
CMPD 8	3	7.62	0.09
CMPD 9	10	6.95	0.18
CMPD 10	3	6.69	0.04
CMPD 11	4	6.05	0.02
CMPD 12	4	7.23	0.10

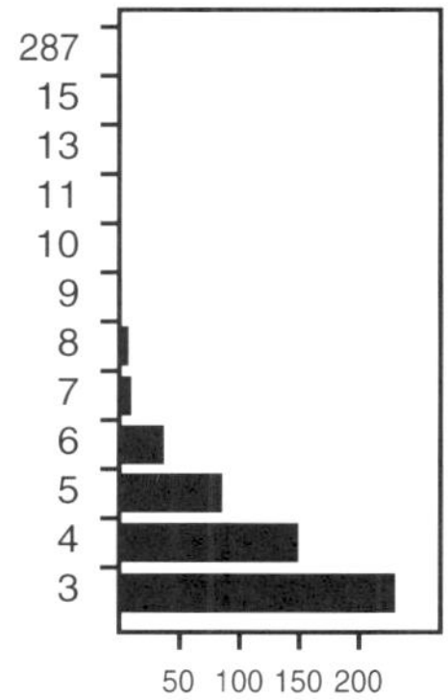

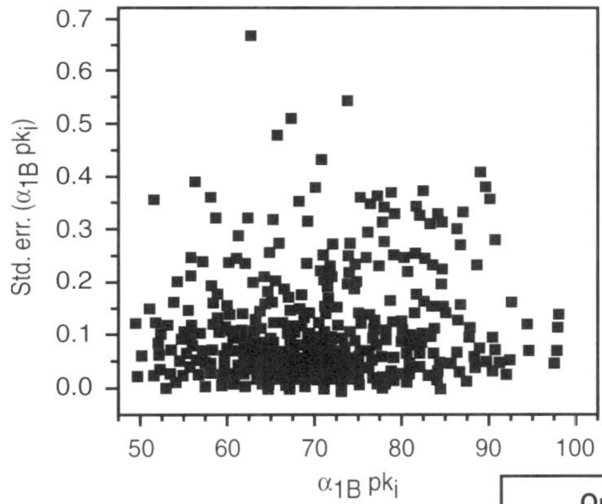

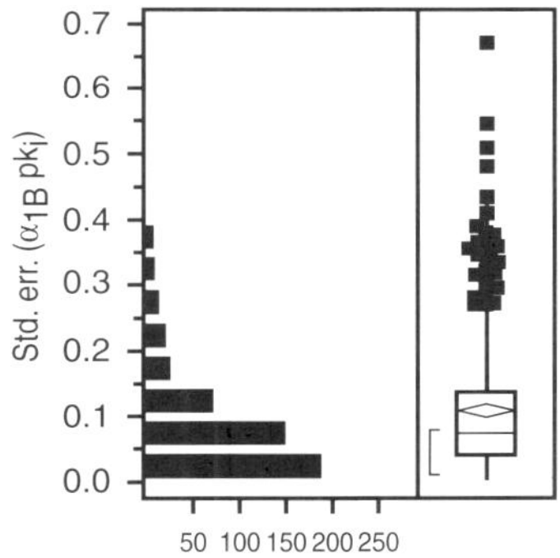

Response: α_{1B} pk$_i$	
RSquare	0.92
RSquare Adj.	0.90
Root Mean Square Error	**0.32**
Mean of Response	7.09
Observations (Sum Wgts.)	2448.00

Quantiles		
maximum	100.0%	0.67
	99.5%	0.52
	97.5%	0.36
	90.0%	0.25
quartile	75.0%	0.13
median	50.0%	0.07
quartile	25.0%	0.04
	10.0%	0.02
	2.5%	0.01
	0.5%	0.00
minimum	0.0%	0.00

Moments	
Mean	0.10
Std. Dev.	0.10
Std. Err. Mean	0.00
Upper 95% Mean	0.11
Lower 95% Mean	0.10
N	543.00
Sum Weights	543.00

Fig. 6.5. (*continued*)

6.2.6 MEASUREMENTS OF CENTRAL TENDENCY AND VARIANCE

Measurements of central tendency and variance are the most common descriptive statistics. The simplest definition of central tendency is the region of a distribution where values tend to congregate, i.e. the peak of the probability distribution. The

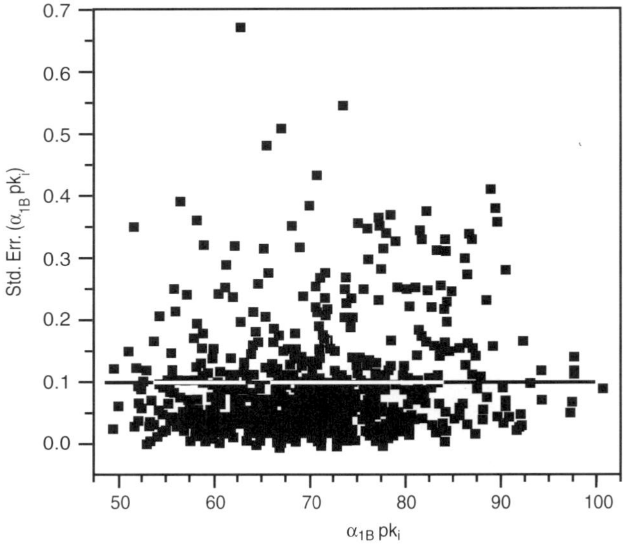

Mean Fit	
Mean	0.104
Std Dev [RMSE]	0.097
Std Error	0.004
SSE	5.095

Fig. 6.6. Plot of variability versus signal. The variability (standard error of α_{1B} pK_i estimates) does not appear to be a function of signal (α_{1B} pK_i estimate). The mean is approximately 0.1 on a pK_i unit scale. Abscissa: mean α_{1B} pK_i from replicate measurements ($n = 3$–10, depending on the compound). Ordinate: standard error of the mean of the α_{1B} pK_i measurements. RMSE = root-mean-square error; SSE = sum of squares error. (Data provided by David Saussy, Glaxo Wellcome.)

mean or sample average is the most common measure of central tendency. The mean is calculated by adding all of the sample elements and dividing by the number of elements. A weighted mean is calculated by computing the sum of products of sample elements and weights and dividing the quantity by the sum of the weights:

$$\bar{x} = \frac{\sum_{i=1}^{n} x_i w_i}{\sum_{i=1}^{n} w_i} \qquad \ldots (6.1)$$

where w_i are the weights and x_i are the observations in the sample.

The median is a measure of central tendency that is defined as the middle value. In a sample that has an odd number of elements, the median is equal to the value that bisects the sample: there are equal numbers of elements greater than the median and less than the median. If a sample has an even number of elements, the median is the average of the two values that are in the middle of the sample. The median is a measure of central tendency that is robust to outliers. Unlike the mean, sample values on the extremes of the distribution have no effect on the value of the median (Fig. 6.7).

The mode is the most common value in the sample. By definition, it is the peak value of the histogram. Like the median, it is resistant to outliers. Additionally, it is not influenced by the shape of distribution of the samples. For example, the mode can be a value at one extreme of the distribution, with the rest of the sample distributed entirely to one side of the mode.

The three measures of central tendency are all equal in the case of a normal distribution. As the distribution becomes skewed, the measures differ. The mode most closely tracks the peak of the distribution. Figure 6.8 shows that the median and mean are influenced more by the skew of the distribution than the mode.

Variation is the extent to which elements in a sample differ among themselves. Although measures of central tendency are associated with the peak of a probability distribution, measures of variation are associated with the width of the distribution. For a population, the variance is the sum of the squared differences between the members of the population and the mean divided by the size of the population. On average, this variance underestimates the variance of a sample drawn from the population. For a sample of a population, variance is calculated by dividing the sum of squared deviations by the number of sample elements minus one. The reason for division by $N - 1$ to compute a sample variance is based on sampling theory. For a population, the mean of the population is considered to be known. For a sample, by definition, the sample mean is an estimate of the population mean. Therefore, before the sample standard deviation can be calculated, the mean must be estimated first at the cost of one degree of freedom. Expressed differently, if the sample mean is calculated, then it and the value of $N - 1$ elements of the sample are sufficient to define the entire sample. This leads to the desired property: that the average value of the sample standard deviation computed for many different random samples is equal to the population standard deviation.

6.2.7 SAMPLING THEORY

Sampling theory is an important part of the framework for the analysis of data. Some aspects of sampling theory apply regardless of the size of the population; others are more critical for the large sample sizes associated with high-throughput screening. Sampling theory is the aspect of statistics that bridges the gap between an observed sample and a theoretical distribution of an infinite number of points. Figure 6.9 illustrates the concept of a sampling distribution in relationship to a

A.

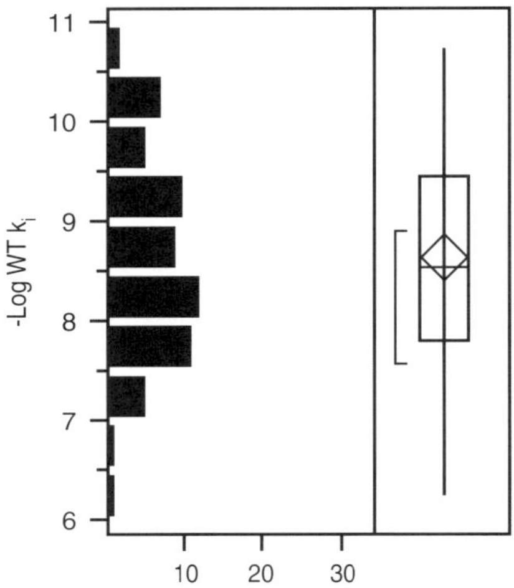

Quantiles		
maximum	100.0%	10.77
	99.5%	10.77
	97.5%	10.74
	90.0%	10.05
quartile	75.0%	9.44
median	50.0%	8.57
quartile	25.0%	7.82
	10.0%	7.43
	2.5%	6.63
	0.5%	6.27
minimum	0.0%	6.27

Moments	
Mean	8.65
Std. Dev.	1.00
Std. Err. Mean	0.13
Upper 95% Mean	8.90
Lower 95% Mean	8.40
N	63.00
Sum Weights	63.00

B.

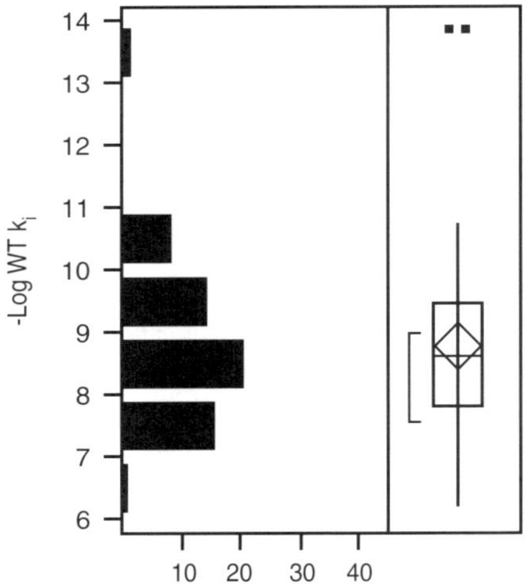

Quantiles		
maximum	100.0%	14.00
	99.5%	14.00
	97.5%	14.00
	90.0%	10.18
quartile	75.0%	9.51
median	50.0%	8.64
quartile	25.0%	7.84
	10.0%	7.44
	2.5%	6.66
	0.5%	6.27
minimum	0.0%	6.27

Moments	
Mean	8.81
Std. Dev.	1.35
Std. Err. Mean	0.17
Upper 95% Mean	9.15
Lower 95% Mean	8.48
N	65.00
Sum Weights	65.00

parent distribution or population. The parent and sample distributions all have the characteristic shape of a normal distribution and have a common mean. The variability of the sample distributions, however, is a function of the size of the sample. The normal distribution sampling theorem states that if a variable is normally distributed with a mean μ and a standard deviation σ, then the sampling distribution of the mean x, based on random samples of size n, also will be normally distributed and have a mean equal to μ and a standard deviation equal to $\sigma/\sqrt{n}$. Note that the standard error of the mean estimates how far off the sample mean is likely to be from the true mean, and not the standard deviation of the population.

The sampling theorem for the normal distribution is an important result. However, the theorem is an extension of a more general aspect of sampling theory: the central limit theorem. This theorem applies to distributions that are not normal and states that if a variable has a distribution with mean μ and standard deviation σ, then the sampling distribution with mean x, based on random samples of size n, will have a mean equal to μ and a standard deviation equal to $\sigma/\sqrt{n}$ and will tend to be normal as the sample size becomes large ($n \geq 30$). This remarkable result applies to the sampling distribution. Although the parent distribution is not normal, or may be of unknown shape, as larger samples are drawn the sampling distributions tend to be normal.

6.2.8 SAMPLE DISTRIBUTIONS

The concept of a sample distribution and the relationship between sample distributions and probability distributions are keys to understanding many statistical methods and, specifically, the framework for interpretation of high-throughput screening data. First, consider a collection of percentage inhibition measurements from a high-throughput screening assay, or a set of IC_{50} values from a secondary assay. The raw data can be tabulated into bins of different widths and a histogram can be plotted. On the histogram, the bins correspond to a range of values for the response and the height of each bar associated with the bin corresponds to the number of compounds with activity in the bin range. The histogram itself is a picture of the sample. If the sample corresponds to a combinatorial library, the peak of the distribution will be centered on the most common activity of the compounds in the library. If the library is a focused, designed library for a specific

---

Fig. 6.7. The impact of outliers on measures of central tendency. (A) A fairly uniform distribution of pK_i values for the potency of a series of inhibitors of HIV protease. The mean and median are similar in magnitude. (B) Two outliers are introduced into the dataset. The outliers, or extreme values in this distribution, have a large impact on the value of the mean. The median and the mean differ in value by almost 0.3 log unit and the outliers are identified on the box plot with values of 14. Outliers have a more substantial effect on the mean than the median or the mode. (Data provided by Eric Furfine, Glaxo Wellcome.)

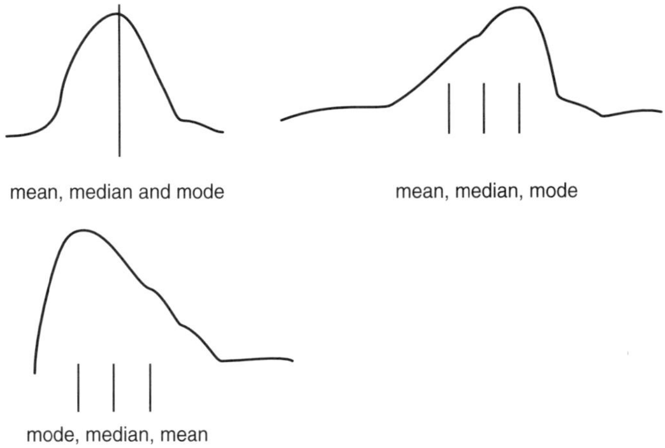

Fig. 6.8. Effect of the shape of a distribution on measures of central tendency. The mean is most affected by the skew of a distribution. The longer tail of the distribution has the effect of pulling the mean toward the tail. The median is less sensitive to the skew of the distribution and the mode is less sensitive than the median

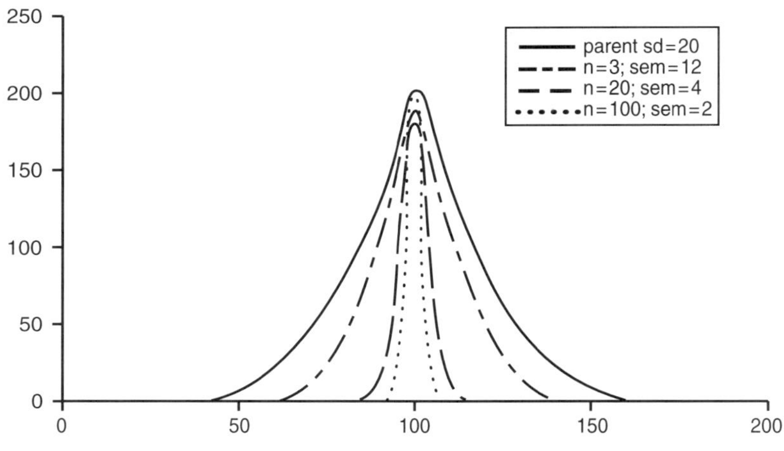

Fig. 6.9. Sampling from a parent distribution. In this figure, the parent distribution has a mean of 100 and a standard deviation (sd) of 20. The other distributions are for samples of different sizes drawn from the parent. As the sample size increases, the variability of the estimate of the mean decreases. The width of the sample distributions is associated with the standard error of the mean(sem). The standard deviation is given only for the parent distribution corresponding to the variability of this distribution. For the samples, a standard error of the mean is reported corresponding to the variability of the estimate of the mean of the parent population based on the sample

target, this activity is likely to be high. Diverse or screening libraries are likely to have a peak at zero, in that most of the compounds are not active against the target. The right-hand tail of the distribution contains compounds that are active against the target. Figure 6.10 shows some of the probability distributions that are commonly encountered in pharmaceutical research.

Sample distributions are drawn for chemical properties. Figure 6.11 shows a set of sample distributions for a 256-member combinatorial library of amides. Sixteen primary amines and sixteen carboxylic acids were reacted in an electronic condensation reaction to construct the library. The profiles were constructed to compare the libraries against known drugs. For example, the molecular weight for the library covers a range 139–429, with an average of 276. As a comparison, for the world drug index of 43 000 compounds, the molecular weight range is 30–42 600. Most orally bioavailable drugs are in the range 180–700. The sample distributions are useful for monomer selection. If it is desirable to keep molecular weight under 500, then the combination of monomers associated with that section of the library product sample distribution is chosen. In order to make the selection, it is necessary to build a data structure that contains both monomers and enumerated products. The products are used to calculate descriptors and selection is made from the lists of monomers. For descriptors that are assumed to be additive, enumeration can be skipped and selection made on the monomers themselves.

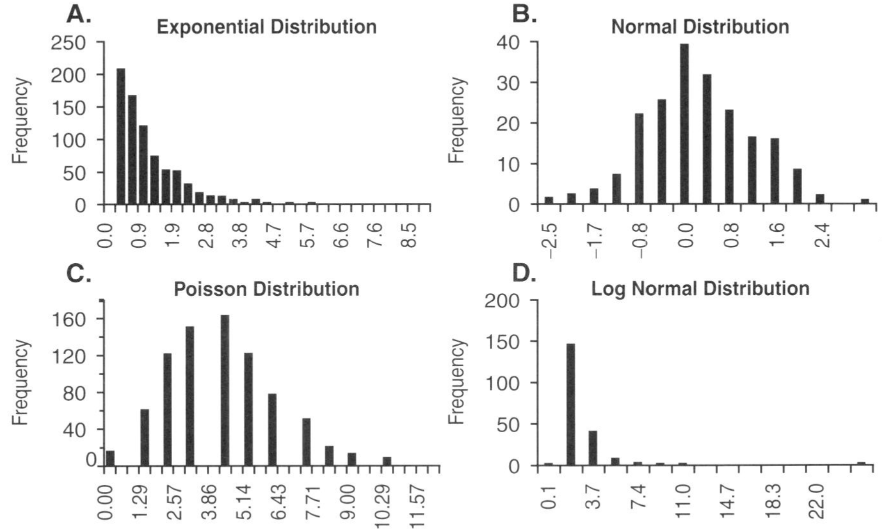

Fig. 6.10. Distributions commonly encountered in drug discovery research: (A) normal; (B) exponential; (C) Poisson (often associated with counts); (D) log normal (often associated with concentrations)

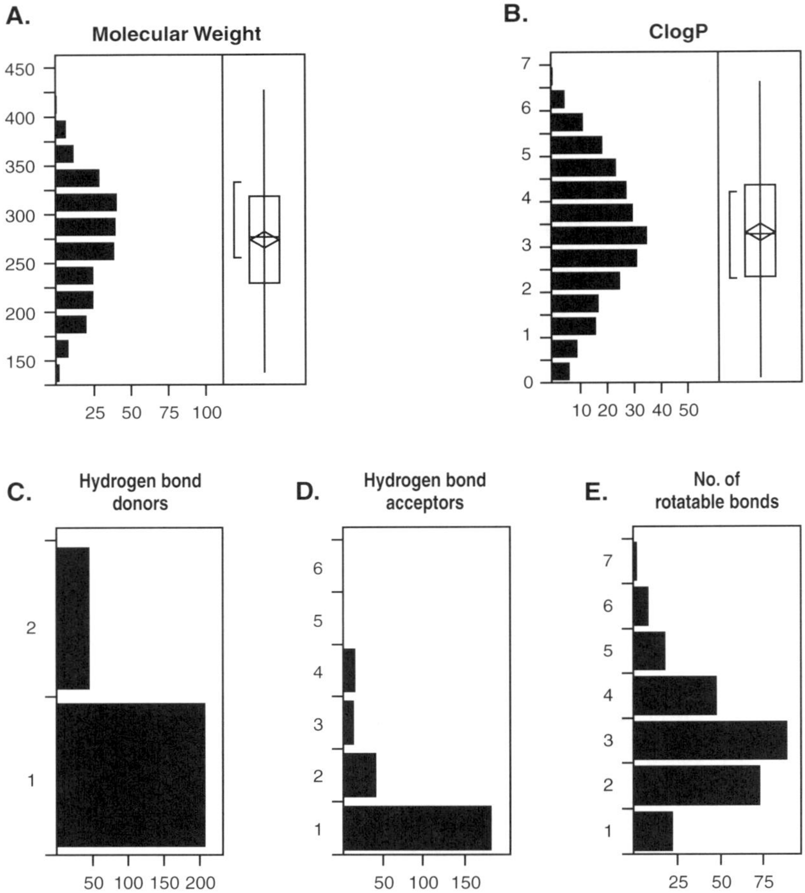

Fig. 6.11. Profile of a combinatorial library. Sixteen primary amines and sixteen carboxylic acids were reacted electronically to create a 256-member combinatorial library of amides. The distributions of chemical properties associated with the library are shown: (A) molecular weight; (B) calculated log *P* (Clog P); (C) number of hydrogen-bond donors; (D) number of hydrogen bond acceptors; (E) number of rotatable bonds. Viewing the distributions of the properties allows selection of monomers or components associated with specific design criteria, e.g. molecular weight or rigidity of the molecule

Computer simulation is used to predict sample distributions for high-throughput screening experiments. The simulations are based on mathematical combinations of probability distributions. Figure 6.12 shows a simulation for a high-throughput screening experiment. A combinatorial library of 500 000 compounds was simulated by calculating ΔG or free-energy values. The ΔG values were assumed

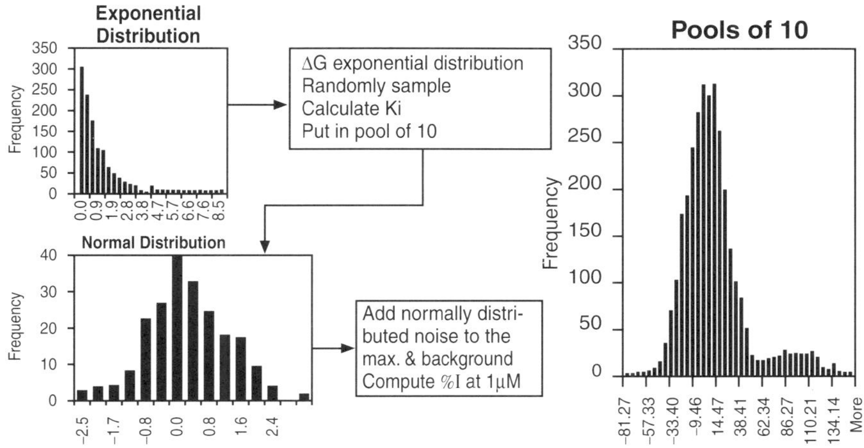

Fig. 6.12. Simulation of a high-throughput screening experiment. A library of 5000 compounds was simulated by assuming an exponential distribution for ΔG values. The screening experiment involved combining the compounds into pools of ten compounds each. Measurements of a control compound defining maximal effect and background were simulated and used to calculate the percentage inhibition at 1 μM. The controls were simulated by assuming a normally distributed error. The simulation shows that the distribution of percentage inhibitions for this experiment is skewed with a heavy right tail

to follow an exponential distribution where most compounds are inactive at 1 μM and a small subset of compounds will have nanomolar activity. The library was sampled to obtain 5000 random compounds with different ΔG values. The ΔG values were converted into inhibition constant (K_i) values. The compounds were put into random pools of ten. The level of inhibition at 1 μM was computed from the K_i values. Positive and negative control values were determined and used to scale the activity. Finally, normally distributed noise was simulated for each control and the percentage inhibition was recorded. The histogram was constructed from the percentage inhibition values. There are several points to make about this simulation. In a theoretical sense, the simulation is able to address questions about screening strategy. Different levels of noise in the controls, libraries having different distributions for ΔG values and different pool sizes and schemes are easily simulated by changing the parameters of the model. The key to the simulation is working with a set of probability distributions: an exponential distribution for the library and normal distributions for the controls. There is statistical theory for working with combinations of probability distributions. This theory gives mathematical models for each combination. For the library simulation, the distributions are simply generated on the computer, sampled and used as a computational model. The model is easily implemented in Microsoft EXCEL®; the program code is shown in Appendix 1.

6.2.9 SAMPLE SIZE ESTIMATION

Sample size estimation is an important part of a research project design if statistics are to be used to make inferences from the data. Sample size estimation addresses the question of how large a sample must be in order to detect a size of some difference. Figure 6.13 shows a schematic for sample size estimation. For the example, measurements of the pK_i values for two compounds are being compared to determine if their difference is statistically significant. The relevant information includes the size of the difference to detect, an estimate of sample variability, the false positive rate and the statistical power. The specific statistical test being used will also affect sample size estimation; in this case, an unpaired *t*-test is used. The schematic of Fig. 6.13 can be generalized to other tests. In Fig. 6.13, it is seen that a 1.0 log unit difference is considered significant and an estimate of the variability is 0.2 log units. The size of the difference is based on the research question being asked. The α level is the probability associated with making a type 1 error. In this case, it would mean indicating that the pK_i for the two drugs were different, when in fact the difference is not statistically significant. An α value of 0.05 indicates that the model assumes that there is a 5% chance of making this error. Statistical power is the probability associated with correctly identifying a difference at least as large as 1.0 log unit. For a given difference, as the desired power increases, more replicates are needed.

In a practical sense, plotting power curves is an effective means to estimate sample sizes. Power curves can be constructed by assuming that some prior estimate of variability has been obtained. Figure 6.14 shows a family of power

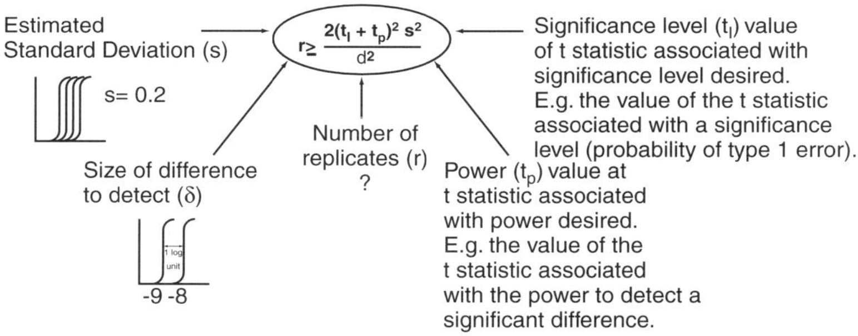

Fig. 6.13. Schematic of calculation of sample size estimates. An estimate must be obtained of the standard deviation of the measurement, e.g. variability of the concentration–response curve location parameter (pK_i or IC_{50}). The size of the difference to detect, the power to detect a significant difference and the significance level of the test must be specified. With these parameters, an equation can be used to estimate the minimum sample size needed for the experiment in order to detect a difference between-groups of the specified magnitude. The equation is modified for tests other than a *t*-test, but has the same general form of a numerator that has a component for variability and statistical power and a denominator based on the size of the difference to detect

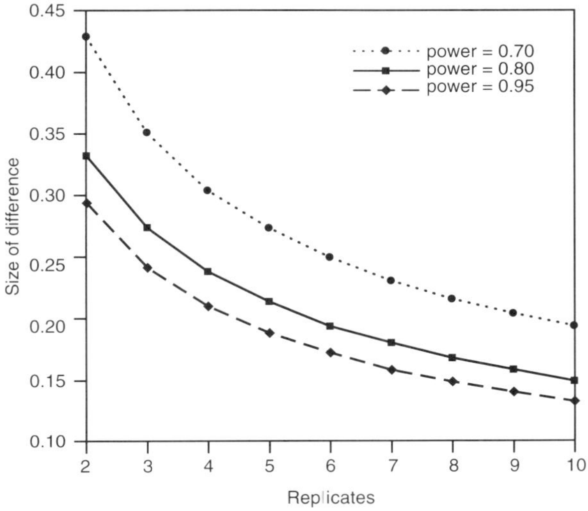

Fig. 6.14. Sample size estimation curves or power curves. These plots are used to estimate the size of a sample for an experiment based on the schematic in Fig. 6.13. Abscissa: the number of replicates needed in the sample. Ordinate: size of the difference that can be detected assuming a false-positive rate (α) of 5% ($P = 0.05$). The curves show this relationship for different levels of statistical power (β). To determine a sample size, choose the statistical power and size of difference to detect and read off the corresponding minimum sample size from the abscissa

curves using the schematic of Fig. 6.13. The curves are generated by assuming that a t-test is to be done to compare two independent samples. The plot shows the number of replicates estimated to be necessary to detect various differences in response. To use the plot, a difference is chosen on the ordinate scale and the corresponding estimate of the number of replicates is located on the abscissa. Note that the different curves are drawn for different levels of statistical power. The effect of rather small changes in the estimate of variability (σ) is substantial on the curves, as shown in Fig. 6.15.

The power curves are easily calculated using statistical software such as JMP®. Appendix 2 includes the equation used in JMP® to draw the curves. Other statistical software includes algorithms to estimate sample sizes for a variety of statistical tests.

6.2.10 DESCRIPTIVE AND INFERENTIAL STATISTICS

Statistics are used to summarize and to make inferences about data. Descriptive statistics summarize data in terms of measures of central tendency, variability and shape. Because data can be expressed in different scales, different statistics

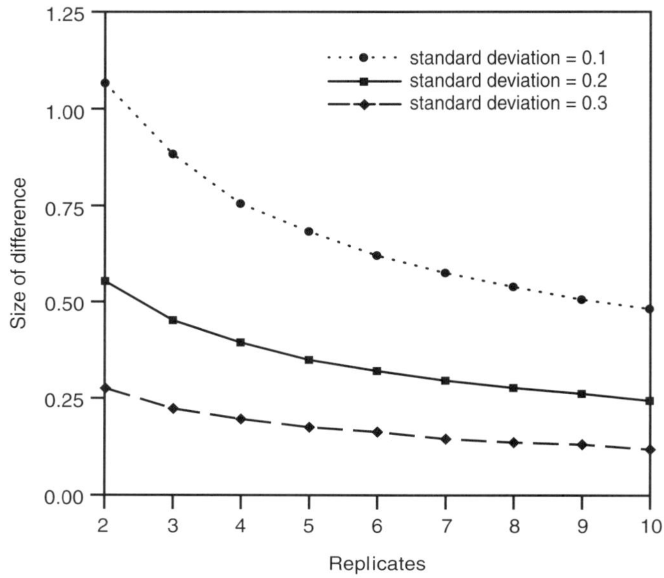

Fig. 6.15. Effect of variability on sample size estimates. Abscissa: the number of replicates needed in the sample. Ordinate: size of the difference that can be detected assuming a false-positive rate (α) of 5% ($P = 0.05$) and a power of 80%. The sample size curves are drawn for different levels of variability to show how the size of the difference increases as variability increases

for summarizing data are needed, including the mean and standard deviation for continuous and ratio data and contingency tables for counts of nominal data.

Hypothesis testing is the domain of inferential statistics. The objectives of hypothesis testing are to compare groups, taking into account variability, and to determine if differences between groups are larger than expected by chance. Hypothesis testing is part of a process that includes experimental design and sampling. Inferential statistics is a more general term that includes hypothesis testing and other methods for making inferences about a population from a sample. Using a sample mean or sample standard deviation to estimate the corresponding parameters of a population is a type of inference.

The framework for hypothesis testing is to set up mutually exclusive questions about the true value of a population parameter or vector of parameters. Data are collected to sample the population and statistics are used to determine if the questions about the parameter values are true. The tests put a level of significance on their conclusion that gives a relative measure of certainty. The question of certainty is related to the probability of drawing a conclusion by chance alone. Whenever a sample is drawn, there is always some chance that the inferences made about the population from the sample are incorrect. Important distinctions must be made about the difference between statistical significance and size of effect. The question of a biochemically relevant effect is different from a statistically significant effect. Statistical significance is an indication that the certainty of a population estimate or hypothesis test, e.g. the difference between the means of two groups, is larger than would be expected by chance alone. Significance depends not only on the size of the difference but on the variability of the two sample means, the number of observations in each sample, the sampling method and the number of tests. Small differences can be significant if the variability is low, whereas large differences can test to be not significant if the variability is high. This is one reason why graphs of mean effects versus time or concentration can be very misleading without an indication of variability of the groups. Whether two compounds are pharmacologically different if their $\mathrm{p}K_{\mathrm{i}}$ differs by 0.5 log unit, or if a blood pressure lowering effect is important if it is greater than 20 mmHg, are scientific questions that are not directly address by statistics. Statistics addresses the questions of whether 0.5 log unit differences between a set of compounds are statistically different, taking into account the variability of the $\mathrm{p}K_{\mathrm{i}}$ estimate, or whether the blood pressure difference is significant for the number of subjects tested and the natural variability. If the research goal is to detect these differences, then the sample size and variability must be optimized in order to make these tests.

6.2.10.1 Hypothesis Testing and *t*-Tests

Hypothesis testing is based in probability theory and some of the terminology commonly used can appear to be arcane. Box 1 defines many of the terms used in hypothesis testing. If the research question can be framed as two mutually exclusive hypotheses, then they are identified as the null (H_0) and alternate (H_1) hypotheses. The hypotheses can involve any parameter or combination of parameters, including differences between means, proportions and paired measures. The analysis compares the two groups. Take, as an example, measurements of heart rate in anesthetized animals where one group receives no treatment and another group receives a test compound. If the mean of the control group is designated μ_c and the treatment group μ_t, the null hypothesis of no difference in mean heart rate can be written as:

$$H_0\colon \mu_{\mathrm{c}} = \mu_{\mathrm{t}} \qquad \ldots(6.2)$$

Box 1

Terms used in Hypothesis Testing

Hypothesis testing: Use of sample statistics (means, standard deviations) to support or reject statements or conjecture about the true value of population parameters.

Null hypothesis: An assumption about the parent population, e.g. a drug treatment does not significantly change heart rate over control levels. In symbolic terms: $\{H_0 :$ HR (treated) $=$ HR (control)$\}$. This allows a test statistic to be formulated with a specific probability distribution if the null hypothesis is true. The sample (observed data) is used to test whether this assumption is supported by the data or not.

Alternative hypothesis: The alternative statement associated with the null hypothesis. For the example: $\{H_1 :$ HR (treated) $\neq$ HR (control)$\}$. If the null hypothesis is not supported by the data or sample, the alternative hypothesis is considered to be supported by the data. If the combination of the null hypothesis and the alternative hypothesis defines all possible values of the population parameters, or all states, then the hypotheses form an exhaustive set. If the two hypotheses do not cover all conditions, they form a truncated set of hypotheses.

One-tailed test: One-tailed tests are a result of using a truncated hypothesis set. For example, if the null hypothesis is that drug treatment does not significantly increase heart rate over control levels $\{H_0 :$ HR (treated) $=$ HR (control)$\}$, then the alternative hypothesis is that treated heart rate is increased (not simply unequal), e.g. $\{H_1 :$ HR (treated) $>$ HR (control)$\}$. The "one-tailed" nomenclature comes from the concept of the distribution of the test statistic. Envision a probability distribution of the test statistic (a t-distribution or normal distribution). If the null hypothesis is true, it is most likely that the observed values are in the center of the distribution. Extreme values, in the tails of the distribution, argue against supporting the null hypothesis. If, based on theoretical, practical or other prior knowledge, the direction of the comparison is restricted to direction (e.g. only high values or only low values of the test statistic), then only one tail of the distribution is used in the test.

Two-tailed test: Two-tailed tests result from using an exhaustive hypothesis set. For example, if the null hypothesis is that drug treatment does not significantly change heart rate over control levels $\{H_0 :$ HR (treated) $=$ HR (control)$\}$, then the alternative hypothesis is that treated heart rate is different, e.g. $\{H_1 :$ HR (treated) $\neq$ HR (control)$\}$. As with the one-tailed tests, it is most likely that the observed values are in the center of the distribution. Unlike the one-tailed tests, both tails of the distribution are used for the comparison, e.g. there is no direction on the magnitude of the change from control levels.

Significance level α: This is a threshold level of probability assigned to a hypothesis test. If the observed significance level is greater than α, the data are considered to support the null hypothesis. Because it is a probability, this number will range from 0 to 1; α is the probability of making a type I error (see definition below).

Power: Technically, the probability of correctly rejecting a false null hypothesis. In symbolic terms, power is $1 - \beta$. The term is derived from the concept of the power of a test to detect a significant difference. The power of a statistical test depends on the size of the difference between the true value of the population parameter and the hypothesized value, the significance level of the test (α), whether a one-tailed or two-tailed test is used, the standard deviation of the sample and the sample size. The power of a test: increases as the difference between the population parameter and the hypothesized value of this parameter increases; increases as the standard deviation of the sample decreases; increases as sample size increases; increases as α increases; and will be larger for a one-tailed test.

***P* value:** Observed significance level. This is the probability that is calculated from the sample distribution. The form of the P value is determined by the probability statement being made, e.g. $P = \text{prob}(x < y)$ is the probability that the observed value of a test statistic (x) is less than a random variable that is associated with a distribution of a given form and a number of degrees of freedom. In order to calculate a P value, the form of the distribution must be specified (t, normal, F), the number of degrees of freedom must be known and it should be specified whether one or both tails of the distribution are to be used. Many statisticians caution against overinterpretation of P values. P values indicate statistical significance, not necessarily scientific or clinical significance. Some analysts prefer to report statistics with confidence intervals rather than P values to allow a graphical interpretation of the sample statistics.

Type I error: This type of error occurs when the null hypothesis is rejected but the true state (population) is that the null hypothesis is true. The probability of making a type I error is often indicated by the symbol α, which is set in advance of making the hypothesis test, e.g. an investigator sets the probability of making a type I error by chance. If α is set at 0.05, for example, when the test statistic is computed and the associated P value is calculated, the null hypothesis is accepted if $P \geq 0.05$ and rejected if $P < 0.05$. The probability of accepting the null hypothesis based on the observed sample when it is in fact true is $1 - \alpha$. Note that α is a criterion for an unlikely but random event.

Type II error: This type of error occurs when the null hypothesis is accepted when the true state (population) is that the null hypothesis is false. The probability of making a type II error is often indicated by the symbol β. The probability of a type II error is linked to power (see definition above). The probability of a type II error decreases as the difference between the hypothesized and true values of the population parameter increases, decreases as α

is increased, decreases as the standard deviation decreases and decreases as the sample size gets larger.

Sample (Test Result)	Population (True State): Null True	Population (True State): Null False
Accept Null	Correct Result Pr (1−α)	Type II Error Pr (β)
Reject Null	Type I Error Pr (α)	Correct Result Pr (1−β)

or

$$H_0\colon \mu_c - \mu_t = 0 \qquad \ldots(6.3)$$

The alternative hypotheses are:

$$H_1\colon \mu_c \neq \mu_t \qquad \ldots(6.4)$$

or

$$H_1\colon |\mu_c - \mu_t| > 0 \qquad \ldots(6.5)$$

The direction of the difference is not important in these tests: the treatment group mean can be larger or smaller than the control group mean and be significantly different. For many research problems, the hypothesis tests can be restricted to one direction. These are known as one-sided, one-tailed or curtailed hypothesis tests. If the test of interest for the example is only in whether the compound increases heart rate, we could write the hypothesis tests as:

$$H_0\colon \mu_c = \mu_t \qquad \ldots(6.6)$$

$$H_1\colon \mu_c < \mu_t \qquad \ldots(6.7)$$

These tests are mutually exclusive but not exhaustive, because the possibility of a control group mean being greater than the treatment group exists. To use the one-tailed test we would have to be certain, based on scientific grounds, that the probability that the compound can lower heart rate is zero.

Figure 6.16 shows pictorial examples of some simple hypothesis tests and their interpretations. Figure 6.16A shows a comparison of a sample mean to a reference value or a test of the mean. Figure 6.16B is a variation on this test, where the

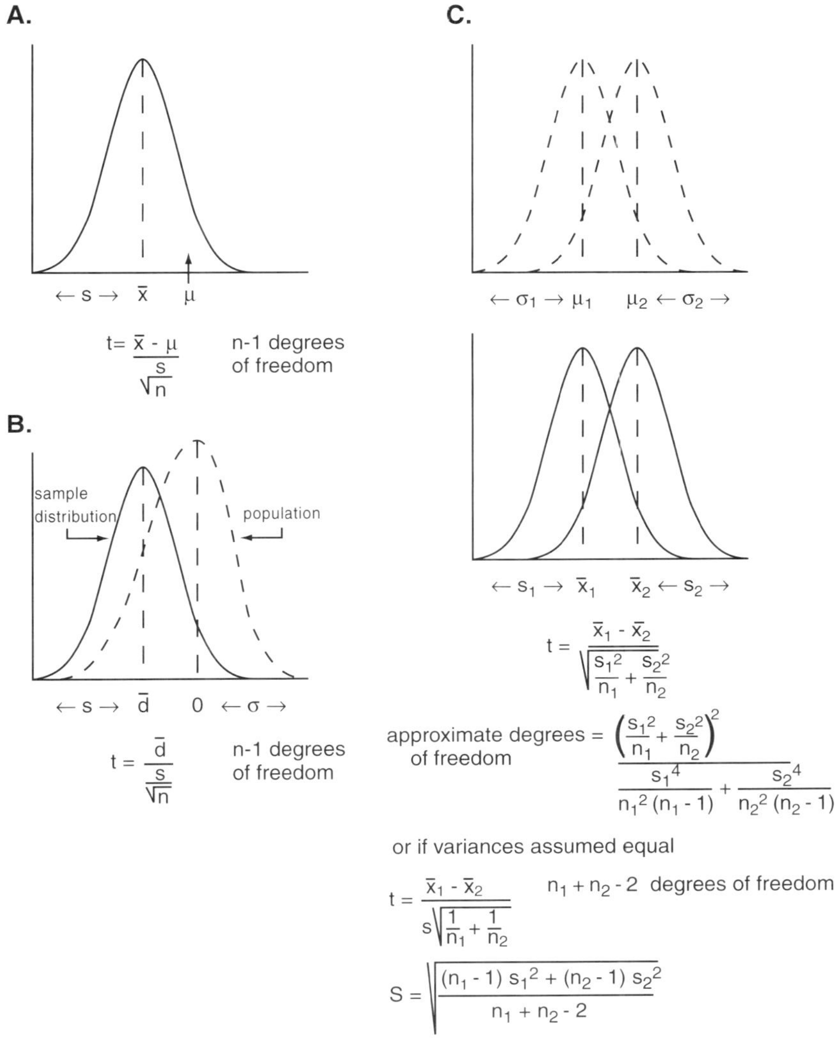

A.

$$t = \frac{\bar{x} - \mu}{\frac{s}{\sqrt{n}}} \quad \text{n-1 degrees of freedom}$$

B.

$$t = \frac{\bar{d}}{\frac{s}{\sqrt{n}}} \quad \text{n-1 degrees of freedom}$$

C.

$$t = \frac{\bar{x}_1 - \bar{x}_2}{\sqrt{\frac{s_1^2}{n_1} + \frac{s_2^2}{n_2}}}$$

$$\text{approximate degrees of freedom} = \frac{\left(\frac{s_1^2}{n_1} + \frac{s_2^2}{n_2}\right)^2}{\frac{s_1^4}{n_1^2 (n_1 - 1)} + \frac{s_2^4}{n_2^2 (n_2 - 1)}}$$

or if variances assumed equal

$$t = \frac{\bar{x}_1 - \bar{x}_2}{s\sqrt{\frac{1}{n_1} + \frac{1}{n_2}}} \quad n_1 + n_2 - 2 \text{ degrees of freedom}$$

$$S = \sqrt{\frac{(n_1 - 1)\, s_1^2 + (n_2 - 1)\, s_2^2}{n_1 + n_2 - 2}}$$

Fig. 6.16. Pictorial examples of hypothesis tests and their interpretations. (A) Comparison of a sample mean to a reference value or a test of the mean. The reference value is μ (population mean or standard) and the sample mean is $\bar{x}$. (B) Testing the difference between paired measurements and zero (e.g. a paired *t*-test). The population mean is centered on zero. The sample distribution (difference of paired measurements) has a mean of $\bar{d}$. (C) Testing the difference of two independent means. The population means are μ_1 and μ_2 and the sample means are $\bar{x}_1$ and $\bar{x}_2$

difference between paired measurements is tested against zero. Figure 6.16C is the test of two independent means. In this case, the comparison is to a difference of zero, but because the means are independent, the variability of each group must be considered separately.

Hypothesis tests are commonly reported in literature studies. Indeed, $P < 0.05$ for a statistically significant test is well established in the literature of many scientific fields. The P value is a probability: it can assume values in the range 0–1. For a two-sample test of equality of mean, the P value is the probability that random sampling of the two populations would lead to a difference in sample means as large or larger than observed. The $P < 0.05$ criterion indicates that there is a 5% chance of observing a difference as large as the difference observed in the experiment even if the population means are the same. Therefore, the P value is a function of both the difference between the means and the variability of the group. As an alternative to reporting P values confidence intervals are calculated to put bounds on a statistic. Confidence limits can be calculated for parametric statistics: sample means, differences between means and parameters of a linear model. The concept of a confidence interval is best described by a simple equation:

$$\text{Confidence interval} = \text{Sample statistic} \pm \text{Standard error} \times \text{Percentile of the distribution for the statistic} \quad \ldots (6.8)$$

Subtraction and addition of the product on the right-hand side of the equation define the boundaries for the statistic. Standard error is the standard error of the sample statistic. The percentile of the distribution for the statistic is calculated using functions from a statistical software program, spreadsheet or table. The percentile is a function of the degrees of freedom for the statistic and the desired confidence level (e.g. 90%, 95% or 99%). Confidence intervals define a range of reasonable values for the statistic or parameter. Technically, the interpretation of a confidence interval is the distance around a population parameter associated with a sample statistic. In other words, there is a given level of confidence that the population parameter (mean or other parameter) is in the range defined by the confidence interval. Pragmatically, confidence limits are useful because they are easy to interpret and can be displayed graphically to show the results of analysis associated with a sample.

6.2.10.2 Analysis of Variance

Analysis of variance (ANOVA) is a very general method for extending hypothesis testing to multiple groups. In analysis of variance, means and variances are calculated for different groups that correspond to a treatment structure or set of factors. If the treatment structure involves a single factor, the analysis of variance is termed a one-way ANOVA. Two factor treatment structures are two-way ANOVAs. Designs that involve larger numbers of factors are termed n-way ANOVAs. The result of the analysis is an indication of whether there are significant differences between the groups for the factors and whether there is a significant interaction between the factors. The design for any analysis of variance experiment must include a sufficient number of observations in each group to estimate both the effects of each factor and the interactions. To determine an estimate of variance, at least two observations are needed in each group. The ANOVA is that part of the statistical framework used to test hypotheses about the presence of relationships between predictor and response variables. Regression analysis and mathematical models describe the nature of the relationship between the variables (linear, non-linear, polynomial) and the associated r^2 statistic in regression measures the strength of the relationship between the variables.

For a standard ANOVA, a statistical test known as the F test is used to test for the significance of effects and interactions. The F test and associated F distribution are based on ratios of variances. If there are two sample variances, s_1^2 and s_2^2, the F statistic is the ratio:

$$F = \frac{s_1^2}{s_2^2} \qquad \ldots(6.9)$$

If s_1^2 and s_2^2 are two estimates obtained from the same normally distributed parent population, then the ratio will be approximately equal to one, with sampling variation accounting for differences. For an F statistic there are two associated degrees of freedom — one for the numerator and one for the denominator — where the degrees of freedom are equal to one less than the sample size. The F distribution has different shapes for different combinations of the degrees of freedom. For samples larger than 40 or 50 with equal degrees of freedom in the numerator and denominator, the distribution is approximately normal with a mean of 1. The distribution becomes narrower as sample sizes increase. For different degrees of freedom in the numerator and denominator, the distribution is skewed.

The basis of analysis of variance methods is to compare sample variability *between* groups with sample variability *within* groups. At first, it may be difficult to see why variances are compared and not means. The explanation is a function of the statistical model used in the method. The concepts of group means and a grand mean of all the data are easy to understand. Assume that each observation has several components that determine its value: a contribution from two factors labeled α and β, a possible interaction between the factors labeled γ and a random error term labeled ε. Building an additive statistical model, the equation for an observation becomes:

$$y_{ijk} = \mu + \alpha_i + B_j + \gamma_{ij} + \varepsilon_{ijk} \qquad \ldots(6.10)$$

The observations are all relative to a grand mean μ, or mean all of the data. The factors are the independent variables or controlled variables. Examples include dose, time, compound, treatment or any factor to which levels can be associated. The subscripts i and j index the different levels or settings of the factors. The interaction term γ is associated with levels for the two different factors and captures whether there is a contribution to the sum from a specific combination of levels of factors. The error term ε is the random component. Each observation is indexed by the two main factor levels and the subscript k for replicates. For more than two factors, the equation generalizes to include more main effect terms and interaction terms.

Once the group means and interactions are defined, the next step in understanding ANOVA methods is to partition the variability in the data. For an ANOVA, there must be several observations for each level of all factors in the model and interactions. Variability is expressed as a mean square (MS), which is a ratio of the sum of squares divided by the corresponding degrees of freedom. Sums of squares (SS) are the sum of squared differences between observations and a mean. Therefore, we can define total variability (*Total SS*) as the sum of squared differences between each observation and the grand mean. Within-group variability (*Within SS*) is the sum of squared differences between observations and their respective group mean, e.g. if there is a contribution to the within-group variation from each observation it is simply paired with the appropriate group mean. The *Within SS* is often called a pooled variance estimate. Finally, the between-group variance (*Between SS*) is the sum of squared differences between the group means and the grand mean multiplied by the number of observations in each group. Because the sums of squares are assumed to be additive, the primary equation for ANOVA is

$$\textit{Total SS} = \textit{Between SS} + \textit{Within SS} \qquad \ldots(6.11)$$

Between SS can be a complicated statistical model involving interactions, however it can always be reduced to one sum. The sums of squares must be additive to use an ANOVA model. Failures in the assumptions for ANOVA and non-linear effects lead to non-additivity of the sums of squares. Degrees of freedom are the number of independent pieces of information associated with a statistic. Whenever a constraint is added to a problem or equation, the degrees of freedom decrease by the number of parameters that are estimated. For the sample mean, the number of degrees of freedom is equal to the number of points minus one.

The null hypothesis for an ANOVA is that the group means are equal. If the null hypothesis is true, the between-group variance and the within-group variances are also equal because they both estimate the variance of a homogeneous population, which is why we use the F statistic: a ratio of the variances that is $\sim$1 one if the group means are equal. If the group means are not equal, the between-group variation is larger than the within-group variance (F statistic > 1) because at least one group mean is larger than the grand mean. Rather than comparing all means in a combinatorial sense, analysis of variance reduces the problem to a single F test. Therefore, the problem of finding a chance significant effect by simply running many tests is avoided. Analysis of variance indicates that at least one of the group means is different from the remainder, but it does not identify which variance is different. For most problems in drug discovery, it is also important to identify which group mean is different. Multiple comparison procedures or *posthoc* tests are done after analysis of variance to determine which group means are different by comparing pairs.

Because the analysis of variance method compares variances, it is important to test whether the variances of the different groups are equal. If variances are unequal, groups with means that are different in magnitude will not show significant differences due to the differences in variability.

Analysis of variance is a versatile and common method for data analysis. Many types of statistical analysis (hypothesis tests, repeated measures analysis of variance, multiple comparisons) are based in the theory of analysis of variance or are subsets of the more general method. The assumptions and equations used in analysis of variance calculations are found in statistical texts (Snedecor and Cochran 1989). Standard statistical software (SAS®, JMP®) have the capability to perform analysis of variance. To plan and interpret results from analysis of variance, it is important to understand the principal assumptions of the method and the concept of interactions and to develop a decision path for the different tests that are done when using the method.

The three assumptions that are made for analysis of variance are independence, equal variances and normality. Technically, observations in an analysis of variance are independent, variances between groups are equal and the observations are obtained from a population that has a normal distribution. These assumptions rarely are satisfied in practice.

Independence implies that each observation does not depend on any other observation (e.g. is uncorrelated). In technical terms, this means that the errors associated with the observations are independent of each other. The errors must also be independent of the treatment effect, e.g. the error in an observation does not depend on the associated treatment group. Tosses of a fair coin and rolls of dice are the classic examples of independent phenomena. Measurements made on individual observational units (cell culture, tissue, animals) are independent if they are considered to be drawn from a truly random population. Measurements made on the same observational unit under different conditions (time, dose) are not independent. Measurements made on observational units drawn from a stratified population are also not independent. e.g. if sets of 96-well plates are prepared at the same time using a set of reagents and compared to another set of 96-well plates prepared at a different time using the same set of reagents, there could be a time effect that must be tested in the model. Or, if different sets of reagents are used, there could be a reagent effect associated with each block of data.

Independence is usually known from the design of the study. Simple statistical tests based on analysis of variance can be done to assess the significance of a blocking factor. A simple plot of group variances versus group means is a good test of whether the errors are independent of treatment group. If the group variances are independent, they will be randomly scattered around the grand mean (mean of the group means). A pattern in the data indicates that there is a dependence on treatment group. Most importantly, if a study design is based on repeated measurements, then there are methods to include the correlation structure in the analysis of variance and perform more appropriate and often more powerful statistical tests.

Equality of variances is the most important assumption in ANOVA. Because the method compares group means using their variance, if the variances are unequal then the statistics are not valid. Figure 6.17 shows the problem with unequal variances. In Fig. 6.17A the group means have roughly equal error bars and ANOVA indicates significant differences between the groups. In Fig. 6.17B, although the means are different, the large variance in group C contributes to a non-significant result. It is important to point out that the *data* themselves are valid, as is the interpretation of the graph. Many biological phenomena show increased variance as a function of some factor. The ANOVA, however, is not valid in that the F statistic assumes equality of variance between the groups.

There are several tests for equality of variance. Four common tests, all implemented in JMP® software, are: O'Brien's test, the Brown–Forsythe test, Levene's test and Bartlett's test. O'Brien's test is simply an ANOVA on the group variances themselves. A significant ANOVA for an O'Brien's test indicates differences between the group variances. The Brown–Forsythe test and Levene's test are

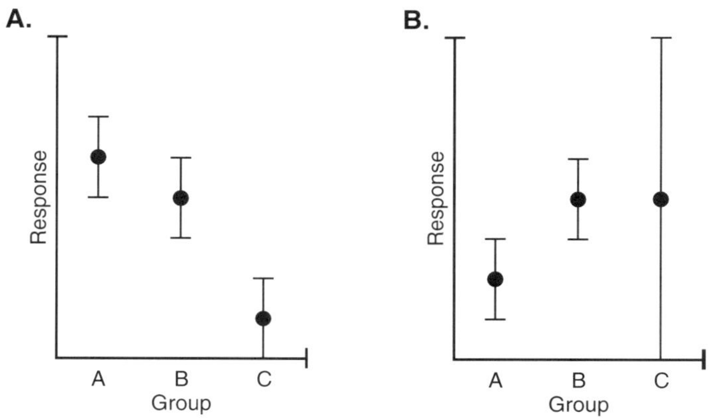

Fig. 6.17. Effect of unequal variances on analysis of variance. (A) The variances between-groups A, B and C are approximately equal. It is clear that the difference between the mean of group C and groups A or B are different because the error bars do not overlap. (B) The variance of group C is considerably larger than the variances of groups A and B. In this case, there are no significant differences between the group means because the high variability of group C dominates the comparisons

both analyses that compare the absolute value of the differences between observations in a group and a measure of the central tendency of each group: group medians for Brown–Forsythe and group means for Levene's test. A significant result indicates unequal variances. Bartlett's test is based on the group variances and a transformation to allow a chi-squared statistic to be calculated. In practical terms, the tests should be done prior to running an ANOVA and if significant results are obtained either transformation of the variables or an ANOVA that does not make the assumption of equal variances is used. The ANOVA methods for unequal group variances usually include some correction to the group means. A Welch ANOVA, for example, weights the group means by the reciprocal of the sample variances of the group means. Non-parametric methods comparing medians and median absolute deviations are also used.

Sample sizes are important to consider. The ANOVAs are considered balanced if there are equal numbers of observations in each group. If there are unequal numbers of observations, it is more likely that there will be differences in variance between the groups. There are two important considerations about unequal group sample sizes. First, if the differences in sample sizes is great, e.g. several (2–4) times as many observations, interpretation of statistics related to interactions becomes difficult. The second consideration is to make sure that the software used for the ANOVA will work with unbalanced group sizes. Some software has different options for balanced and unbalanced ANOVA.

Normality of observations in the treatment groups is an assumption in ANOVA. Normality can be assessed by plotting histograms of the data organized by treatment group and determining if the distributions are symmetrical. Tests of normality can be performed. The implications of the assumption of normality are tied to sample sizes. For small sample sizes, if the distributions are symmetric then the normality assumption is generally satisfied as long as the variances are approximately equal. For large sample sizes, even if the distributions are not normal, the ANOVA is usually valid by virtue of the central limit theorem.

Interactions are important in statistical analysis and experimental design. Presence of an interaction implies that the effects that factors have on response are not independent. The simplest interactions involve two factors. Figure 6.18 shows the interaction between two factors. In Fig. 6.18A there is no apparent interaction between the factors, whereas in Fig. 6.18B the response to the first factor (plotted on the abscissa) depends on the level of the second factor. Interactions can be tested in an ANOVA model if there are observations made at each level for each factor.

Figure 6.19 shows a flow chart for performing an analysis of variance experiment. A quick plot of means and variances provides detailed diagnostics about the assumptions. First, a plot of standard deviations of each group mean versus the group mean helps to determine whether variances are equal and whether the errors are independent of treatment effect. Unequal variances are detected by

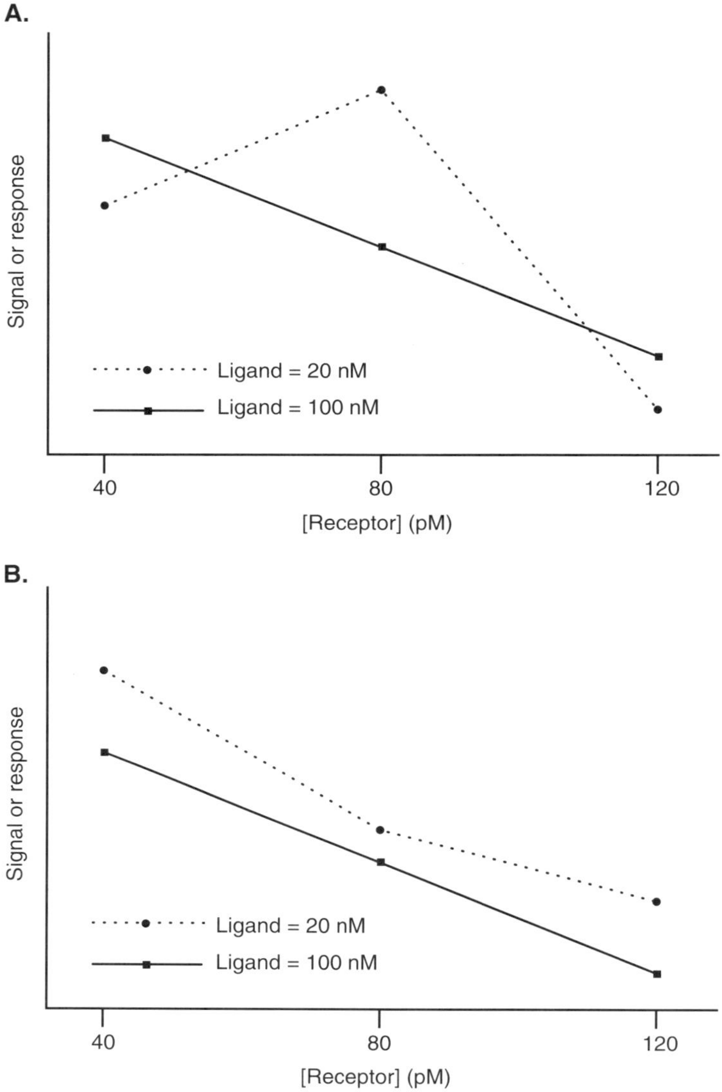

Fig. 6.18. Interactions in an analysis of variance model. (A) No interaction. The lines do not intersect, showing that there is not an interaction between ligand concentration and receptor concentration for these data. (B) Interaction. The intersection of the lines shows an interaction between ligand concentration and receptor concentration for these data. Abscissa: receptor concentration. Ordinate: signal or response. The curves are drawn for different ligand concentrations

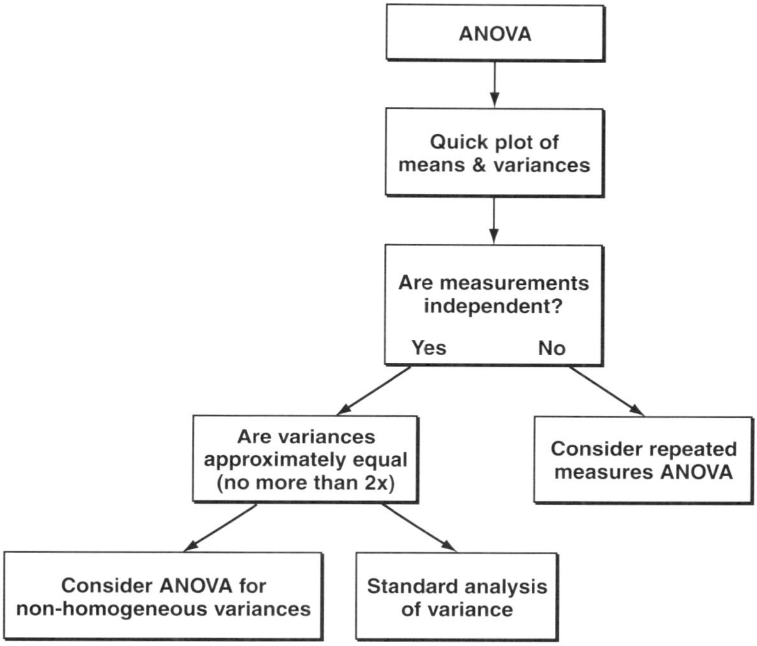

Fig. 6.19. Schematic for analysis of variance. The first step involves plotting means and variances to check assumptions about equality of variances and to define the samples that comprise each group in the analysis. Decisions are made based on whether the observations are independent and whether the variances are equal

comparing the sizes of the error bars, independence is checked by determining if there is a pattern to the data. Plots must be made for all factors and interactions.

If measurements are not independent, repeated-measures ANOVA should be considered (see next section). If the measurements are independent, equality of variances is checked using the tests mentioned earlier. Depending on the outcome of these tests, either a standard ANOVA or an ANOVA for non-homogeneous variances is done.

Results of an analysis of variance are reported using an ANOVA table (Table 6.1). This table partitions the variance of the data into the different sources. For the example, the data are counts observed from a set of five microtiter plates with 80 observations on each plate. The sources of variation are row and column. It is also possible for there to be an interaction between row and column, so that is coded as a source of variation as well. Each row of the table gives the sum of squared deviations and degrees of freedom for the source of variation. Because a mean is calculated for each group, the number of degrees of freedom for an effect is simply the number of levels associated with the effect minus one for

Table 6.1. Format of an analysis of variance (ANOVA) table

Source	Degrees of freedom	Sum of squares	Mean squares	F ratio	$P > F$
Model	79	10086251	127674	3.87	0.0001
Error	320	10548699	32965		
Corrected total	399	20634950			
Row	7	4555016.0	650716.6	19.74	0.0001
Column	9	1302351.0	144705.7	4.39	0.0001
Row × Column	63	4228883.7	67125.1	2.04	0.0001
Summary					
R^2	0.49				
Adjusted R^2	0.39				
Root mean square	181.56				
Mean of response	4174.12				
No. of observations	400.00				

the group mean. For interactions, the degrees of freedom are a product of the degrees of freedom for the component effects. From the table we see that the plates have eight rows and ten columns. For the row × column interaction, there are 63 degrees of freedom corresponding to (rows − 1) × (columns − 1). The F ratios are constructed by dividing the mean square errors for the factors (sum of squares divided by degrees of freedom) by the error mean square. The error mean square is the residual error after all of the factors have been included in the model. In the example, the 79 degrees of freedom for the model account for all levels for all factors plus the interaction (e.g. every combination of row and column). Subtraction from the total number of observations (400) gives 321, and one degree of freedom is needed for the grand mean, giving 320 error degrees of freedom. The F ratio is then compared to the critical value of the F statistic to determine if the effect is statistically significant. In the example, both main effects (row and column) and the interaction are significant ($P < 0.0001$). The row effect is most significant, with an F ratio of >19. Variations on this table report different components of the variance, e.g. mean squared error in addition to sum of squares.

6.2.10.3 Repeated Measures Analysis of Variance

In some experimental designs, multiple measurements are made in the same subject. The advantage to these designs is that it is statistically more powerful to detect differences within a subject than between subjects, because the variability between measurements within a subject is much lower than that between subjects. This is the reason why a paired t-test has more power to detect significant differences than an unpaired t-test. It is important to recognize that the measurements made in the same subject are correlated. Tests that assume independence are

Table 6.2. Example of a repeated-measures analysis of variance table

Source	Wilks lambda	F ratio	Degree of freedom (numerator)	Degrees of freedom (denominator)	$P > F$
Between subjects (sum of responses)					
All tests	0.71	2.34	4	23	0.086
Intercept	0.06	373.51	1	23	<0.0001
Treatment	0.71	2.34	1	23	0.09
Contrast control and compound 1	0.95	1.18	1	23	0.29
Within subjects					
All tets (all within subject interactions)	0	1.26	76	22.09	0.275
Time	0.01	36.44	19	5	0.0004
Time × Treatment	0	1.26	76	22.09	0.275
Time × Treatment (contrast control and compound 1)	0.1	0.1	19	5	0.16

not appropriate for these data, e.g. comparison of response at two different time points with an unpaired t-test would not be correct.

Repeated measures data are analyzed using multivariate statistical methods. The variability is partitioned into a within-subject variance and a between-subjects variance. The correlation structure of the repeated measure is included in the analysis. Statistical software, including SAS® and JMP,® support repeated measures designs. Table 6.2 shows the analysis of a repeated measures experiment where the amount of food consumed by mice was measured over a timecourse.

Another approach to repeated measures analysis that is used in practice is to transform the multivariate problem into a univariate problem. Timecourse data can be characterized in terms of peak response, area under the curve, magnitude of a derivative response or some other univariate statistic calculated from the timecourse. The metric is calculated for each experimental unit and used as the dependent variable for analysis of variance, regression or other statistical method. Figure 6.20 shows the result of analyzing the data from Table 6.2 by calculating the area under the curve for each subject. The areas are compared with a one-way analysis of variance, and Dunnett's test is used to compare the treatment groups to a control. The advantages to this approach are that the metric captures the relevant biochemical response and the analysis is simpler. The primary disadvantage is that the metric may not capture details about the timecourse. The accuracy of the area under the curve calculation is compromised if it is based on an inadequate number of time points.

6.2.10.4 Multiple Tests

Some statistical analyses involve multiple hypothesis tests. If multiple drug interventions are to be compared to a control, then multiple hypothesis tests are

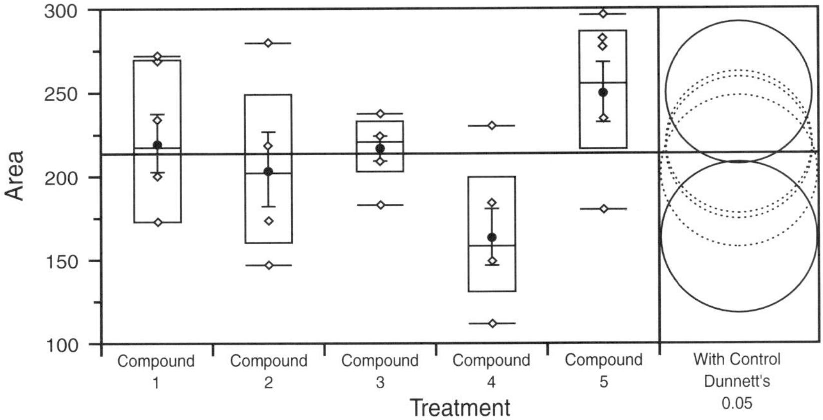

Means comparisons

Dif=Mean[i]–Mean[j]	Control	Compound 1	Compound 2	Compound 3	Compound 4
Control	0.00	30.72	33.49	46.51	86.72
Compound 1	–30.72	0.00	2.77	15.80	56.00
Compound 2	–33.49	–2.77	0.00	13.02	53.23
Compound 3	–46.51	–15.80	–13.02	0.00	40.20
Compound 4	–86.72	–56.00	–53.23	–40.20	0.00
Alpha=	**0.05**				

Comparisons with a control using Dunnett's Method [d] 2.62872

Abs (Dif)–LSD	Control
Control	–61.24
Compound 1	–30.52
Compound 2	–27.75
Compound 3	–17.71
Compound 4	22.49

Positive values show pairs of means that are significantly different.

Fig. 6.20. Area under the curve analysis for a repeated-measures analysis of variance. The area under the timecourse curves is calculated and used as a response variable. For each treatment (control and four different compounds), the average area and variability are plotted. The error bars show 95% confidence limits around the mean. The rectangle shows the 25th and 7th quartiles, with the median at the center of the box. Dunnett's test is used to compare the four compounds with the control response. A significant difference ($P = 0.05$) is found between the control and compound 4. In the graphic, if circles do not overlap, the differences between the means are significantly different at the level set by the analyst. The variances of the different groups are approximately equal

done. It is well known that if enough hypothesis tests are performed, a test will show significance just by chance alone. Multiple comparison methods are used to correct for the number of tests being performed. These methods are often called post hoc analysis because they are done after an analysis of variance is found significant. In the case of the multiple drug examples, an analysis of variance would be done on all of the data and if it were not found significant then no further testing would be done. If the analysis of variance is significant, then multiple comparisons are done to identify which groups are significantly different.

In terms of the framework, it is essential to check the experimental design to ensure that multiple comparison methods are needed in the first place. If the groups that are being compared are correlated, e.g. drug response at different time points in the same group of experimental units or subjects, then a repeated measures analysis is a more appropriate and powerful test.

One of the simplest methods to correct for multiple tests is the Bonferroni adjustment. For this adjustment, the α value for a given test is divided by the corresponding number of tests. In other words, when setting the critical values from the distribution, rather than using $\alpha/2$ the Bonferroni method uses $\alpha/2k$, where k is the number of tests. This has the result of setting the experiment-wise error rate (chance of making an error for the experiment overall) less than or equal to α and the comparison-wise error rate (chance of making an error for each comparison or test) equal to α/p. The chance of making an error for these tests of means is the error in finding that there is a significant difference in means (technically, reject the null hypothesis of no difference in means) when in fact there is no difference in the population means.

Other methods for multiple testing often bear the name of the statistician that first proposed them. Dunnett's test is a method that is often used when a number of groups are compared to a control. This method has the advantage of considering fewer comparisons because the only comparisons are between control and treatment groups. Other tests include Tukey's test and the Student–Newman–Keuls tests. Algorithms for the tests are different and often give different results. A difference between groups that is significant with one test is not significant with another. The tests are characterized as being conservative or liberal. Conservative tests require larger differences between groups to show significant differences. Choice of a test depends on several factors. If the variances between the groups are different, some tests should not be used. For an exploratory analysis, where the goal is to identify significant differences to be followed by additional designed experiments, a conservative test is often used.

An alternative approach to multiple comparisons that is useful in drug discovery is to use planned comparisons or linear contrasts. The methods discussed so far are unplanned comparisons, in that they were done after the experiment and after an ANOVA is shown to be significant. If a comparison between two group means is planned before the experiment, e.g. control treatment against drug treatment, a linear contrast or one degree of freedom comparison can be performed. In setting up a linear contrast, the sum of the coefficients in the contrast must be zero and the sum of the products of the corresponding coefficients in any two contrasts must be zero. This has the effect of partitioning the between-group sums of squares into additive components that are tested using a one degree of freedom (in the numerator) F test. Often, when contrasts are done they are reported in the ANOVA table directly under the row that contains the between-group mean square.

Linear contrasts are easily performed using statistical software. The programs SAS® and JMP® support linear contrasts in addition to other specialized contrasts such as testing for polynomial effects or comparing adjacent levels of a factor. The difficult part of using linear contrasts is that the experiment must be set up in such a way that the contrasts can be defined before the experiment is done. For example, if three drugs and a control were tested in the same experiment, it would be possible to set up the orthogonal linear contrasts. The contrasts are set up by assigning coefficients to the components of a linear model:

Control	Drug 1	Drug 2	Drug 3
−1	1	0	0
0	0	−1	1

in order to estimate the effects by comparing drug 1 to control and drug 2 against drug 3. The coefficients in a row all add up to zero and the sum of the products of the coefficients (columnwise) also add up to zero. Weighted contrasts involving several treatments at once can also be constructed if the experimental design corresponds to the test. Testing all of the drugs against control would not be an orthogonal contrast, e.g.:

Control	Drug 1	Drug 2	Drug 3
−1	1	0	0
−1	0	1	0
−1	0	0	1

In this case, although the sum of the coefficients in each contrast is zero, the sum of the products of the coefficients is one for each pair of contrasts. If there are k treatment groups, each set of contrasts can contain, at most, $k - 1$ orthogonal contrasts. To handle this situation, three sets of orthogonal contrasts similar to the first one would have to be constructed to make the relevant comparisons.

6.2.11 NON-PARAMETRIC STATISTICS

Statistics that are based on theoretical probability distributions, including the t statistic, are often termed parametric statistics. When these statistics are used, it is assumed that the sample has the same characteristics as the distribution that is the mathematical basis for the statistic. It is also assumed, when using most parametric statistics that compare groups (e.g. a treatment and control group), that the variances of the groups are similar. Many parametric statistics are based on the normal distribution. For these statistics, the central limit theorem shows that as sample sizes get larger the distributions of the *sample statistics*, such as the mean and standard deviation, are approximately normal even if the distribution of the population is not.

Statistical methods are available that are not based on specific probability distributions. These methods are often termed non-parametric or distribution-free methods. Although the majority of statistical analysis in the literature is based on parametric tests (e.g. t-tests, F tests), in the process of drug discovery there are situations where the use of non-parametric tests is appropriate. In practice, these situations involve cases where the sample distributions are markedly different from the probability distributions used to derive the parametric statistics or cases where there are large numbers of outliers.

In cell-based assays where samples are replicated three to six times, it is often standard practice to calculate the mean of the replicates for use in analysis. The presence of an outlier will have a significant impact on the mean, particularly for such a small number of replicates. Use of the median instead of the mean alleviates this problem and often is a better estimate of central tendency for these assays.

For measurement of central tendency, the median is a non-parametric analog to the mean. For measurement of variation, the median absolute deviation (MAD) is a commonly used non-parametric analog. To calculate the median absolute deviation, the median value is subtracted from all values. For example, if the values are designated v_i, then the median absolute deviation is:

$$MAD = \frac{Median(|v_\mathrm{i} - Median(v_\mathrm{i})|)}{0.6745} \qquad \ldots (6.12)$$

Division by 0.6745 is done to scale the MAD to standard deviation units.

6.2.12 OUTLIERS

The presence of outliers or extreme values and appropriate methods to handle outliers are issues that arise in many analytical problems in drug discovery. Outliers are an old and difficult problem in statistics. Multiple outliers and outliers in multivariate space are a particularly difficult problem. Theoretical treatment of outliers can be found in the statistics literature and in several texts (Barnett and Lewis 1978; Beckman and Cook 1978; Hawkins 1980), but it is often difficult to apply a method in an automated sense.

Outliers result from a number of causes: errors in conducting the experiment, systematic effects that have an impact on the measured response; and biological variability. Particularly for small (under ten experimental units) studies, outliers can affect measures of central tendency and variation, estimates of parameters from regression and correlation analysis and hypothesis tests. Determining whether an outlier should be removed from the analysis is a question often posed to statisticians, but one that should be addressed first by investigating the conditions under which the data point was generated. If experimental error is known to be the source of the outlier, then the point should not be included in the analysis. Likewise, if there are experimental reasons why the point is erroneous, e.g. an IC_{50} value that is outside the range of the concentration tested, then the point should be eliminated. More troubling are the cases where the reason for an outlier cannot be identified; these are cases where an objective rather than a subjective determination to remove the outlier must be made. When mathematical models are fited to data, handling outliers is particularly vexing. Outliers can significantly change estimates of the parameters of the model, but they also may be the most interesting cases in the data-set. In this sense, an outlier is not an aberrant point but a point that does not satisfy the theoretical or scientific model, perhaps because of a different biochemical mechanism.

A difficulty with outliers is that the statistics used for detection of outliers are themselves affected by the outlier or outliers. For example, if all points exceeding twice the standard deviation are to be deemed an outlier, in practice few points are ever identified because the standard deviation increases due to the outlier (Fig. 6.21). Possible solutions to this problem are to use historical estimates of variability for the comparison or to use robust statistics such as the MAD.

The Bradu–Hawkins method (Bradu and Hawkins 1982) is an outlier detection algorithm that is useful for large datasets and is amenable to automation. The

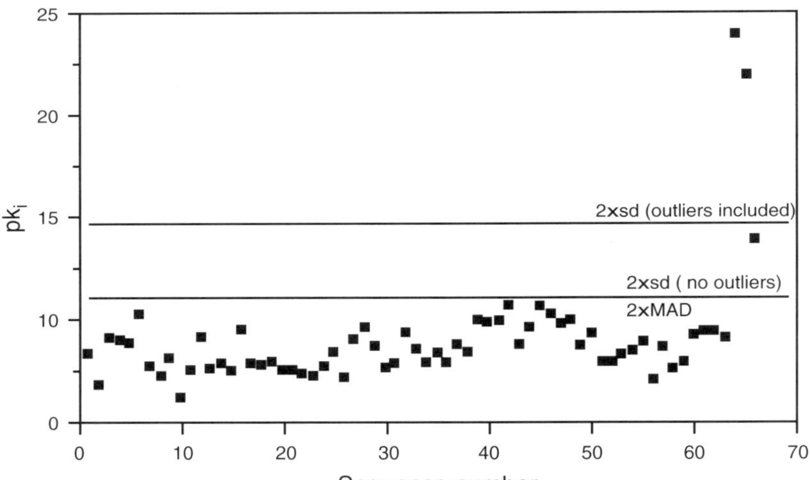

Fig. 6.21. Outlier detection. The presence of three outliers inflates the value of the standard deviation(sd). If all points greater than twice the standard deviation are declared to be outliers, in this example the outlier with the lowest value (12.5) is not identified because the standard deviation used in the calculation is unusually large. A non-parametric estimate of variation, the median absolute deviation (MAD), has the property that it is robust to outliers. If all points greater than twice the MAD are declared outliers, then all outliers in this example are correctly identified. The estimate of $2 \times$ MAD is approximately equal to the standard deviation of the sample if all of the outliers are excluded. Abscissa: sequential number. Ordinate: response (pK_i from a binding assay)

method is particularly well suited for data arranged by rectangular coordinates, such as a 96-well plate. The method will work for plates that contain concentration–response data or single determination data. It is based on a standard cell-means model where it is assumed that there are no interactions between the rows and columns:

$$Y_{ij} = N(\mu_{ij}, s^2) \quad \ldots(6.13)$$

$$\mu_{ij} = \mu + \alpha_i - \beta_j + \delta_{ij} \quad \ldots(6.14)$$

This notation, which is very similar to the ANOVA model discussed earlier, includes a concise description of the model and its assumptions. In statistical notation, eq. 6.13 lists the assumptions about the observations from the wells in a plate or set of plates that are identified as Y_{ij} (the row coordinate is i and the column coordinate is j). These observations are assumed to be normally distributed with a mean of μ_{ij} and a variance of s^2. Equation 6.14 is the model form for the expected value of the cell, μ_{ij}. The data point is estimated to be the sum of a grand mean μ, a row effect, a column effect and an interaction term. In this model, α_i are row effects, β_j are column effects and δ_{ij} are interaction terms where, if $\delta_{ij} = 0$, the cells are table inliers (no interaction) and, if $\delta_{ij} <> 0$, the cells are outliers. Tetrads, or the sum of cells adjacent to a given cell, are calculated as:

$$T_{ijkl} = Y_{ij} - Y_{kj} - Y_{il} + Y_{kl} \quad \ldots(6.15)$$

Tetrads are portions of the table based on the sum of four adjacent cells. For any cell Y_{ij}, a median of all tetrads that include the cell can be listed. The outlier detection algorithm is to save the median tetrad for each cell Y_{ij}. The median tetrads then can be plotted on a half-normal plot as shown in Fig. 6.22. One way to visualize this algorithm is to think about a set of four squares moving around the data matrix examining whether there are row and column interactions (Fig. 6.23).

As published, the algorithm required inspection of a half-normal plot to identifiy outliers. To automate the algorithm, it is possible to standardize the median tetrads, rank the resulting values and compare them to expected values obtained from the distribution of the order statistics (Stan Young, Glaxo Wellcome, personal communication). Order statistics give the probability of changes in ranked values for a sequence of ordinal numbers, they are based simply on ranking the sample values. The modified algorithm is easily implemented in a statistical spreadsheet. The algorithm is a useful tool for outlier identification during data reduction of plate-oriented assays, where samples are replicated of the order of three to four times and for fast identification of outliers in tables of data from high-throughput screening. In the latter example outliers stand out from the remainder of the data as a signal in a matrix of noise, whereas in the former application the data are organized so that columns represent different

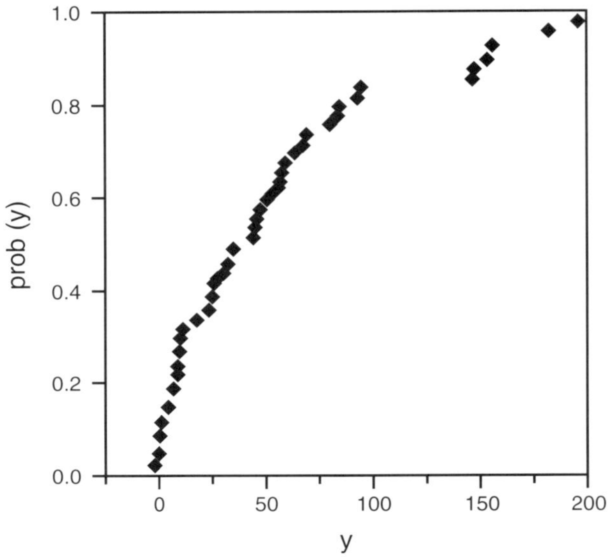

Fig. 6.22. Half-normal plot for tetrad outlier identification algorithm. Outliers are suspected when the curve breaks from linearity. Abscissa: median value of all tetrads associated with an individual cell. Ordinate: associated probability score for the median tetrad value.

concentrations of compound, rows are replicates and a significant interaction between the two indicates an outlier.

Comparison to the order statistics is not limited to use with tetrads. If a vector or list of numbers can be scaled and ranked, the same method can be used. Rather than computing tetrads, the median of a vector is calculated and subtracted from all the values and scaled using the MAD. The resulting numbers are ranked and compared to the order statistics.

All outlier identification methods must deal with the problem of multiple outliers. Although one outlier in a data set is easily identified, the presence of multiple outliers makes the problem more difficult because the outliers mask their presence. First, any estimates of variability used to set outlier thresholds are affected by the presence of multiple outliers. Second, the effect of an outlier in one direction can be masked by an outlier in the other direction. These problems are usually resolved by iterative approaches, e.g. leaving out several points at random and recalculating the statistics like variance. Alternatively, the statistics used to determine whether a point is an outlier are calculated without using the point itself.

Multivariate outliers are associated with problems where the outlier can be in one or several of the dimensions of the problem. For example, a compound in a

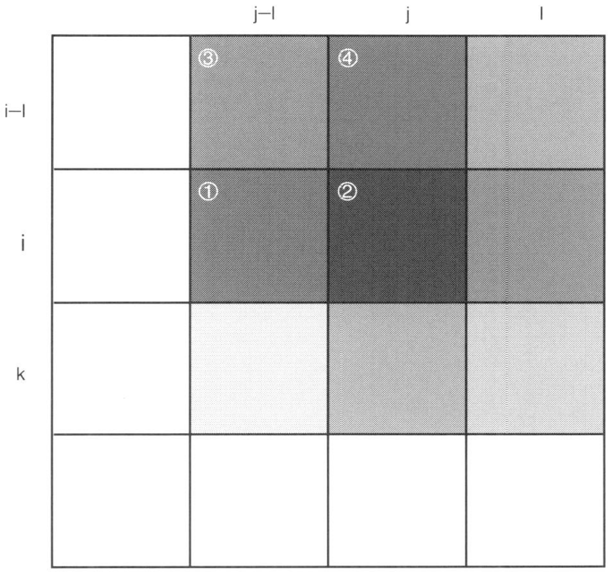

Tetrads for cell i, j

Cell i, j	
right upper	①
left upper	②
right lower	③
left lower④	

tetrad ① Yi, j–1 – Yi, j –Yk, j–1 + Yk, j
tetrad ② Yi, j – Yk, j – Yi, l + Yk, l
tetrad ③ Yi, j–1 – Yi–1, j – Yi, j–1 + Yi, j
tetrad ④ Yi–1, j– Yi–1, l – Yi, j + Yi, l

Fig. 6.23. Illustration of the tetrad algorithm. A tetrad is a collection of four cells. The tetrad is constructed via equations for analysis of variance models using the three cells adjacent to a given cell. For the cell with row coordinate *i* and column coordinate *j*, the equations for calculating the four tetrads are given

homologous series that is tested on three different receptor subtypes appears as an outlier for one subtype (Fig. 6.24A) or all three (Fig. 6.24B). A Mahalanobis distance plot is a convenient method to identify multiple outliers. In this plot, the Mahalanobis distance from the multivariate mean is plotted against some independent variable or simply the case (row) number (Fig. 6.25). If the data are organized by rows where each column is a different multivariate response, the Mahalanobis distance can be calculated for each row. The Mahalanobis distance takes into account the correlation structure of the multivariate responses

A.

Variable	Correlations		
	α_{1B} (pk_i)	α_{1C} (pk_i)	α_{1D} (pk_i)
α_{1B} (pk_i)	1.0000	0.6054	0.7925
α_{1C} (pk_i)	0.6054	1.0000	0.8398
α_{1D} (pk_i)	0.7925	0.8398	1.0000

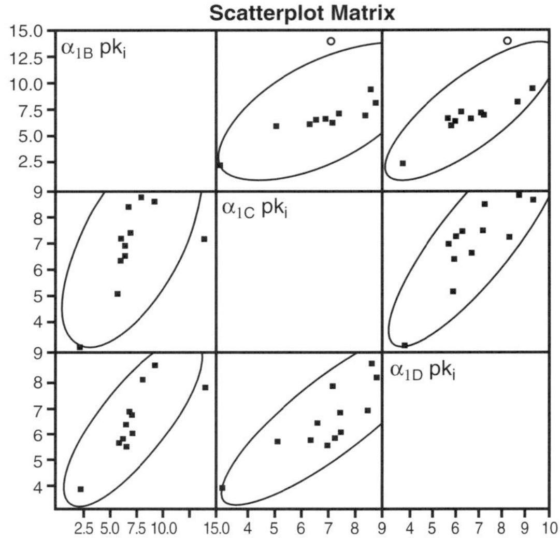

B.

Variable	Correlations		
	α_{1B} (pk_i)	α_{1C} (pk_i)	α_{1D} (pk_i)
α_{1B} (pk_i)	1.00	1.96	1.96
α_{1C} (pk_i)	0.96	1.00	0.94
α_{1D} (pk_i)	0.96	0.94	1.00

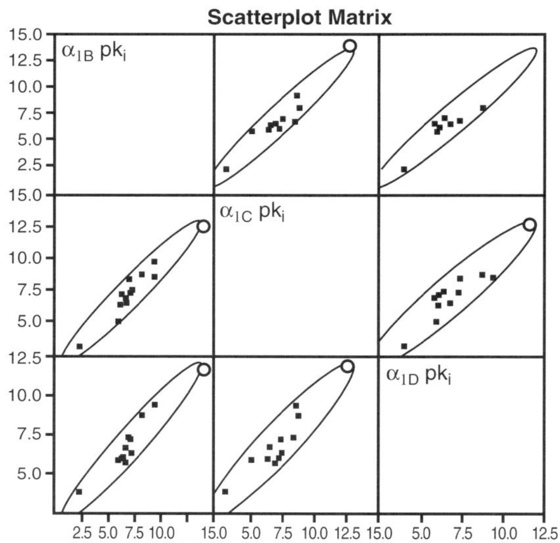

by including the covariance matrix:

$$d_i = \sqrt{(y_i - \overline{y})' S^{-1} (y_i - \overline{y})} \qquad \ldots (6.16)$$

In this equation, y_i is a vector of data for row i and $\overline{y}$ is the vector of means for the variables, the 'prime' indicates transposition (rows become columns and columns become rows for the multiplication to take place), S is the covariance matrix (variances of each variable on the diagonal and covariances on the off diagonals) and the S^{-1} indicates the inverse of the covariance matrix.

Larger distances on the plot indicate a larger deviation from the multivariate mean. A reference distance or threshold is calculated based on the number of variables in the multivariate analysis. This distance is a function of the number of data points and variables (SAS ,1994):

$$Threshold = \sqrt{F \bullet v} \qquad \ldots (6.17)$$

where v is the number of variables and F is the value of the quantile for a given significance level (e.g. 95%) from the F distribution where the numerator has v degrees of freedom and the denominator has $n - v - 1$ degrees of freedom (n is number of rows). Points that have a distance greater than the reference line are identified as possible multivariate outliers.

The problems with univariate outliers are amplified with multivariate outliers. Correlations between multivariate responses are a specific problem and can easily cause masking. It is possible for a point not to appear as an outlier in any single variable but be identified as an outlier in a multivariate sense due to high correlations between the variables.

6.2.13 RESAMPLING METHODS

Probability theory is the statistical basis for many of the tests described in this chapter. The mathematics for these methods require the use of closed-form equations for the statistics. For example, the standard error of the mean has a closed-form solution. These statistics also require considerations of the assumptions that are relevant for a particular statistic, such as independence or correlation structure. Biological factors, such as the inability to wash a drug from tissue,

---

Fig. 6.24. Multivariate outliers. Multivariate outliers can be very difficult to detect. Each plot is a graphical correlation matrix for the affinities of a set of twelve compounds for three subtypes of the α_1 adrenergic receptor (α_{1B}, α_{1C} and α_{1D}). (A) An outlier in only one dimension (α_{1B}) is easily identified because it appears as an outlier in all plots that involve α_{1B} . The presence of the outlier also decreases the apparent correlation between the affinities. (B) The multivariate outlier, in this case in all three dimensions, is harder to identify. All of the correlations are >0.9 and the scatterplots do not identify the outlier, except as a possible large value (12.5)

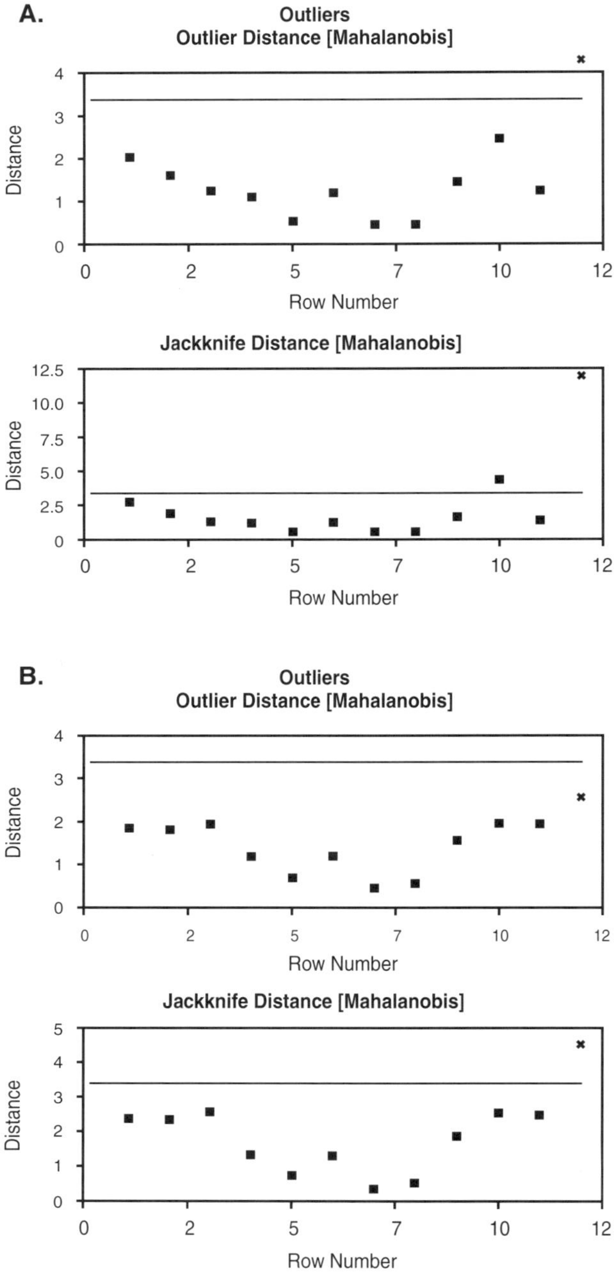
A.
Outliers
Outlier Distance [Mahalanobis]
Distance
Row Number
Jackknife Distance [Mahalanobis]
Distance
Row Number
B.
Outliers
Outlier Distance [Mahalanobis]
Distance
Row Number
Jackknife Distance [Mahalanobis]
Distance
Row Number

may preclude several types of traditional experimental designs and consequently invalidate the assumptions of conventional statistical methods.

Resampling is a non-parametric statistical method that utilizes the fast computational capabilities of digital computers to estimate statistics, particularly for cases where a closed-form algorithm does not exist (Diaconis and Efron 1983; Efron and Tibshirani 1993). Resampling has many applications in biology and chemistry where the application precludes using statistics with a closed-form solution or the experimental protocol invalidates the assumptions of a statistical analysis. Resampling is a powerful method to have in the framework of data analysis tools.

Bootstrap resampling (Efron 1982; Efron and Gong 1983; Hinkley 1988) is based on the use of bootstrap samples, which are samples of some given size drawn with replacement from the original dataset. The bootstrap samples are used to calculate the statistic of interest, often standard error. The name bootstrap refers to the use of the original dataset to generate new datasets (Efron and Tibshirani 1991). A principal advantage of the bootstrap is that it can be applied to any statistic (Efron and Gong 1983). The bootstrap has been applied to a wide variety of statistical problems, including linear regression models (Freedman 1981; Freedman and Peters 1984; Wu 1986), non-linear regression models (Alkuzweny 1986), confidence intervals (DiCiccio and Romano 1988), *P* values for multiple comparisons (Westfall and Young 1989), quasi-likelihood estimators (Simonoff and Tsai 1988) and parameters obtained from pharmacokinetic modeling (Niedzwiecki and Simonoff 1990). Meng *et al* (1989) utilized the bootstrap method to calculate standard errors and confidence limits for pA_2 estimates in a model-free analysis of competitive antagonism.

The theory of bootstrap resampling has been reviewed by Efron (1982) and Efron and Tibshirani (1993). Assume that a statistic $\hat{\theta}(x_1, x_2, \ldots x_n)$ is defined symmetrically in $X_1, X_2, \ldots, X_n$ where X_i are independent and identically distributed according to some probability distribution F. The statistic can have any functional form. It is a calculation based on the sample. From the previous discussion about sampling, it is known that the standard deviation (σ) of $\hat{\theta}$ depends on the sample size (n), the functional form of $\hat{\theta}$ and the probability

---

Fig. 6.25. Mahalanobis plots for outlier detection. Mahalanobis plots are a useful tool for outlier detection. Larger distances on the plot indicate a larger deviation from the multivariate mean. (A) Regular Mahalanobis distances. Abscissa: row number or case order. Ordinate: Mahalanobis distance. (B) Jackknife distances. Mahalanobis distances are calculated by using a resampling method known as the jackknife. The distance for any observation is calculated by using estimates of the mean, standard deviation and correlation matrix that do not include the observation itself. Jackknife resampling is, by definition, without replacement, e.g. without replacing the observation in the sample. Bootstrap resampling allows the observation to be replaced in the sample distribution F. The sample size and functional form of $\hat{\theta}$ are known, the distribution F is unknown or:

$$\sigma = \sigma(F, n, \hat{\theta}) = \sigma(F) \qquad \ldots(6.18)$$

and F can be approximated by an empirical probability distribution $\hat{F}$ where a weight of $1/n$ is placed on each data point. The bootstrap estimate of $\sigma(\hat{F})$ is:

$$\hat{\sigma}_{\text{Boot}} = \sigma(\hat{F}) \qquad \ldots(6.19)$$

For most statistics, it is not possible to write $\sigma(F)$ in closed-form. In bootstrap resampling, $\hat{\sigma}_{\text{Boot}}$ is determined by drawing a large number of bootstrap samples from the data. The basic algorithm is:

(1) Draw bootstrap samples $x_1^*, \ldots x_n^*$ from the original data. This sample is drawn, with replacement. As each member of the sample is drawn, it is replaced and can be chosen again.
(2) Compute the statistic $\hat{\theta}^*$ for each bootstrap sample.
(3) Repeat step 1 and step 2 a large number (R) of times; R is usually of the order of several thousand.
(4) Approximate $\hat{\sigma}_{\text{Boot}}$ by:

$$\hat{\sigma}_{\text{Boot}} = \sqrt{\frac{\sum_{r=1}^{R} (\hat{\theta}^* - \hat{\theta})^2}{R - 1}} \qquad \ldots(6.20)$$

and

$$\hat{\theta} = \frac{\sum_{r=1}^{R} \hat{\theta}^*}{R} \qquad \ldots(6.21)$$

After the bootstrapped statistic is calculated, confidence intervals can be calculated. Several methods are used to calculate confidence intervals. One approach is to use the bias-corrected percentile method (Efron and Gong 1983). Define $\Phi(z)$ as the cumulative distribution function of the standard normal distribution (returns the zth quantile from the standard normal distribution, where z is some probability) and $\Phi^{-1}(q)$ is the inverse function of the standard normal cumulative distribution function (returns the probability that an observation from the standard normal distribution is less than or equal to q). Let $\hat{C}(\hat{\theta})$ be obtained by drawing a normal probability plot of some variable. Define:

$$z_0 = \Phi^{-1}[\hat{C}(\hat{\theta})] \qquad \ldots(6.22)$$

where $\hat{\theta}$ is the estimated mean value of the statistic. The confidence interval can be calculated as $[\hat{C}^{-1}(\alpha_1), \hat{C}^{-1}(\alpha_2)]$, where:

$$\alpha_1 = \Phi(2z_0 - z_\alpha) \qquad \ldots(6.23)$$

$$\alpha_2 = \Phi(2z_0 + z_\alpha) \qquad \ldots(6.24)$$

This method is illustrated graphically in Fig. 6.26. In practice, the following algorithm is used to compute the confidence interval (reference is made to functions provided in the statistical analysis software program JMP®, SAS Institute, Cary NC, version 2.0):

(1) A histogram is made for each statistic. The histogram is used to compute the probability scores $\hat{C}(\hat{\theta})$ for each value of the statistics; this is equivalent to computing an empirical cumulative distribution function (normal quantile plot in JMP®). Using the plot of $\hat{C}(\hat{\theta})$ versus $\hat{\theta}$, $\hat{C}(\hat{\theta})$ is obtained for $\hat{\theta}$: the mean estimate of the variable obtained from the analysis.

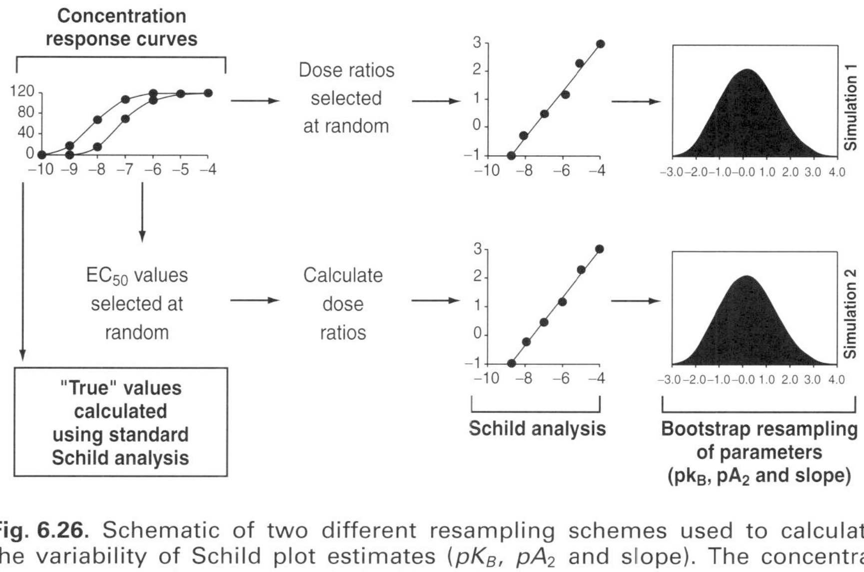

Fig. 6.26. Schematic of two different resampling schemes used to calculate the variability of Schild plot estimates (pK_B, pA_2 and slope). The concentration–response data are sampled with replacement: either raw EC_{50} values or dose ratios. These are used to calculate Schild plots using standard regression analysis. Histograms are drawn of the resulting analysis for a large number of bootstrap samples and are used to estimate the variability. (Reprinted from *J Pharmacol Toxicol Methods*, Vol. 34, Lutz MW, Menius JA, Choi TD, *et al*, Use of resampling techniques to estimate the variance of parameters in pharmacological assays when experimental protocols preclude independent replication: an example using Schild regressions, pp. 37–46. Copyright 1995, with permission from Elsevier Science.)

(2) Probability z_0 is computed using the equation (Φ^{-1} is the normQuant function in JMP®):

$$z_0 = \Phi^{-1}[\hat{C}(\hat{\theta})] \quad \ldots(6.25)$$

(3) Values α_1 and α_2 are calculated using the equations (Φ is the normDist function in JMP®):

$$\alpha_1 = \Phi(2z_0 - z_\alpha) \quad \ldots(6.26)$$

$$\alpha_2 = \Phi(2z_0 + z_\alpha) \quad \ldots(6.27)$$

For 95% confidence intervals, $z_\alpha = 1.645$.

(4) Using the values of α_1 and α_2, the plot of $\hat{C}(\hat{\theta})$ versus $\hat{\theta}$ is used to estimate the actual values of the confidence interval (this is done in JMP® using the normal quantile plot and the measuring tool). Confidence limits can also be calculated using the percentile-*t* method. In this method, the bootstrap estimates of the variables are standardized:

$$t_{\mathrm{r}}^* = (\hat{\theta}_{\mathrm{r}}^* - \hat{\theta})/\hat{\sigma}_{\hat{\theta}} \quad \ldots(6.28)$$

The standardized variables are plotted and the appropriate 95% confidence intervals are obtained by reading off the quantiles and by making the calculations:

$$\hat{\theta}_{\mathrm{lowerci}} = \hat{\theta} - t_{\alpha/2}^* \hat{\sigma}_{\hat{\theta}} \quad \ldots(6.29)$$

$$\hat{\theta}_{\mathrm{upperci}} = \hat{\theta} + t_{\alpha/2}^* \hat{\sigma}_{\hat{\theta}} \quad \ldots(6.30)$$

The standard deviation used in these equations ($\hat{\sigma}_{\hat{\theta}}$) is obtained by another round of resampling and has been termed the 'double bootstrap' (Mooney and Duval 1993).

The algorithm as described provides estimates of statistics and their confidence limits. The estimates obtained by bootstrapping, however, are biased. The bootstrapped estimate of bias is the difference between the analytical point estimator and the bootstrapped point estimate of the parameter (Efron and Tibshirani 1993; Mooney and Duval 1993):

$$\mathrm{Bias}(\hat{\theta}) = \hat{\theta} - \hat{\theta}^* \quad \ldots(6.31)$$

$$\hat{\theta}^* = \sum_{r=1}^{R} \frac{\hat{\theta}_r}{R} \quad \ldots(6.32)$$

Lutz *et al* (1995) described a general resampling-based method to estimate the variance of parameters from pharmacological assays when the protocol precludes independent replication and used Schild regressions to illustrate the method. The motivation was to use this method for cases where a dose ratio could not be obtained from a single biological sample. Assuming that a single EC_{50} is obtained per tissue, the steps in the method are:

(1) Estimate EC_{50} values for all dose–response curves.
(2) Construct a table of EC_{50} estimates indexed by antagonist concentration (each row contains an antagonist concentration and EC_{50} calculated for the tissue).
(3) Choose rows from the table at random N times (N should be approximately equal to the number of tissues tested in the experiment).
(4) Ensure that at least one control curve and one treatment curve (any non-zero antagonist concentration) are present in the sample.
(5) Calculate the geometric mean of the EC_{50} values for the control curves to use as the denominator for the dose ratios (a modification would use each control curve as a possible denominator; the same type of approach also can be used for scaling parameters e.g. maximal response).
(6) Calculate the dose ratio for each treatment curve and use to construct the sample dataset of log[antagonist] versus log (dose ratio − 1).
(7) Calculate a linear regression for the data set. Record slope and y intercept and compute pA_2 (y intercept/slope).
(8) Calculate the pK_B using the equation:

$$pK_B = \log(\text{dose ratio} - 1) - \log[\text{antagonist}] \qquad \ldots(6.33)$$

Record pK_B (note that until the analysis is complete and the confidence band around the slope is estimated, the interpretation of pK_B as a binding constant is not valid).
(9) Repeat steps 3–8 a large number of times (800).
(10) Utilize the histograms of estimated statistics: (slope, pA_2 and pK_B) to determine statistics estimates and to calculate confidence intervals.

The results of this method as applied to atenolol antagonism of isoproterenol in guinea pig left atria are shown in Fig. 6.27. The histograms show the results for 800 bootstrap samples. Table 6.3 compares the results of the bootstrap estimates to conventional Schild analysis.

Statistics based on resampling-based methods have a role in pharmaceutical research applications. These methods are not a panacea: care must be taken to calculate the bias of the estimates and understand that the sampling strategy is appropriate for the problem. Statistically, good practices of experimental design, sample determination and randomization, identifying and minimizing sources of variability and clearly stating the research questions apply to resampling methods. The advantage to the methods is that they can be used for statistics where a closed-form solution cannot be determined analytically.

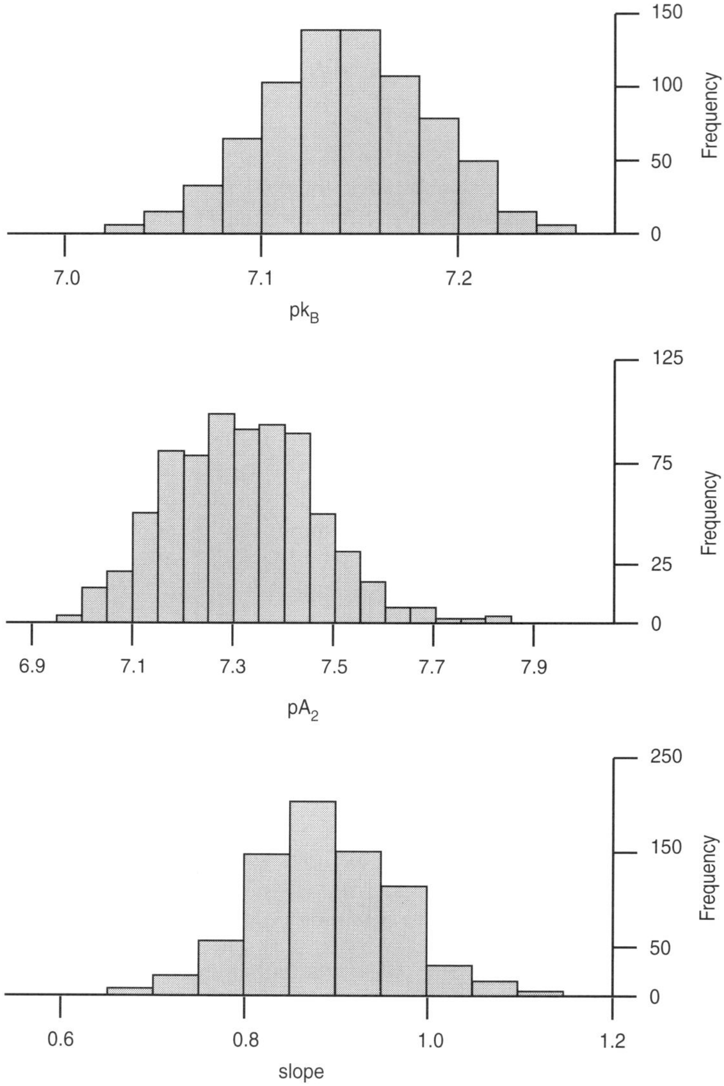

Fig. 6.27. Histograms of pK_B, pA_2 and Schild plot slope for 800 bootstrap samples where the pairing of the control and treatment tissues was retained. Data were obtained for atenolol antagonism of isoproterenol in guinea pig left atria. (Reprinted from *J Pharmacol Toxicol Methods*, Vol. 34, Lutz MW, Menius JA, Choi TD, *et al*, Use of resampling techniques to estimate the variance of parameters in pharmacological assays when experimental protocols preclude independent replication: an example using Schild regressions, pp. 37–46. Copyright 1995, with permission from Elsevier Science.)

Table 6.3. Comparison of pA_2, Schild plot slopes and pK_B and associated 95% confidence intervals obtained by conventional and bootstrap resampling methods

Method	pA_2	Slope	pK_B
Conventional	7.30 (7.04–7.69)	0.89 (0.71–1.07)	7.14 (7.04–7.24)
Bootstrap, paired-dose ratios	7.32 (7.09–7.58)	0.88 (0.75–1.03)	7.14 (7.06–7.21)
Bootstrap, random pairing of dose ratios	7.29 (7.06–7.54)	0.90 (0.75–1.03)	7.13 (7.06–7.20)

6.3 MATHEMATICAL MODELS

6.3.1 MODEL COMPLEXITY

Mathematical models are used in biology and chemistry for a variety of purposes including calculation of descriptive statistics, simulation of complex biochemical systems and testing of alternative models of a biochemical process. Types of models can be characterized by the complexity of the model and the ability to estimate parameters from experimental data (Fig. 6.28). The complexity of a model is a function of the number of parameters that specify the model. If a biochemical system is well characterized and parameters are identified for the components of the system, a complex model is constructed.

The second axis of model characterization is parameter estimability. Parameters are estimated from laboratory experiments, assuming that sufficient data are available for the estimation process. To estimate parameters there must be; an experimental procedure that allows estimation of the unique parameter, in isolation from other parameters; sufficient data to estimate the parameter with statistical confidence; and the parameter must be identifiable in a statistical sense.

6.3.2 DATA-FITTING MODELS AND CONCENTRATION–RESPONSE CURVES

Data-fitting models, by definition, involve fitting mathematical equations to data. These models are important for many areas of drug discovery. The relationship between concentration and effect is modeled in order to calculate values for potency and efficacy. For enzyme kinetics studies, mathematical models are fitted to timecourse data. Some measurements, such as calcium transients and the contraction of cardiac muscle, yield timecourse data that are fitted to equations.

The best example of a data-fitting model is the concentration–response or dose–response curve. These models have low complexity (two to four parameters) and have a simple structure based on the Langmuir adsorption isotherm (Chapter 2). Experimental procedures for fitting concentration–response models

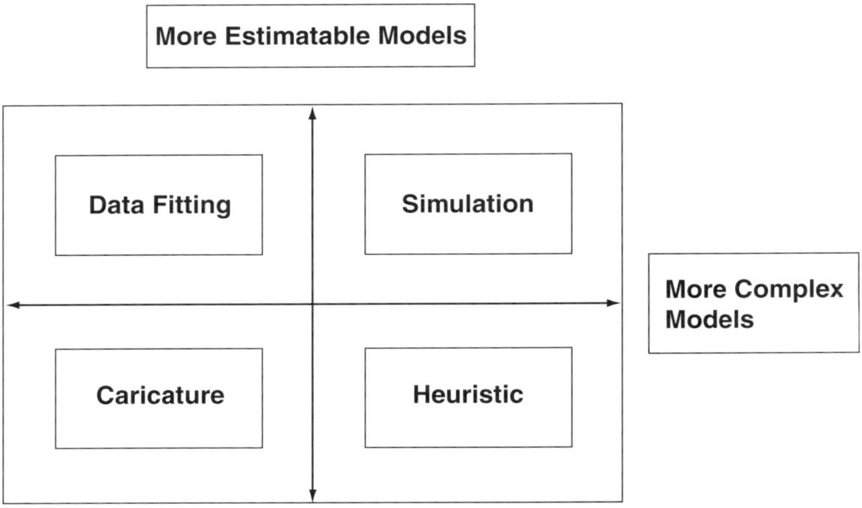

Fig. 6.28. Types of mathematical models. The matrix defines four types of mathematical models in terms of model complexity (number of parameters) and the ability to estimate parameters from experimental data.

involve recording the response at several different concentrations. The concentrations are chosen to cover all regions of the concentration–response surface: the maximum and minimum asymptotes (upper and lower plateaus of the curve) and the region near the IC_{50} or EC_{50}. If there are three or four data points per parameter and if the concentration range covers all regions of the surface, the IC_{50} or EC_{50} value can be estimated. Concentration–response models fit data that have the shape of a sigmoid curve. These models are used to fit data for pure enzyme systems, isolated tissues, cell-based assays and *in vivo* experiments. The assumptions involved in reporting an IC_{50} or an EC_{50} are statistical, e.g. the assumptions in the regression analysis and goodness of fit. As defined, these parameters are concentrations that are associated with a half-maximal response. These parameters do not have the theoretical meaning of K_i or K_d, which require specific assumptions about kinetics to be valid. Statistical assumptions about fitting equations to data are discussed in the section on regression and correlation (section 6.3.7).

6.3.3 SIMULATION MODELS

Simulation models come closest to simulating reality. These models often have large numbers of parameters (eight or more) that can be determined from experimental data. Enzyme kinetic models are a good example of simulation models. When these models are used, the mathematical model is derived directly from the rate equations of the enzyme system. Several experiments are performed to estimate all of the parameters of the system. It is important to recognize that all of the parameters in a model may not be estimated from a single experiment or estimated simultaneously. Depending on the structure of the equation, parameters may be unidentifiable or highly correlated with other parameters, precluding estimation by regression methods. This situation is often encountered with simulation models, when there are more than four parameters to estimate. If data can be collected to estimate the parameters, the behavior of the biochemical system can be simulated.

6.3.4 CARICATURE MODELS

All biochemical models approximate reality. Caricature models offer a very crude approximation of a biochemical system. Physiological models of an entire organism are examples of caricature models. Compartmental models of the human cardiovascular and gastrointestinal systems are constructed to model the pharmacokinetics of drug absorption and distribution. These models approximate complex organ systems with a exponential equations, however they are useful tools for drug metabolism studies. Provided that estimates can be obtained for the rate constants associated with compartments, these models are used to simulate the metabolism of compounds.

6.3.5 HEURISTIC MODELS

Heuristic models have the complexity required to capture the essential features of the biochemical system being studied, but independent experimental estimates for all of the parameters are not available. Heuristic models are used to design experiments and test alternative theoretical models of biochemical systems. The evolution of models of G-protein-coupled receptor systems is a good example of the development of a heuristic model.

6.3.6 MODELS OF G-PROTEIN-COUPLED RECEPTOR SYSTEMS

The data-fitting model most commonly applied to concentration–response data is a simple equilibrium-binding model that involves a ligand and a protein or receptor. The signal transduction pathway for many receptors involves an accessory protein, or G protein. The receptor–ligand species binds with a membrane-bound G-protein to form a ternary complex that, in turn, causes dissociation of the alpha and beta/gamma subunits of the G-protein and subsequent activation of more downstream parts of the signal transduction cascade. Although the simple equilibrium-binding model is often fitted to data from systems that involve activation of G-proteins, this model clearly does not adequately capture the complexity of the system.

A natural extension to the simple equilibrium-binding model is to add the equilibrium between the receptor–ligand complex and the G-protein. This model

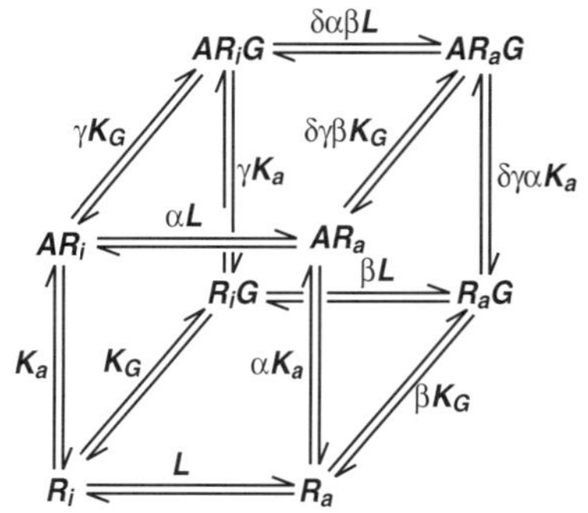

Fig. 6.29. The cubic ternary complex model. This model shows different forms of receptor species that allow the receptor to exist in either an inactive or active state, bound or unbound to G-protein and bound or unbound to ligand. Equilibrium association constants are put on each reaction step. These parameters define the characteristics of the simulation model and, in turn, the system that the model represents. By changing the parameters, the concentration–response behavior of a model system can be studied

is known as the ternary complex model. The ternary complex model is considered to be a heuristic model because it is not possible to obtain independent estimates of the two equilibrium dissociation constants (K_a and K_g). A more recent extension of the ternary complex model is the cubic ternary complex model (CTC) shown in Fig. 6.29. This model was developed to explain observations that the α_{1B}-adrenoceptor can exist in active (R_a) and inactive (R_i) forms. For completeness, both forms can bind to ligand and G protein, forming active (AR_aG) and inactive (AR_iG) ternary complexes. As with the ternary complex model, although estimates cannot be obtained for all of the parameters of the model, it does capture many essential features of the interaction of ligands with G-protein-coupled receptors. The model is useful to simulate specific problems in receptor theory, e.g. overexpression of a receptor species or an abundance of one of the ternary complexes. The CTC model is easily generalized to construct models that involve multiple ligands, G-proteins and receptors.

6.3.7 REGRESSION AND CORRELATION

Many statistical textbooks place a discussion about regression and correlation before analysis of variance. Linear relationships between variables are intuitive to understand and the assumptions and methods for fitting a line to data allow the reader to understand concepts such as the sum of squared errors, calculation of the coefficients in the model equation and estimates of error in the coefficients. The analysis of variance model is introduced as a general type of linear model. Estimation of coefficients and their errors are extensions of the theory introduced from the linear model.

In this book, ANOVA models are introduced as part of the statistical framework based on theory from probability distributions and sampling. Regression and correlation are discussed as tools for mathematical modeling. This presentation is based on how these tools are used in biochemical research. The ANOVA models are used to test questions about whether statistically significant relationships exist between variables, whereas regression is used to model mathematically the relationship between variables.

When working with data to be fitted to an equation, the first steps involve selecting an equation or model of the data and understanding the variability of the data. The important second step is sometimes overlooked: estimates of the variability of the data are needed to make decisions about whether to use weighted curve fitting and to assess the quality of the fit. In terms of the statistical framework, at least two variability patterns are considered: random biological or biochemical variability and variability that is a function of some aspect of a system. Natural biochemical variability is random variation that is measured by the standard deviation of measurements taken under similar conditions for different experimental units. These measurements are made on independent experimental units: different animals or tissues, different 96-well plates on different days, different runs of a biochemical assay or true replicate measurements in an experimental run.

Variation is often a function of some aspect of the test system. For example, variation often increases as a function of signal strength or time. Data analysts plot variation, usually expressed as the standard deviation from replicate measurements, as a function of all independent variables and time in order to detect whether there are systematic patterns. Lack of a trend or pattern indicates that the variance is homoscedastic as opposed to heteroscedastic. If data are heteroscedastic, the regression model is changed to account for the variation. The simplest changes involve weighting the data by a function that adjusts the weights in order to account for variation. For example, if variation increases with signal, the weighting function $1/x$ or $1/x^2$ is used. When the weight is multiplied by the data points, the larger values of signal effectively down-weight the observations with higher variability. The most elegant approach to heteroscedastic data is to use a variance function model (Davidian and Carroll 1987; Carroll and Ruppert 1988; Giltinan and Ruppert 1989). These models involve models for both the deterministic and random parts of the data. In other words, parameters are estimated during the fitting process for both the mean and signal and the variability.

The assumptions in regression analysis are shown pictorially in Fig. 6.30. Each data point is assumed to be independent of the other points and to be measured without error in the independent (X) variables. Independent variables are also known as explanatory variables, predictors or regressors. Error in the dependent (Y) variable (or variables) is assumed to be normally distributed and consistent across the range of independent variables. The dependent variables are the

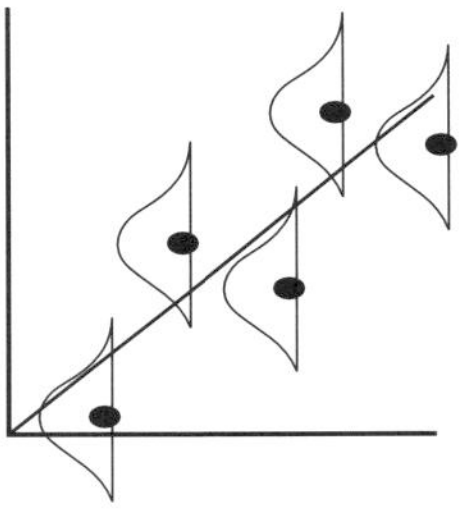

Fig. 6.30. Assumptions in regression analysis

responses that are measured. The assumptions in regression are similar to the assumptions of ANOVA methods. As with ANOVA, they are rarely fully satisfied in real data. If there is error in some of the independent variables and the regression equation is needed for prediction or hypothesis testing, the analyst must determine whether to proceed with classical regression methods or to use a method that accounts for variation in both the independent and dependent variables, often termed an error in the predictors approach. If the experimental objective is to measure the strength of the linear relationship between the variables, correlation analysis can be used if there are errors in both of the variables. For correlation analysis, the terms independent and dependent variable are not used because all variables are sampled and contain sampling variability. Generally, if the measurement error in the predictors is negligible and if the errors are randomly distributed, many analysts use regular least-squares regression because these methods are most commonly available in statistical software. However, calculations to perform regressions with error in the independent variables can be set up using statistical software that supports matrix manipulation, such as SAS®.

There are two definitions of linearity that are needed to categorize different regression methods. Relationships between variables can be linear or non-linear. Linear relationships between variables indicate that they are best fitted with a straight line: there is no curvature in the relationship. An equation is linear in its parameters if the partial derivatives for the equation with respect to each of the parameters contain only the predictor (X) variables and constants. If the partial derivatives include parameters, the equation is not linear in its parameters. An equation that is linear in its parameters can model non-linear response, polynomial equations are an example. Because the power terms on a polynomial are constants, the partial derivatives of the equation with respect to the parameters only contain constants. An equation that has a parameter as an exponent, or an exponential equation, is non-linear in its parameters.

The equations used for linear regression, to fit an equation to data and to estimate parameters and associated standard errors are simple and can be set up easily using a spreadsheet program. There is only one unique solution to any equation that is linear in its parameters. Solving non-linear equations is more complex. Because there are multiple solutions to solve the equation, iterative methods are necessary. Non-linear equations start with initial estimates for the parameters and use different algorithms to refine the fit in order to minimize the sum of squared differences between the equation and the data. The mathematics for non-linear regression are beyond the scope of this book and are covered in texts on regression (Draper and Smith 1981). Statistical software is used to fit both linear and non-linear equations. Understanding the mathematical basis for linear regression is helpful in order to understand the assumptions inherent in the method.

The simplest form of the linear model has three components: intercept, slope multiplied by an independent variable and a random error term. This form is written:

$$y = \alpha + \beta X + \varepsilon \qquad \ldots(6.34)$$

where α is the intercept, β is the slope and ε is the error term. This is similar to the ANOVA model, with the exception that X is a continuous variable multiplied by the slope. The general model contains these components in addition to a terms that are raised to a specific powers and potentially several independent variables, each multiplied by a corresponding slope. Fitting equations to data involves minimizing the sum of the squared differences between the line or curve that defines the equation and the data points. For eq. 6.34 the slope is:

$$\beta = \frac{\sum xy - \dfrac{\sum x \sum y}{n}}{\sum x^2 - \dfrac{\left(\sum x\right)^2}{n}} \qquad \ldots(6.35)$$

and the formula for the intercept is a function of the slope and the means of x and y:

$$\alpha = \overline{y} - \beta\overline{x} \qquad \ldots(6.36)$$

Simple algebra shows that these values of the parameters minimize the sum of squares. More generally, the regression problem is written in matrix form, e.g.:

$$Y = X\beta + \varepsilon \qquad \ldots(6.37)$$

where Y and X are vectors for the dependent and independent variables and β is a vector of parameters. The product of X and β accommodates different forms of the independent variables (power terms, interaction terms, multiple independent variable) along with a coefficient from the parameter vector β as a multiplier. Matrix manipulation is used to solve the equation for the parameters in order to solve the resulting equation. Regardless of the complexity of the equation, a regression model partitions the sum of squares into a model-dependent term and an error term. The total sum of squares is simply the sum of squared differences from the mean, e.g.:

$$SS_{\text{total}} = \sum_{i=1}^{n} \left(Y_i - \overline{Y}\right)^2 \qquad \ldots(6.38)$$

If $\hat{Y}$ is the response predicted from the model, then the total sum of squares is:

$$SS_{\text{total}} = \sum_{i=1}^{n} \left(\hat{Y}_i - \overline{Y}\right)^2 + \sum_{i=1}^{n} \left(Y_i - \hat{Y}\right)^2 \qquad \ldots(6.39)$$

The first summation (model response) gives the sum of squared differences between the predicted observations and the grand mean; the second summation (error term) is the difference between the observed response and the predicted response.

Diagnostic plots are used to determine whether the data satisfy the assumptions of a regression model. The first step in making these plots is to calculate the residuals, i.e. the differences between the predicted response and the observed response:

$$r = y - \hat{y} \qquad \ldots(6.40)$$

where y is the observed response and $\hat{y}$ is the predicted response. The residuals are plotted as a function of both the independent and dependent (predicted) variables. Patterns in the residual plots indicate some departure from the assumption. If the residuals increase in magnitude as a function of the independent variables, it is an indication that variance is not constant. If the terms of the equation model the data, the residuals will be randomly scattered around a mean of zero. If the residuals have curvature, this indicates that the data have a non-linear component that is not included in the model. A histogram of the residuals indicates whether the errors (ε) are normally distributed. Finally, to check for independence, the residuals are plotted as a function of case order, time, concentration or other variable. A pattern in the residuals (curvature or periodicity) indicates that the observations are not independent. Residual plots also indicate the presence of outliers and points that are influential on the regression. An outlier appears on any plot as a point that has a large magnitude. Points that have high leverage or are influential are identified by looking at the residuals plotted against the independent variables. If the residuals cluster around certain values of an independent variable and there are large values of the residuals at relatively few values of the independent variable, the latter are the influential points of the regression.

Patterns or trends in the residual plots as a function of time or case order indicate lack of independence. If there is a periodic trend (similar to a sine wave with different amplitudes for the peaks and troughs) where each residual tends to be followed by residuals having the same sign (positive or negative difference from zero) and the estimate of the error variance appears to be small compared to the variability of the process (e.g. small residuals), the residuals are considered to have a positive autocorrelation. On the other hand, if the trend is alternating signs (positive residual followed by negative residual in a long periodic series), the residuals are said to have a negative autocorrelation. The error variance is usually high for negatively correlated residuals. The Durbin–Watson statistic is a convenient test for autocorrelation and is calculated as:

$$d = \frac{\sum_{i=2}^{n}(r_i - r_{i-1})^2}{\sum_{i=1}^{n} r_i^2} \qquad \ldots (6.41)$$

Lack of correlation between the residuals is associated with a value of ~2 for this statistic. Values of d that are ~4 indicate negative autocorrelation and values of d that are near zero indicate positively correlated residuals. Calculated values of d can be compared to a table that lists critical values of the Durbin–Watson statistic for different numbers of independent variables in order to assess whether the errors are correlated.

Depending on the application of the regression model, statistics are calculated to assess the overall fit of the equation to the data and to make inferences about the parameters. The overall fit to the data is assessed using the coefficient of determination, R^2:

$$R^2 = \frac{\sum_{i=1}^{n}(\hat{Y}_i - \overline{Y})^2}{\sum_{i=1}^{n}(Y_i - \overline{Y})^2} \qquad \ldots (6.42)$$

which is the ratio of the variance explained by the model to the variability in the data, and varies between 0 and 1. Expressed as a percentage, R^2 is the percentage of the variation in the data explained by the model. Note that the grand mean is subtracted from each observation in the numerator and denominator. One way of thinking about the fit of a linear model is to compare the model with the simplest model, i.e. the grand mean of the data. For the R^2, the variance of the model and the data around the grand mean are compared. The full model is the linear model, with two parameters; the reduced model is simply the mean (one parameter). These models are easily compared with an F test:

$$F = \frac{MSE_{\text{reduced}} - MSE_{\text{full}}}{MSE_{\text{full}}} = \frac{\left(\dfrac{SS_{\text{reduced}} - SS_{\text{full}}}{df_{\text{reduced}} - df_{\text{full}}}\right)}{\dfrac{SS_{\text{full}}}{df_{\text{full}}}} \qquad \ldots (6.43)$$

where MSE = mean square error and SS = sum of squares. For this F test, the numerator has degrees of freedom equal to the difference in the degrees of freedom between the models, in this case 1 because there is one more parameter in the full model. The denominator has degrees of freedom equal to the number of data points, i.e. 2 for the two parameters.

In writing the equation for R^2 (eq. 6.42), the sum of squares was used. This makes intuitive sense, because the sum of squares reflects the variability of each term of the ratio. However, the general form of the linear model can have many or few terms and often it is desirable to compare the model fits to data. Unfortunately, if R^2 is used, a model with more parameters (more complexity) will always fit better and have a higher R^2. If the ratio of mean squares (sum of squares divided by degrees of freedom) is used instead, the resulting R^2 does not necessarily increase as more parameters are added to model. This R^2 is commonly called an adjusted coefficient of determination or multiple correlation coefficient if expressed as the square root of R^2 (e.g. R). It is easily calculated from the R^2:

$$R^2_{\text{adj}} = 1 - \frac{n-1}{n-p}(1 - R^2) \qquad \ldots(6.44)$$

where n is the number of data points and p is the number of parameters in the model. Note that this statistic can be negative, but will have a maximum value of 1.

In many applications in drug discovery research, it is important to measure how well a model predicts response. For example, a quantitative structure–activity analysis often yields a linear equation to predict biological response as a function of a number of molecular descriptors. Of interest is how well this equation will predict other compounds, either similar to the dataset used to estimate the parameters of the model or compounds that are not structurally similar to the compounds in the test dataset. These equations must always be interpreted with caution. It is wise to examine all values of the predictors used in estimating the equation by plotting histograms. It is also important to determine if there are correlations between the predictors, which is often the case for descriptors calculated from the two-dimensional structure of a chemical compound. Descriptors such as calculated log P and calculated molar refractivity tend to be highly correlated. The easiest cases of collinearity to diagnose are those where there is a pair of variables that are collinear. In some cases, the collinearity is between one predictor and a set of predictors. Unfortunately, collinearity introduces a number of problems for linear regression and model fitting. The variances of parameter estimates can be very large: parameter estimates may not test significant even though there is a mathematical relationship between the predictor variable and the response and prediction for cases that are outside of the design space may not be valid. Often, the first clue that predictors are collinear is a very good fit of the equation to data with extremely large standard errors for the estimates. By plotting the independent variables against each other, it can be determined if there are pairwise correlations or a collinear relationship. Additionally, it is possible to calculate the variance inflation factor for each independent variable. The variance inflation factor is calculated as:

$$VIF = \frac{1}{1 - R^2} \qquad \ldots(6.45)$$

where R^2 is calculated for a regression where each predictor is regressed against all of the other predictors in the model. If there is not a linear correlation between the predictors, R^2 will be close to zero and the corresponding VIF will be ~ 1. Large values of the variance inflation factor (>10, corresponding to $R^2 > 0.9$) indicate problems with collinearity.

If the equation is used for the prediction of compounds that have values of the predictors outside the range, there is no statistic that can estimate the error in this prediction. If the dataset is small, the equation may give a good fit to the data but the samples necessary to estimate the parameters may be too small to cover a useful range of the descriptor space. If there is scatter in the data, the errors in the prediction may be very large. To address questions about prediction, a useful statistic to compute is the PRESS statistic: predicted extra sum of squares. This statistic is associated with a process known as cross-validation. To compute the PRESS statistic, an observation is dropped from the sample before the analysis or fitting is done. A PRESS residual is computed that is the deviation between the deleted observation and the new regression equation, e.g.:

$$r_{\text{PRESS}} = Y_i - \hat{Y}_{(i)} \qquad \ldots (6.46)$$

where Y_i is the observation omitted from the fitting and $\hat{Y}_i$ is the predicted value of this observation. The PRESS statistic is simply the sum of the PRESS residuals:

$$\text{PRESS} = \sum_{i=1}^{n} [Y_i - \hat{Y}_{(i)}]^2 \qquad \ldots (6.47)$$

Smaller values of the PRESS statistic indicate better prediction. If PRESS is appreciably larger than the error sum of squares from the model, it is likely that there are outliers or influential points in the data. The PRESS statistics are easily calculated by computer programs that can repeatedly fit the dataset leaving out observations. Cross-validation methods cover a variety of techniques that involve leaving observations out of a dataset, fitting a model and then predicting the remainder. For example, a large dataset may be split in half at random. The model (QSAR equation, neural network or recursive partitioning model; (see Chapter 9) is fitted to half of the data. The resulting model is used to predict the remainder of the dataset and statistics are calculated to measure the accuracy of the prediction, e.g. standard deviations between predicted and observed responses. These methods assume that the split dataset contains two homogeneous parts and that the split is at random. The method does not indicate the ability of the model to predict cases that are not covered by the original dataset or address questions of biased sampling of the original data, e.g. the entire dataset may be representative of a small number of chemical classes.

Diagnostics used for linear regression are also applicable for non-linear regression. Plots of residuals to detect trends in variation, normality of the error term, homogeneity of variances, non-independence and lack of fit of the model are

Table 6.4. *t*-statistics and confidence intervals for the parameters of a linear model

Parameter	Statistic	Degrees of freedom	Confidence interval
Slope	$t = \dfrac{\beta - \beta_t}{\dfrac{RMS}{\sqrt{s_x^2}}}$	$n - 2$	$\beta \pm \dfrac{t_{\alpha/2}, n - 2 \cdot RMS}{\sqrt{s_x^2}}$
Intercept	$t = \dfrac{\alpha - \alpha_t}{RMS\sqrt{\dfrac{1}{n} + \dfrac{\overline{x}^2}{s_x^2}}}$	$n - 2$	$\alpha \pm t_{\alpha/2}, n - 2 \cdot RMS \cdot \sqrt{\dfrac{1}{n} + \dfrac{x^2}{s_x^2}}$
Predicted y equal to y_p	$t = \dfrac{\hat{y} - y_p}{RMS\sqrt{\dfrac{1}{n} + \dfrac{(x_p - \overline{x})^2}{s_x^2}}}$	$n - 2$	$\hat{y} \pm t_{\alpha/2}, n - 2 \cdot RMS\sqrt{\dfrac{1}{n} + \dfrac{(x_p - \overline{x})^2}{s_x^2}}$

$\hat{y} = \alpha + \beta X$; $s_x^2 = \sum_{i=1}^{n}(X - \overline{X})^2$; $RMS = \sqrt{\dfrac{\sum_{i=1}^{n}(y - \hat{y})^2}{n - 2}}$

useful in determining whether the equation chosen adequately model the variation in the data and if the data satisfy the assumptions of the model.

One specific application of equation fitting or regression is statistical testing of the coefficients or parameters of the model. Are the parameters different from zero? if not, this indicates that the term associated with the parameter does not have a statistically significant effect on the model for the dataset being used? Is a parameter different from a theoretical value (e.g. slope of 1 in a Schild plot)? Does the magnitude of a parameter suggest experiments for further development of the model? Table 6.4 (Table 9.1 of Dowdy and Wearden 1983; see Further reading) shows the test statistics (*t*-tests) and confidence limits for the parameters of a linear model (slope and intercept) and for a predicted response. For the *t*-tests, the values of b_t and a_t are what the regression coefficients are compared to for the slope and intercept, respectively. For example to test for a slope equal to zero, b_t is set to zero. Testing a slope of zero corresponds to a test of whether the linear model offers a significantly better fit over simply fitting the mean of the data (e.g. a horizontal line). This test is constructed as an *F* test:

$$F = \frac{MS_{\text{model}}}{MS_{\text{error}}} = \frac{\dfrac{\sum_{i=1}^{n}(\hat{Y}_i - \overline{Y})^2}{1}}{\dfrac{\sum_{i=1}^{n}(Y_i - \hat{Y})^2}{n - 2}} \qquad \ldots(6.48)$$

This F statistic has one degree of freedom in the numerator and $n - 2$ in the denominator. Note that the numerator is the variability (mean square) of the model around the mean. The denominator is the error term: sum of squared differences between the observed and predicated data. If the linear model does not offer a significantly improved fit to the data, the value of the F statistic is ~ 1.

Non-parametric statistical methods are useful for determination of weights used in regression model fitting. One method for determining the weights is iteratively-reweighted non-linear least squares (IRLS) (Beaton and Turkey 1974). In this method, the weights are determined by looking at the magnitude of scaled residuals and are updated at each iteration of the non-linear curve-fitting algorithm. If the raw residuals are designated r_i, then a scaled residual v_i is defined as:

$$v_i = \frac{\left(\frac{|r_i|}{\sigma}\right)}{B} \qquad \ldots(6.49)$$

where σ scales the residual and B is a tuning constant (default value of 4.685). The weights (w_i) are defined as:

$$\text{If} \quad v_i \leq B \quad \text{then } w_i = (1 - v_i^2)^2 \qquad \ldots(6.50)$$

Or, if $v_i > B$, then the point is identified as an outlier and assigned a weight of zero, effectively excluding the point from the fit: $w_i = 0$.

Note that the tuning constant scales the residual and sets a threshold for outlier identification. One difficulty with this algorithm is the estimation of σ, the standard deviation to scale the residuals. If there are outliers, the estimate of σ will be affected and will be correspondingly larger, causing estimates of the weights to be biased and fewer points to be excluded. As an alternative, σ can be estimated using median values of gaps between the residuals and the median residual:

$$\sigma = \frac{median(|r_i - median(r_i)|)}{0.6745} \qquad \ldots(6.51)$$

Although the double use of the median may appear unusual, the principle is to subtract a robust measure of central tendency from all the residuals (the median residual) and then calculate the median gap. Note that the set of residuals used in this equation come from an initial unweighted fit to the data.

Correlation analysis is often used in research to examine the relationship between two variables and measure the strength of their linear association. The mathematics used in correlation analysis are derived from the principles of linear regression; however, there are important differences between the assumptions and objectives of the two types of analysis. Figure 6.31 is a pictorial representation of the assumptions in correlation analysis. In correlation analysis, measurements are made on observational units chosen at random and pairs of variables are

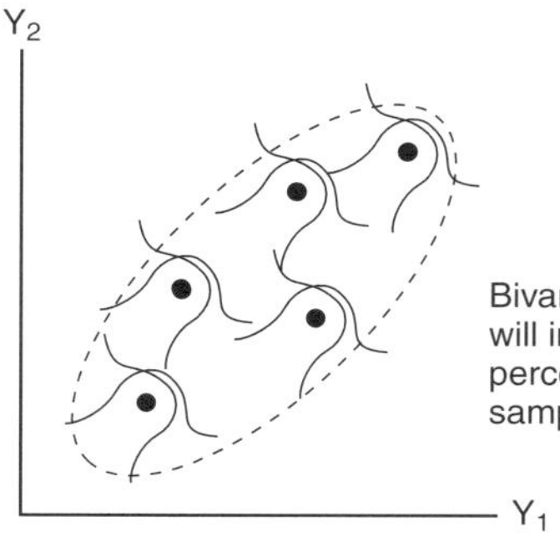

Fig. 6.31. Assumptions in correlation analysis

compared to test whether a linear relationship exists. The strength of the association is measured by the correlation coefficient, usually denoted r:

$$r = \frac{\sum_{i=1}^{n}(x_i - \overline{x})(y_i - \overline{y})}{(n-1)s_x s_y} \qquad \ldots(6.52)$$

In this equation, n is the sample size, $\overline{x}$ and $\overline{y}$ are the sample means and s_x and s_y are the sample standard deviations. Pearson's product moment correlation is another term often used for r, derived from the fact that the numerator is the sum of the product of the first moments of x and y. The values of r are in the range $1 \geq r \geq -1$. As r values approach $+1$, the strength of the linear correlation increases. Likewise, as r values approach -1, the strength of the linear correlation increases; however, the relationship is an inverse, e.g. as one variable increases, the other decreases. An r value of zero indicates lack of linear correlation.

Calculating confidence limits for r is not as straightforward as for a sample mean. Because r has definite bounds, the procedure used to calculate confidence intervals must be changed. Fisher developed the procedure that is commonly used to make this calculation. A transformation can be made to r to give it the properties of a nearly normal sampling distribution:

$$r_i = \ln\sqrt{\frac{1+r}{1-r}} \qquad \ldots(6.53)$$

The standard deviation of this transformed parameter is a function only of the sample size:

$$s(r_t) = \frac{1}{\sqrt{(n-3)}} \qquad \ldots(6.54)$$

To calculate the confidence limits, the normal distribution is used instead of the t distribution to calculate the appropriate critical value for a desired significance level:

$$r_t - z_{\alpha/2}s(r_t) < l < r_t + z_{\alpha/2}s(r_t) \qquad \ldots(6.55)$$

Here, α is the significance level and z is the critical value from the normal distribution. The confidence limits are in the transformed space, hence the use of l to indicate a transformed correlation coefficient. To make the inverse transformation (from the confidence limits or from l to r) a table or function is used.

Other types of correlation analysis are available for problems that do not satisfy the assumptions of standard linear correlation analysis. Instead of using the value for each variable, the variables can be ranked and correlated. The resulting statistic is known as Spearman's rho. This analysis is useful if the variances of the different distributions around the points are markedly different. It is also important to recognize that the correlation coefficient r only applies to linear correlations. If the relationship between the variables is non-linear, the r value can be misleadingly small even if a relationship exists between the variables.

One of the most important points to make about correlation analysis is that correlation does not prove causality. If there is a linear relationship between variables, it may be a result of a causal relationship or there may be an underlying factor that causes both variables to vary in a linear fashion. Only by running designed experiments where factors are set by the investigator can inferences be made about cause and effect.

It is worth noting the similarities and differences between regression and correlation analysis by comparing Figs. 6.30 and 6.31. In regression analysis, the independent variables are set by the investigator and do not have sampling error. At each level of the independent variables, random samples are chosen and measurements of the dependent variable or variables are made. The sampling or measurement error for the points is assumed to be normally distributed and the variance of each distribution is approximately equal for all levels of the independent variables. For correlation analysis, a random sample is chosen and measurements of the variables of interest are made. The variables have an associated measurement error that is assumed to be normally distributed. For each variable, the error distributions are assumed to have the same variance. The joint distribution of any two variables is therefore bivariate normal.

6.3.8 *MODEL SELECTION*

When a theoretical model of the data is used, identification of the equation to fit is dictated by the assumed scientific model or is done by a model selection procedure. If a theoretical model is not used, the equation is chosen from a set of general forms: dose–response equations, polynomials and linear equations. It is important to realize that when linear regression and correlation methods are used, the assumption is made that the data involve a linear relationship between the predictors and response variables. Many biological processes are not linear (concentration–response relationships, enzyme kinetics). It is also important to realize that although the relationship between variables is linear for a range of predictor values, the relationship can be non-linear outside of the range. The octanol/water partition coefficient ($\log P$), for example, is often a parabolic function for esters of increasing molecular weight.

There are several approaches to modeling a non-linear relationship. If the goal is simply to model the relationship with an empirical function, the inclusion of terms that raise the predictors to different powers maintains the linear form of the model while adding the capability of model curvature. A sequential F test is used to determine when the addition of more complexity to the model is no longer justified based on a statistical test.

There are many cases where the relationship between variables is known to be non-linear. In these cases, non-linear regression is used to fit equations to data. Non-Linear curve fitting is preferable to linear transformation in most cases when a theoretical relationship is assumed.

6.3.9 PRACTICAL GUIDELINES FOR FITTING MODELS TO DATA

Box 2 contains a set of practical guidelines for fitting equations to data. At least three to four data points should be available per fitted parameter. This is necessary to estimate the standard errors of the parameters. Additionally, there must be data available for each region defined by the equation. For non-linear equations, this is particularly important. For example, an equation to fit dose–response data defines several regions of space (Fig. 6.32): the upper asymptote (or region without inhibitor), the lower asymptote (region with a high concentration of inhibitor); the IC_{50} (location parameter where the response is decreased 50% of the range); and the Hill coefficient (slope of the linear portion of the dose response curve in the immediate vicinity of the IC_{50}). Without data, the parameters associated with the different regions cannot be estimated. The statistical check points for concentration–response curve fitting are shown in Fig. 6.33.

Box 2

Practical Guidelines for Fitting Equations to Data

- There should be four or five observed data points for each estimated parameter
- All regions in a non-linear function must be defined by some data
- Check for systematic lack of fit (graph or runs test)
- Compare RMS error to your best estimate of variability in the system

- Non-linear curve-fitting standard errors are not estimators of biological variability; they depend on both the model and the observed data. Repeat the experiment to estimate variability
- Simpler models (fewer estimated parameters) are more robust. More complex models will always fit the observed data better
- Strong correlations between parameters indicate problems in determining unique estimates for the parameters
- Check that the equation and parameters make scientific sense or recognize that you are smoothing/filtering
- It is usually better to have more points at different X than replicates at each X, in order to obtain more precise estimates for the parameters
- If variance changes as a function of something, consider using a weighting function
- Scales are important; large differences in parameter scales can cause problems with the estimation algorithms
- If estimates have poor precision, check to see whether a full concentration–response curve has been obtained. Constrain top and/or bottom of curve if data are not available to estimate these regions of the curve. If data from controls are available, use the control estimates to constrain these regions of the curve
- If estimation of the parameters is unstable, make a graph of the data. Use the graph to obtain better initial estimates

Leverage plots (Sall 1990) are an excellent diagnostic tool for linear regression models. A leverage plot has the structure of a scatterplot; the scales are drawn

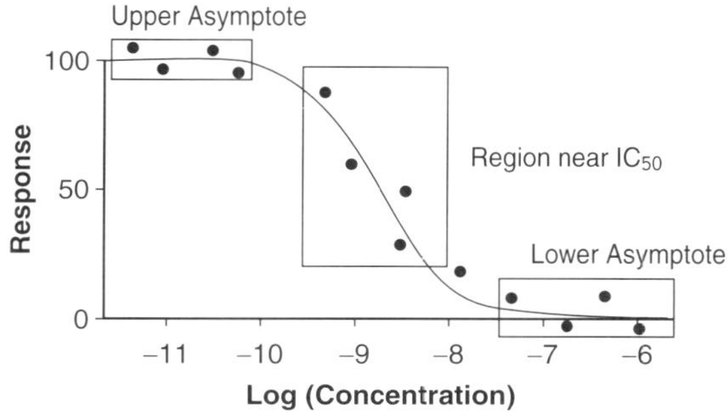

Fig. 6.32. Regions of a concentration–response curve. When fitting equations to data, all of the regions defined by the equation must have data if the corresponding parameters are to be estimated. For concentration–response curve fitting, data must be obtained for each of the regions defining the curve: upper and lower asymptotes and region near the IC_{50}/EC_{50}

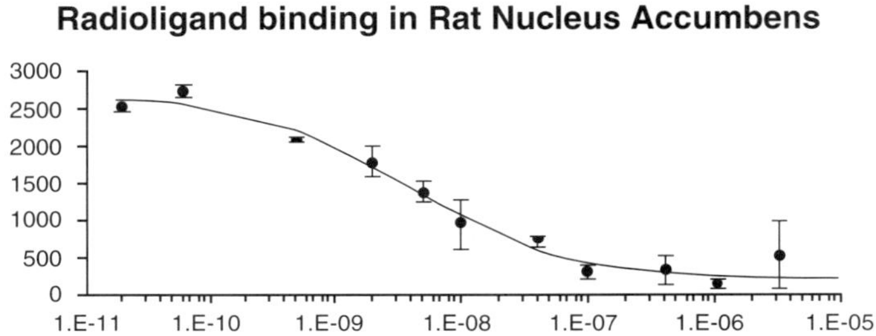

Parameter	Value	SE	Lower CL	Upper CL
Vmax	2473	258	1947	2999
K	3.9E-09	1.5E-09	1.8E-09	8.4-09
Y2	217.0	117.4	−22.2	456.2
n	0.7	0.2	0.3	1.1
	SS	**df**	**MSE**	**F**
Model	75401407	4	18850352	175.3129
Residual	3440770	32	107524.1	
RMS	328			

Fig. 6.33. Checkpoints for concentration–response curve fitting. Data are randomly distributed around the line; all regions of the concentration–response curve are defined by data; the root-mean-square (RMS) error is comparable to the pure error of the biological response. These data are obtained for prazosin binding in the rat nucleus accumbens. (Data provided by Aaron Goetz, Glaxo Wellcome.)

in the units of variables, however the data are plotted to show the effects of each predictor in the model. The data points are plotted to show the leverage of each predictor in the model, with the effect of the other parameters cancelled out. In other words, each plot shows the effect of a predictor or the interaction between predictors after all the other predictors have been taken into account. The confidence intervals on each plot indicate the statistical significance of the parameter. If confidence limits are near the line and do not cross the mean of the response, the parameter is not significant at the confidence level used. The use of leverage plots is described in Chapter 7.

6.4 DATA MINING

Designed experiments involve a high level of planning. Data mining, a relatively new method used in drug discovery, involves exploratory analysis of historical data. In a practical sense, data mining does not have the same degree of planning as the types of experiments described in the previous section. Data mining is often termed exploratory data analysis, with the premise that data mining finds interesting and significant patterns in historical data. Pharmaceutical research laboratories collect data on a variety of biochemical targets, often directing synthesis and testing at one specific target. As a historical database of the testing results grows, data mining is used in the hit identification process to identify compounds to test against new targets that are similar to targets in the historical database. Data mining is used to identify interesting patterns in the data, for example, an interesting pattern of receptor subtype selectivity or a consistent fragment in ligands that appears to be associated with activity against a receptor. Alternatively, data mining is used to partition data, e.g. analyzing high-throughput screening data to build a tree that correlates activity with patterns in chemical structure. Finally, data mining is used to make predictions. An example would be to build a model of a dataset to use for virtual screening of compounds.

The statistical methods used for data mining have some similarities to those used for classical analysis but also have some differences. Although many statistical methods are developed for sample sizes of several hundred observations, data mining often deals with hundreds of thousands or millions of observations. Adaptations are made for larger samples sizes and the potential violation of standard assumptions, e.g. equal sample sizes or large numbers of tests. Data mining generally is concerned with secondary analysis, e.g. finding patterns in data that are unknown but are potentially valuable to the scientific project teams. From an informatics standpoint, many aspects of data mining are considered before the analysis takes place. An informatics view of data mining is shown in Fig. 6.34. Three different databases that span the different types of data collected in drug discovery are shown on the schematic: chemical, biochemical and clinical genetics. The chemical database contains data on the chemical structures and properties of compounds and the combinatorial libraries available for testing, the biochemical database contains assay results from *in vitro* and *in vivo* testing of

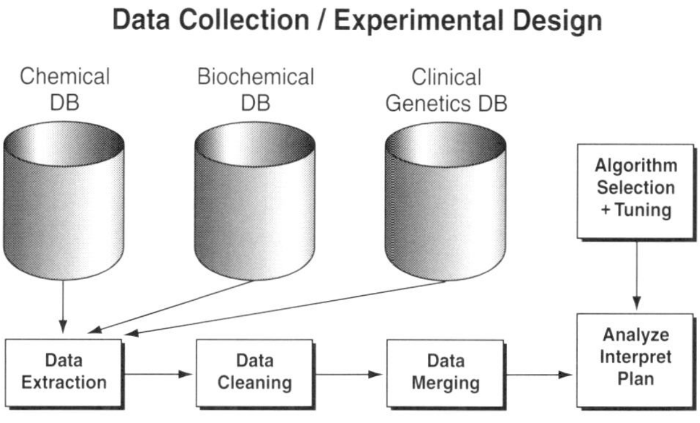

Fig. 6.34. An informatics view of database(DB) mining. This schematic shows the different steps associated with the process of automated pattern recognition from databases

compounds; and the clinical genetics database contains data from clinical genotyping studies for a variety of studies. Before the data can be used, it must be extracted and put into a form suitable for analysis. These steps are time consuming and tedious when the data are collected and stored in a format that is reasonable when considered in the context of a single project, but not for ad hoc queries.

After the data have been put in an acceptable format, the statistical data mining algorithms operate to identify patterns in the data. Current algorithms include tree regression, genetic algorithms and heuristic methods. These methods are covered in detail in Chapter 9. Data mining is an inductive or an exploratory process rather than the confirmatory or hypothesis testing model of classical statistics. Many factors complicate data mining. The underlying populations can change and the data to be analyzed (the sample) may not be random. The assumption that observations are independent, which is intrinsic to many statistical methods, often does not hold for data mining problems. Multiple comparisons are a problem. Most multiple comparison methods were designed for fewer than 20 different groups. In data mining there are often thousands of comparisons to make.

Successful data mining projects involve an intersection of disciplines: the areas of science (chemistry, biology, genetics), statistics and computer science. Computer science provides the specialized methods to handle the storage and data manipulation techniques needed for the large datasets. Statistics brings methods for pattern identification and assessment of probabilities, addressing questions of whether observed patterns are spurious or are representative of relationships in the data. Statistical theory continues to be developed to handle large datasets with multiple variables, dealing with problems of correlation between the variables, lack of independence between the observations and large numbers of comparisons.

6.5 SUMMARY

There are numerous sources of information available for pharmaceutical research. Terms such as bioinformatics and cheminformatics, in their most general sense, refer to the use of information derived from protein, receptor or chemical structures and sequences to build models for structure–activity relationship analysis or to model a biochemical system. Much of this information is quantitative: data from high-throughput screening, enzyme or receptor characterization (kinetics, concentration–response for compounds); *in vivo* experiments; and genetic and genomic studies. This chapter defined a statistical framework for the interpretation and analysis of data. Identifying research questions and objectives, understanding the assumptions made in constructing samples, listing and measuring sources of variability, using appropriate mathematical and statistical models, employing experimental design and presenting results in a clear and concise format are the cornerstones of the framework. Applying the framework to problems such as analysis of data from an *in vivo* experiment or choosing compounds to retest after a high-throughput screen ensures that the information is utilized in the most effective way possible and that future experiments are designed to obtain the maximal amount of information.

6.6 REFERENCES

Alkuzweny BMD (1986) Bootstrap procedures for linear, robust and nonlinear regression: application to pharmacodynamics. *PhD. Thesis*, University of Wyoming.

Barnett V and Lewis T (1978) *Outliers in Statistical Data*. Wiley, Chichester.

Beaton AE and Tukey JW (1974) The fitting of power series, meaning polynomials, illustrated on band-spectroscopic data. *Technometrics* **16**: 147–185.

Beckman RJ and Cook RD (1978) *Outliers. Technometrics* **25**: 119–149.

Bradu D and Hawkins DM (1982) Location of multiple outliers in two-way tables using tetrads *Technometrics* **24**: 103–108.

Carroll RJ and Ruppert D (1988) *Transformation and Weighting in Regression*. Chapman and Hall, New York.

Davidian M and Carroll RJ (1987) Variance function estimation. *J Am Stat Assoc* **82**: 1079–1091.

Diaconis P and Efron B (1983) Computer-intensive methods in statistics. *Sci Am* **248**: 116–130.

DiCiccio TJ and Romano JP (1988) A review of bootstrap confidence intervals. *J R Stati Soc B* **50**: 338–354.

Draper NR and Smith H (1981) *Applied Regression Analysis*. Wiley, New York.

Efron B (1982) The jacknife, the bootstrap and other resampling plans. *CBMS-NSF Regional Conference series in Applied Mathematics*, Vol. 38 Society of Industrial and Applied Mathematics, Philadelphia.

Efron B and Gong G (1983) A leisurely look at the bootstrap, the jacknife and cross validation. *Am Stat* **37**: 36–48.

Efron B and Tibshirani RJ (1991) Statistical data analysis in the computer age. *Science* **253**: 390–395.

Efron B and Tibshirani RJ (1993) *An Introduction to the Bootstrap*, pps. 45–49, 105–121, 168–176, 178–199, Chapman and Hall.

Freedman DA (1981) Bootstrapping regression models. *Ann Stat* **9**: 1218–1228.

Freedman DA and Peters SC (1984) Bootstrapping a regression equation: some empirical results. *J Am Stat Assoc* **79**: 97–100.

Giltinan DM and Ruppert D (1989) Fitting heteroscedastic regression models to individual pharmacokinetic data using standard statistical software. *J Pharmacokin Biopharm* **17**: 601–614.

Hawkins DM (1980) *Identification of Outliers*. Chapman and Hall, London.

Hinkley DV (1988) Bootstrap methods. *J R Stat Soc B* **50**: 321–337.

Lutz MW (1995) Use of resampling techniques to estimate the variance of parameters in pharmacological assays when experimental protocols preclude independent replication: an example using Schild regressions. *J Pharmacol and Toxicol Methods* **34**: 37–46.

Meng C, Nelson D and Andersen S (1989) A new approach to pA_2 estimation in drug–receptor studies. *1989 Proceedings of the Biopharmaceutical section of the American Statistical Association* pp. 199–204.

Mooney CZ and Duval R (1993) Bootstrapping: a nonparametric approach to statistical inference. *Sage University Paper Series on Quantitative Applications in the Social Sciences*, 07-095, pps. 15–22, 30–44 Sage, Newbury Park, CA.

Niedzwiecki D and Simonoff JS (1990) Estimation and inference in pharmacokinetic models: the effectiveness of model reformulation and resampling methods for functions of parameters. *J Pharmacokin Biopharm* **18**: 361–377.
Sall J (1990) Leverage plots for general linear hypotheses. *Am Stat* **44**: 308–315.
SAS (1994) *JMP Statistics and Graphics Guide*, Version 3 SAS Institute Cary, NC.
Simonoff JS and Tsai CL (1988) Jacknifing and bootstrapping quasi-likelihood estimators. *J Stat Comput Simul* **30**: 213–232.
Snedecor GW and Cochran WG (1989) *Statistical Methods*, 8th edn. Iowa State University Press, Ames, IA.
Westfall PH and Young SS (1989) p value adjustments for multiple tests in multivariate bionomial models. J Amer Statist Assoc **84**: 780–786.
Wu CFJ (1986) Jacknife, bootstrap and other resampling plans in regression analysis. *Ann Stat* **14**: 1216–1350.

6.7 FURTHER READING

Statistics, Regression, General Texts

Bard Y (1974) *Nonlinear Parameter Estimation*. Academic Press, New York.
Bates DM and Watts DG (1988) *Nonlinear Regression Analysis and its Application*. Wiley, New York.
Dowdy S and Wearden S (1983) *Statistics for Research*. Wiley, New York.
Huber P (1981) *Robust Statistics*. Wiley, New York.
Kachigan SK (1986) *Statistical Analysis*. Radius Press, New York.
Milliken GA and Johnson DE (1992) *Analysis of Messy Data*. Chapman and Hall, New York.
Montgomery DC and Peck EA (1981) *Introduction to Linear Regression*. Wiley, New York.
Myers RH (1986) *Classical and Modern Regression with Applications*. Duxbury Press, Boston.
Netter J, Wasserman W and Kutner M (1985) *Applied Linear Statistical Models*. Richard D Irwin, Illinois.
Rosseeuw PJ and Leroy AM (1987) *Robust Regression and Outlier Detection*. Wiley, New York.
Searle SR (1971) *Linear Models*. Wiley, New York.
Steel RGB and Torrie JH (1980) *Principles and Procedures of Statistics*. McGraw Hill, New York.

Outliers

Barnett V and Lewis T (1993) *Outliers in Statistical Data*. Wiley, New York.

Data Mining

Hand DJ (1997) *Construction and Assessment of Classification Rules*. Wiley, Chichester.
Hand DJ (1998) Data mining: statistics and more? *Am Stat* **52**: 112–118.

Mathematical Models and Cubic Ternary Complex Model

Abramson SN, McGonigle P and Molinoff PB (1987) Evaluation of models of radioligand binding data. *Mol Pharmacol* **31**: 103–111.

Feldman HA (1972) Mathematical theory of complex ligand-binding systems at equilibrium: some methods for parameter fitting. *Anal Biochem* **48**: 317–338.

Novak B and Tyson JJ (1993) Numerical analysis of a comprehensive model of M-phase control in *Xenopus oocyte* extracts and intact embryos. *J Cell Sci* **106**: 1153–1168.

Uthurusamy R (1996) From data mining to knowledge discovery: current challenges and future directions. In *Advances in Knowledge Discovery and Data Mining*, ed by UM Fayyad, G Piatetsky-Shapiro, P Smyth and R Uthurusamy, pp. 506–569. AAAI Press: Menlo Park, CA.

Weiss JM, Morgan PH, Lutz MW and Kenakin TP (1996) The cubic ternary complex receptor-occupancy model I. Model description. *J Theor Biol* **178**: 151–167.

Weiss JM, Morgan PH, Lutz MW and Kenakin TP (1996) The cubic ternary complex receptor-occupancy model II. Understanding apparent affinity. *J Theor Biol* **178**: 169–182.

Other Topics

Lai TL and Zhang L (1994) Statistical analysis of ligand binding experiments. *Biometrics* **50**: 782–797.

6.8 APPENDIX 1: CODE FOR COMBINATORIAL LIBRARY SIMULATION

```
Sub pool_sampler()
' limits for the exponential to simulate delta G
' these are the asymptote values for the distribution
yi = -2600
y0 = -15000
' k is the rate constant for the exponential.  Larger values of k
' will make the exponential decay faster
k = 10
' constants for the Ki calculation (temperature and R)
T = 294
R = 1.987
'initial ligand concentration (no inhibitors present)
l0 = 0.000001
' size of the pools
pool_size = 10
For i = 1 To 5000
   accum = 0
   For j = 1 To pool_size
      x = Rnd()
      'calculate delta G
      delta_g = yi + (y0 - yi) * (Exp(-k * x))
      'calculate association constant
      kassoc = Exp(-delta_g / (R * T))
      'calculate Ki
      ki = 1 / kassoc
      'accumulate the sum of the reciprocal Ki for the pool
      accum = l0 / ki
      Next j
```

```
   'random values for the control scaling parameters
   MaxValue = Application.NormInv(Rnd(), 1, 0.2)
   MinValue = Application.NormInv(Rnd(), 0, 0.2)
' calculate a percent inhibitor from the pool Ki and the controls
PercentI = MinValue + (MaxValue - MinValue) / (1 + accum)
   PercentI = 100 * (1 - PercentI)
   'fill in the results in the worksheet
   Worksheets("sheet1").Cells(i, 1).Value = delta_g
   Worksheets("sheet1").Cells(i, 2).Value = ki
   Worksheets("sheet1").Cells(i, 3).Value = PercentI
   Next i
End Sub
```

6.9 APPENDIX 2: JMP FORMULA TO CALCULATE POWER CURVES

$$sd \bullet \sqrt{\frac{2[\text{normQuant }(0.975) + \text{normQuant }(0.95)]^2}{\text{Replicate}}}$$

6.10 APPENDIX 3: STATISTICAL FORMULAE

Sample Mean ($\overline{x}$)

$$\overline{x} = \frac{1}{n}\sum_{i=1}^{n} x_i$$

where x_i are observations and n is sample size.

Weighted Mean

$$\overline{x} = \frac{\sum_{i=1}^{n} x_i w_i}{\sum_{i=1}^{n} w_i}$$

where x_i are observations, w_i are weights and n is sample size.

Geometric Mean

$$\overline{x} = \left(\prod_{i=1}^{n} x_i\right)^{\frac{1}{n}}$$

where x_i are observations, n is sample size and $\prod$ indicates product of terms rather than sum.

Harmonic Mean

$$\bar{x} = \frac{1}{\frac{1}{n}\sum_{i=1}^{n}\frac{1}{x_i}}$$

where x_i are observations and n is sample size.

Sample Variance

$$\text{var}(x) = \frac{1}{n-1}\sum_{i=1}^{n}(x_i - \bar{x})^2 = \frac{\sum_{i=1}^{n} x_i^2 - n\bar{x}^2}{n-1}$$

where x_i are observations and n is sample size.

Sample Standard Deviation

$$sd(x) = s_x = [\text{var}(x)]^{\frac{1}{2}}$$

Standard Error of the Mean

$$SEM = \frac{S_x}{\sqrt{n}}$$

t-Test Comparing Sample Mean to a Constant

$$t = \frac{\bar{x} - k}{\frac{s_x}{\sqrt{n}}}$$

where $\bar{x}$ is sample mean, s_x is sample standard deviation, n is sample size and k is a constant for the test (e.g. $\bar{x} = k$). The test has $n - 1$ degrees of freedom.

t-Test Comparing Two Samples: Variances of Samples are Assumed Equal

$$t = \frac{\bar{x}_1 - \bar{x}_2}{s\sqrt{\frac{1}{n_1} + \frac{1}{n_2}}}$$

where $\overline{x}_1$ and $\overline{x}_2$ are sample means and n_1 and n_2 are sample sizes. There are $n_1 + n_2 - 2$ degrees of freedom for this test. The pooled standard deviation (s) is:

$$s = \sqrt{\frac{(n_1 - 1)s_1^2 + (n_2 - 1)s_2^2}{n_1 + n_2 - 2}}$$

Paired *t*-Test

$$t = \frac{\overline{d}}{\dfrac{s_\mathrm{d}}{\sqrt{n}}}$$

where $\overline{d}$ is the mean of differences, s_d is the standard deviation of differences and n is the number of differences. The test has $n - 1$ degrees of freedom.

t-Test Comparing Two Samples: Variances of Samples are not Assumed to be Equal

$$t = \frac{\overline{x}_1 - \overline{x}_2}{\sqrt{\dfrac{s_1^2}{n_1} + \dfrac{s_2^2}{n_2}}}$$

where $\overline{x}_1$ and $\overline{x}_2$ are sample means, n_1 and n_2 are sample sizes and s_1 and s_2 are sample standard deviations. Degrees of freedom are approximated by:

$$df = \frac{\left(\dfrac{s_1^2}{n_1} + \dfrac{s_2^2}{n_2}\right)}{s_1^4\{n_1^2(n_1 - 1) + s_2^4(n_2^2(n_2 - 1)\}}$$

Confidence Intervals

$$\overline{x} \pm t_{n-1\mathrm{df},1-\alpha/2} \bullet \frac{s_x}{\sqrt{n}}$$

where $\overline{x}$ is sample mean, s_x is sample standard deviation, n is sample size, $t_{n-1\mathrm{df},1-\alpha/2}$ is $100(1 - \alpha/2)$ percentage point of the t distribution for $n - 1$ degrees of freedom and α is the significance level for the test.

Slope and Standard Error of Slope for Linear Regression Line ($y = mx + b$)

$$m = \frac{\sum_{j=1}^{n}[(x_i - \overline{x})(y_i - \overline{y})]}{\sum_{i=1}^{n}(x_i - \overline{x})}$$

$$se(m) = \frac{\dfrac{\sum_{i=1}^{n}\{[y_i - (b - mx_i)]\}}{n - 2}}{\sqrt{\sum_{i=1}^{n}(x_i - \overline{x})^2}}$$

where x_i and y_i are observations, $\overline{x}$ and $\overline{y}$ are means, n is sample size, m is slope and b is intercept of line.

Intercept and Standard Error of Intercept for Linear Regression Line ($y = mx + b$)

$$b = \overline{y} - m\overline{x}$$

$$se(b) = \frac{\sum_{i=1}^{n}\{[y_i - (b - mx_i)]^2\}}{n - 2} \bullet \left[\frac{1}{n} + \frac{\overline{x}^2}{\sum_{i=1}^{n}(x_i - \overline{x})^2}\right]$$

where x_i and y_i are observations, $\overline{x}$ and $\overline{y}$ are means, n is sample size, m is slope and b is intercept of line.

Correlation Coefficient

$$r = \frac{\sum_{i=1}^{n}[(x_i - \overline{x})(y_i - \overline{y})]}{\sqrt{\sum_{i=1}^{n}(x_i - \overline{x})^2 \bullet \sum_{i=1}^{n}(y_i - \overline{y})^2}}$$

where x_i and y_i are observations, $\overline{x}$ and $\overline{y}$ are means and n is sample size.

Non-linear Equation to Fit to Sigmoidal Concentration–Response Curves

$$y = nsb + \frac{(B_0 - nsb)}{1 + \left(\frac{[I]}{10^{-pIC_{50}}}\right)^n}$$

where y is observed response (dependent variable), $[I]$ is inhibitor or ligand concentration (independent variable in the equation), pIC_{50} is $-\log_{10}$ of the concentration of inhibitor $[I]$ giving half-maximal response, B_0 is maximal effect (for binding experiments, concentration bound with no inhibitor ($[I] = 0$)), nsb is background effect (for binding experiments, non-specific binding) or effect at a very large (infinite) value of $[I]$ and n is the Hill coefficient or slope in the region of the IC_{50}.

Back-calculation for Sigmoidal Concentration–Response Curves (calculate *x* from *y*)

$$[I] = 10^{-pIC_{50}} \bullet \left(\frac{y - nsb}{b_0 - y}\right)^{\frac{1}{n}}$$

where y is observed response (dependent variable), $[I]$ is inhibitor or ligand concentration (independent variable in the equation), pIC_{50} is $-\log_{10}$ of the concentration of inhibitor $[I]$ giving half-maximal response, B0 is maximal effect (for binding experiments, concentration bound with no inhibitor ($[I] = 0$)), nsb is background effect (for binding experiments, non-specific binding) or effect at a very large (infinite) value of $[I]$ and n is the Hill coefficient or slope in the region of the IC_{50}.

Calculation of Different Values of IC_v from IC_{50} (e.g. IC_{95} from IC_{50})

$$f = \frac{100 - v}{v}$$

$$\mathrm{IC}_{50}^n = f \bullet \mathrm{IC}_v^n$$

$$\mathrm{IC}_v = \left(\frac{1}{f} \bullet \mathrm{IC}_{50}^n\right)^{\frac{1}{n}}$$

where v is the value of response to use in calculation of the effective concentration (e.g. v is 95 to estimate IC_{95}) and n is the Hill coefficient or slope in the region of the IC_{50}.

Non-linear Equation to Fit IC_v

$$y = \frac{b\text{IC}_v^n + mx^{n/f}}{\text{IC}_v^n + x^{n/f}}$$

where f is fraction definition from equation above, x is independent variable (concentration of ligand or inhibitor), y is dependent variable (effect), b is baseline response (response at $x = 0$), m is maximal response (response at large value of x) and n is the Hill coefficient.

Back-calculation for IC_v Equation (Calculate x from y)

$$x = \text{IC}_v \bullet \left[f \bullet \frac{(y - b)}{(m - y)} \right]^{\frac{1}{n}}$$

where f is fraction definition from equation above, x is independent variable (concentration of ligand or inhibitor), y is dependent variable (effect), b is baseline response (response at $x = 0$), m is maximal response (response at large value of x). IC_v is the ligand or inhibitor concentration associated with half-maximal response and n is the Hill coefficient.

7 Statistical Methods for Target Identification and Validation

7.1 INTRODUCTION

The challenge of target identification and validation is choice of therapeutically relevant biochemical targets. The target becomes the basis of a research program that includes testing to identify and optimize drugs. Choice of a biochemical target for drug discovery research is a process that involves interaction between clinical and non-clinical disciplines, including genetics, genomics, biochemistry, enzymology, chemistry and physiology. Publication of the human genome and high-performance bioinformatics tools add new considerations for target selection.

Genetics and genomics are scientific fields with extensive literature about methods and application studies. It is not possible to list all of the current methods being used in these areas or to discuss even one method in detail. Instead, the goal of this chapter is to present material that will enable scientists to understand how these methods are used in the process of target validation and identification and to provide the background to read papers that assess the association between genes and disease. Three general types of analysis are considered: linkage analysis, association analysis and gene expression analysis.

Proteins are the biochemical targets for most drugs. This general classification can be refined to include receptors, enzymes, ion channels, kinases, phosphatases and other types of proteins. These biochemical entities are the building blocks for the chemical reactions that sustain life: regulation of function, signal transduction, metabolism and other processes. Signal transduction is of particular importance for drug discovery because each step in the signal transduction pathway is a potential site for action of a drug, from cell surface receptors to intracellular signaling elements. Drugs bind to proteins with the effect of activating a response (agonists), blocking a response to an endogenous ligand (antagonists) or modifying the response of the protein (allosteric modifiers). Proteins are synthesized from amino acids and form a characteristic three-dimensional structure. The structure of each protein is fully determined by its DNA sequence, which is in effect the blueprint for a protein. The molecular processes involved in the translation from DNA to protein are shown in Fig. 7.1. The DNA sequence for each protein is part of the genome for each species. Not all of the base pairs on a gene encode information about the protein. Coding information is contained as part of the exons, whereas the introns are the non-coding part of the gene. The presence of

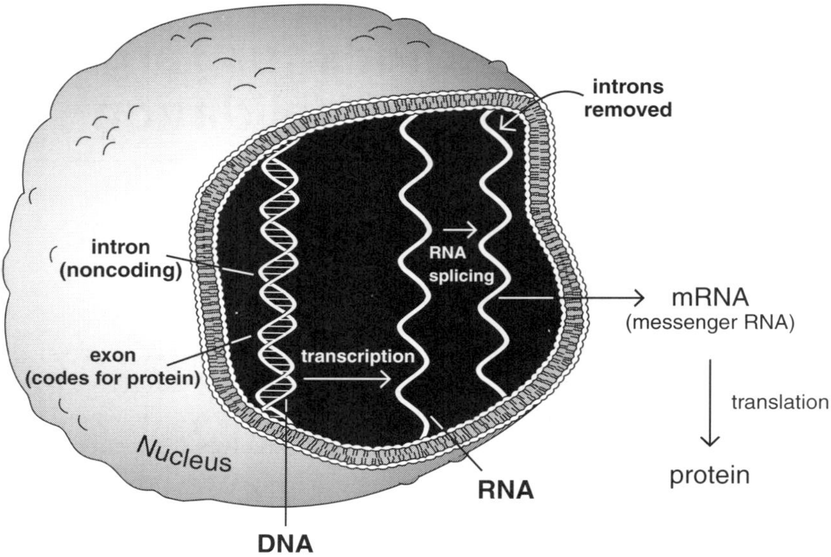

Fig. 7.1. Processes involved in translation from DNA sequence to protein. The genetic codes for making proteins are contained in double-stranded DNA in the nucleus of a cell. The DNA contains segments that code for the protein (exons) and non-coding regions (introns). The DNA code is transcribed to single-strand RNA. The process of splicing removes the introns from the sequence. Messenger RNA contains the code for the protein and is translated into the amino acids during protein synthesis

the introns adds versatility to the process of protein synthesis. Different proteins can be produced by the same gene in different cells, because different splicing locations are used during transcription and RNA splicing. Although most genes are converted into messenger RNA (mRNA) and are used to direct the synthesis of proteins, the product of some genes is the RNA itself.

Although proteins are the predominant targets for most drugs, genes themselves are the targets of potential therapeutic interventions using antisense technology. The antisense strategy involves making a mutant gene that will produce an RNA sequence complementary to the usual sequence during transcription. The presence of the complementary (or antisense) RNA at a high concentration causes hybridization to occur with the normal RNA and theoretically inhibits protein synthesis.

7.2 PRINCIPLES OF MOLECULAR GENETICS

Genetics is the study of heredity and variation in inherited traits, including disease. For target validation, data from genetic study trials are collected and analyzed in order to identify susceptibility genes that are associated with the risk of a disease.

After the relevant genes are identified, the biochemical mechanism and proteins associated with the genes are studied to determine where there are targets for drugs. Each step of this process takes time on the scale of months to years. For a complex genetic study, four years is not an unusually large amount of time to collect and analyze the data. Identifying cause and effect relationships in genetic studies and understanding the assumptions in the studies are critical. In most cases for human genetic studies, qualifications are made for other genetic and/or environmental risk factors for a disease. Increased susceptibility to disease is a critical concept to keep in mind. Genetic factors may increase the chance that an individual will develop a specific disease in their lifetime but an individual genetic defect is often not the direct cause of the disease. Multiple genetic factors can create a cumulative increase in risk, and susceptibility can increase as a function of time. The genetics of Alzheimer's disease illustrates many of these concepts and problems; to date, it is the best example of the identification of a genetic susceptibility factor for a common polygenetic disease (Corder *et al* 1993; Roses 1994). There is general agreement that a specific apolipoprotein E (ApoE) allele, $\varepsilon4$, in the absence of the $\varepsilon2$ allele is a genetic risk factor for a common late-onset form of Alzheimer's disease (Hellman *et al* 1998). However, there are patients with Alzheimer's disease that do not have the $\varepsilon4$ allele and there are patients with the $\varepsilon4$ allele that are unaffected. Results from different studies give conflicting results for some diseases and gene/disease associations. For Alzheimer's disease, a protective effect of allele 2 of the precursor protein of the non-amyloid component of plaque (human α-synuclein) was reported for a sample of elderly individuals who have at least one ApoE $\varepsilon4$ allele (Xia *et al* 1996). A subsequent study using a similar protocol did not support a protective effect for allele 2 (Hellman *et al* 1998). Reasons for the different results are suggested by the statistical framework of the studies. Because of the low frequency of allele 2 (the candidate allele) in the general population, the power to detect significant differences between groups is small, particularly for the relatively small sample sizes (control and treatment groups were samples of approximately 150 patients). Sampling difficulties, discussed in Chapter 6 and later in this chapter, make it difficult to match treatment samples with controls. With diseases that occur late in life, it can be difficult to obtain data on parents to use as controls. Finally, there can be differences between the two populations, such as stratification and admixture.

Chapter 6 described the importance of estimating sample sizes for experiments, understanding the assumptions of the statistical methods used to analyze data and rigorously defining the sampling scheme or experimental design. These principles are critical for genetic studies as well.

7.2.1 INHERITANCE AND DEFINITION OF GENETIC TERMS

In Chapter 1, some basic terms used in genetics are defined. Further definition and use of these terms is described in this section to understand the principles for analysis of genetic data.

The rules for inheritance and the basis for much of genetic theory were described in 1865 by Gregor Mendel with the publication of Mendel's laws. Although this was a full century before the discovery of DNA, Mendel's experiments with peas established concepts for the inheritance of characteristics and ultimately the role of genes. In addition to coding for RNA and protein synthesis, a gene is the basic unit for transmission of inheritable characteristics and recombination in offspring. Genes exist in different forms known as alleles. Alleles are the genetic material at a given location (locus) on a gene. Each offspring carries two copies of each gene, one from the mother and one from the father. Genes are arranged within cells on structures known as chromosomes. Each human cell contains 23 pairs of chromsomes (46 total), each parent contributing one member of the pair. Of the 23 pairs, 22 are autosomes that are the same for male and female offspring. The additional pair of chromosomes are the sex chromosomes, of which female offspring have two X-chromosomes and male offspring have one X-chromosome and one Y-chromosome. If a characteristic shows sex-linked inheritance, the X and/or Y-chromosomes are involved.

7.2.2 GENOTYPES AND POLYMORPHISM

Genotype is determined by using blood samples to identify the specific alleles present in an individual's genome. The alleles are represented by different nomenclature, perhaps capital letters to indicate dominance (the expression of a dominant allele occurs even if a recessive allele is inherited from the other parent) or they may be represented using integers or other codes. Because an individual inherits an allele from each parent, if there are K alleles then there are $K(K+1)/2$ possible genotypes for the gene. If the alleles inherited from both parents are the same, then the genotype is described as being homozygous; if the alleles are different, then the genotype is heterozygous. When multiple genes are considered, the distribution of alleles is considered a haplotype. For a set of genes, a child inherits one haplotype from their mother and another from their father. Geneticists estimate the frequencies of different alleles in the population known as gene frequencies. Theoretically, genotype frequencies in a population depend only on gene frequencies if random mating is assumed and mutation does not occur; this is known as the Hardy–Weinberg equilibrium. In other words, the frequency of a genotype of an individual with two dominant homologous alleles $P(\mathrm{A}|\mathrm{A})$ is equal to the squared frequency of the allele (p_{A}^2), and likewise for two recessive alleles. For a heterozygous individual $P(\mathrm{A}|\mathrm{a})$, the frequency is equal to the product of the gene frequencies $2p_{\mathrm{A}}p_{\mathrm{a}}$. It is common in scientific papers to determine if the distribution of genotypes of patients in a study sample follows the Hardy–Weinberg law.

Data from a family that is analyzed for genetic studies are graphically represented as a pedigree or tree diagram, as described in Chapter 1. Depending on the distribution of individuals expressing a trait or disease, the pedigree can be classified as simplex or multiplex. Simplex pedigrees show a single affected offspring with parental information. Sibling pairs (sib-pairs) involve pairs of affected siblings, whereas sibling sets (sib-sets) extend the pedigree to more than two offspring. The pedigree can be extended to include relatives, and the most general case is a multiplex pedigree involving multiple-related families.

Many alleles are very common when their frequency is estimated across the entire population e.g. most of the population has one specific form of the allele. The exception are polymorphic (multiple forms) genes where the most common allele has a frequency of less than 95%. Although some polymorphic genes have no observable effect, others are interesting for drug-target gene finding because they represent a location on genes that may be associated with disease. There are several classes of polymorphism, shown diagrammatically in Fig. 7.2. There are single base changes, insertion and deletion of parts of the sequence, and tandem repeats where a simple pattern of bases is inserted into the gene.

One method used by biochemists to study DNA sequences and test for polymorphism is to use enzymes that cut the DNA into fragments at specific sequences. For a set of chromosomes, if the sequence at a site is different then the cutting will occur at a different location and fragments of different length will result. Gel electrophoresis shows these different length fragments as different bands on the gel. This method of identifying polymorphism is referred to as restriction fragment length polymorphism (RFLP).

For use in statistical models, it is important to quantify polymorphism of markers. Elston (1995) provides guidelines for quantifying polymorphism that are independent of the mode of disease inheritance. If the genotypic frequencies of a marker are in Hardy–Weinberg equilibrium and p_i is the estimated frequency of the ith allele in a random sample of n people, then the heterozygosity (measure of polymorphism) is:

$$H = 1 - \sum_{i=1}^{n} p_i^2 \qquad \ldots (7.1)$$

Single nucleotide polymorphism

CTACCTTAGACT
CTACATTAGACT

Short tandem repeat or microsatellite

GTACTCACACACATCG
GTACTCACACACACACATCG
GTACTCACACACACACACACATCG

Fig. 7.2. Classes of polymorphism. These include single nucleotide changes, tandem repeats and microsatellites. In single nucleotide polymorphism, only a single base pair is changed. Tandem repeats and microsatellites are short repeating segments of a simple motif or combination of nucleic acids, here CA. In many cases, the polymorphism is simply a different number of repeats, allowing detection using the polymerase chain reaction

An unbiased estimate is:

$$H = 1 - \frac{\left(2n \sum_{i=1}^{n} p_i^2 - 1\right)}{(2n - 1)} \qquad \ldots (7.2)$$

These estimates are the probability that a random individual is heterozygous for any two alleles at a gene locus, given the corresponding frequencies (p_i). The variance of these statistics is (Nei and Roychoudhury, 1973):

$$V(H) = \frac{(2n - 1)}{4n^3}\left[(3 - 4n)\left(\sum_{i=1}^{n} p_i^2\right)^2 + 4(n - 1)\sum_{i=1}^{n} p_i^3 + \sum_{i=1}^{n} p_i^2\right] \qquad \ldots (7.3)$$

for the biased estimate and:

$$V(H) = \frac{1}{n(2n - 1)}\left[(3 - 4n)\left(\sum_{i=1}^{n} p_i^2\right)^2 + 4(n - 1)\sum_{i=1}^{n} p_i^3 + \sum_{i=1}^{n} p_i^2\right] \qquad \ldots (7.4)$$

for the unbiased estimate.

Another estimate of polymorphism that is commonly used is the polymorphism information content (PIC) value as proposed by Botstein *et al* (1980). This value was derived for rare dominant diseases occurring in a family with a codominant marker. The PIC value is defined as the probability that a marker genotype of an offspring will allow identification of which of the two marker alleles it received from the parent with the disease. In that sense, it is a measure of the informativeness of a marker in an offspring. One parent is assumed to be affected and thus be heterozygous for the disease allele. Using the notation of p_j for allele frequencies, the PIC is defined as:

$$PIC = 1 - \sum_{i=1}^{n} p_i^2 - \sum_{i=1}^{n-1}\sum_{j=i+1}^{n} 2p_i^2 p_j^2 \qquad \ldots (7.5)$$

The PIC will always be smaller than the measure of heterozygosity; however, as the number of alleles becomes larger, the two numbers become similar. Genes are considered polymorphic when these measures are greater than or equal to 0.10.

7.2.3 RECOMBINATION

From the description of the inheritance laws and inspection of the pedigrees, it is clear that each parent contributes one allele to each gene. When several genes are studied at the same time, linkage between the alleles and recombination are important considerations. Figure 7.3 shows a simple, two-gene pedigree for three generations of a family. Each row shows the alleles for a single gene; A and B for

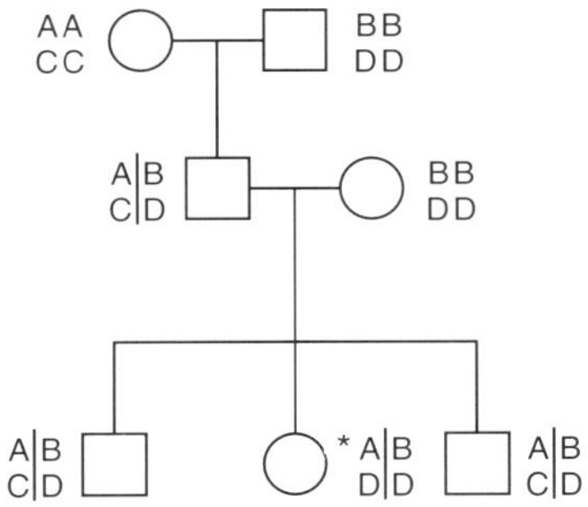

gene 1- alleles A, B
gene 2- alleles C, D
female* is recombinant

Fig. 7.3. Illustration of recombination using a simple pedigree. Three generations of a family are shown. Squares indicate males and circles indicate females. Each small (2 × 2) table adjacent to a symbol shows the alleles for two genes. The rows show the alleles for each different gene: A and B for gene 1; C and D for gene 2. In the parental generation (middle of pedigree), information about the mother's parents is missing; however, the father received alleles A (gene 1) and C (gene 2) from his father and alleles B (gene 1) and D (gene 2) from his mother. If the child inherits a haplotype that is different from his or her parents, the new haplotype is a recombinant haplotype, in other words the alleles recombined from an earlier generation, the grandparents. Here, the female child is recombinant

gene 1, C and D for gene 2. Squares indicate males and circles indicate females. In the parental generation, information about the mother's parents is missing but the father received alleles A (gene 1) and C (gene 2) from his father and alleles B (gene 1) and D (gene 2) from his mother. The assumption is made that the alleles are inherited independently from each other, therefore the father could pass four different haplotypes to his children, all with a probability of $\frac{1}{4}$: A,C; A,D; B,C; or B,D. If the genotype of a child is identical to a parental genotype, then the child's genotype is called parental. The two male children have parental genotypes. If the child inherits a haplotype that is different from his or her parents, then the new haplotype is a recombinant haplotype, i.e. the alleles recombined from an earlier generation, the grandparents. In Fig. 7.3, the female child is recombinant. In some cases, the alleles of several genes appear to be linked, e.g. genetically coupled. This is termed genetic linkage. For the example of two genes, if a parent has a heterozygous genotype and does not produce recombinant offspring, then the genes are considered to be linked. Note that the linkage is between the alleles, and not the genes themselves. Recombination can be determined only if at least one pair of parental genes are both heterozygous, otherwise it is not possible to determine whether the alleles are linked. Parents are considered informative for linkage analysis if a person has a pair of heterozygous genes.

A doubly heterozygous person has two possible ways to receive their alleles, depending on how the alleles are coupled. These are known as phases. Often, the phase is unknown because it is unclear which haplotype came from each parent. Linkage analysis is facilitated if the phase can be determined. Recombination occurs primarily during formation of the gametes (meiosis). The probability that a gamete produced by a parent is recombinant is known as the recombination fraction. Lower numbers indicate linkage (less recombination) and will be less than $1/N$, where N is the number of genes. If recombinations are counted, the recombination fraction is simply:

$$\theta = \frac{\text{No. of observed recombinations}}{\text{No. of opportunities for recombination}} \qquad \ldots(7.6)$$

The recombination fraction varies from 0 to 0.5, 0 indicating linkage and 0.5 indicating random segregation of alleles.

The phenomenon of linkage between alleles of genes is not surprising when you consider the structure of genes on chromosomes. Genes located closely together on a chromosome have a higher probability of being linked (e.g. observing linkage between alleles). This observation is critical for studies that rely on linkage between markers and disease genes.

7.2.4 LINKAGE DISEQUILIBRIUM AND ALLELIC ASSOCIATION

In order to understand the context of using genetic studies for the identification of valid targets, it is critical to understand the nature of the associations being tested, the genetic or inheritance models and the statistical models underlying any tests (Ott 1991). Additionally, good statistical practice requires that the characteristics of the sample be understood. For example, a chi-squared test assumes that all observations are independent, which will not be the case if family members are included in the same analysis.

Associations can be between phenotype and genotype (next section) or between phenotypes. Remember that, by definition, a phenotype is an observable feature or trait. The first evidence for an association between a pair of phenotypes is that they appear in individuals more often than expected if they are independent. For example, assume that individuals receiving a drug treatment have a higher incidence of a disease than the population at large. Using the notation of n_{ta} and n_{tu} for the number of affected and unaffected individuals in the treatment group and n_{ca} and n_{cu} for the number of affected and unaffected individuals in a control group, then an odds ratio is computed:

$$\text{Odds ratio} = \frac{n_{\text{ta}} n_{\text{cu}}}{n_{\text{tu}} n_{\text{ca}}} \qquad \ldots(7.7)$$

It is clear that a higher proportion of affected individuals in the treatment group increases the odds or relative risk for the disease. Alternatively, a correlation coefficient can be calculated:

$$\text{Correlation coefficient} = \frac{Nn_{\text{ta}} - (n_{\text{ta}} + n_{\text{ca}})(n_{\text{ta}} + n_{\text{tu}})}{\sqrt{(n_{\text{ta}} + n_{\text{ca}})(n_{\text{tu}} + n_{\text{cu}})(n_{\text{ta}} + n_{\text{tu}})(n_{\text{ca}} + n_{\text{cu}})}} \quad \ldots(7.8)$$

where N is the total sample size.

As discussed in Chapter 6, sampling protocols can inflate correlations. If the treatment group consists of related family members, the correlation coefficient will be artificially high due to clustered sampling. Admixture is another sampling problem that can inflate apparent correlations. In admixture, a mixture of two populations shows an apparent correlation between a specific factor (genotype or characteristic) and disease when in fact they are independent. Examining the individual populations shows that although there is no correlation between the factor and disease, in one population there is a higher frequency of both the factor and disease. When the populations are combined, there is an apparent correlation that is an artificial consequence of the different frequencies between the groups.

Admixture is a sampling problem that is not unique to genetics. However, admixture is useful in order to understand the basis of linkage disequilibrium and allelic association. The basic concept of linkage disequilibrium is that marker and disease alleles are inherited as a block. Marker alleles are determined during genotyping and are considered observations. Disease alleles can be known or unknown, e.g. an allele that is not necessarily observed. Ott (1991) and Hartl (1988) provide examples that illustrate these concepts. First, assume that there are two populations and that two loci are in Hardy–Weinberg equilibrium before admixture. For the mixed population, after random mating between the two populations, let p_i and q_i be the gene frequencies at the two loci. Let P_{ij} be the frequency of a haplotype that has allele i at the first locus and allele j at the second. If there are two alleles at each locus and the alleles occur independently, the haplotype for allele 1 at both loci occurs with the frequency:

$$P_{11} = p_1 q_1 \quad \ldots(7.9)$$

It may take several generations of random mating for equilibrium between the loci to be re-established. Linkage disequilibrium measures the difference between the frequencies and is defined as:

$$D = P_{11} - p_1 q_1 \quad \ldots(7.10)$$

When the loci are in equilibrium, D is equal to zero. For the example, linkage disequilibrium resulted from admixture, e.g. the apparent linkage between the alleles will decrease over time. The form of the disequilibrium equation is important to understand: P_{ij} is a population frequency for a specific haplotype, and quantities p_i and q_i are gene frequencies. If linkage disequilibrium exists, then there are non-random occurrences of haplotypes. There are other causes for linkage disequilibrium, including tight linkage between the loci themselves. Allelic association refers to differences from equilibrium without regard to the cause.

Later parts of this chapter discuss the differences between linkage analysis and association analysis. Elston (1995) states that association analyses are statistically more powerful provided that there is linkage disequilibrium between the marker and disease loci; however, only linkage analyses have power in the absence of linkage disequilibrium.

7.2.5 THE RELATIONSHIP BETWEEN PHENOTYPE AND GENOTYPE

A phenotype is an observable expression of a specific genotype: presence or absence of a disease or a physical characteristic. Phenotypes are recorded as binary variables (presence or absence of a disease) or as a real number for quantitative phenotypes (height and weight, for example). Using the Mendelian laws of inheritance, the distribution of phenotypes for families can be predicted and compared to observed frequencies using a chi-squared test. However, observed characteristics and diseases are often associated with the interaction of multiple genes, different allelic forms of the same gene and environmental factors. Sorting out the complex interactions and temporal order of events is a difficult task and is the study of research in genetics. In genetic studies the genotypes of individuals, obtained from sequencing DNA from blood samples, are compared to their phenotypes: whether or not they have a disease or trait.

In the remainder of this chapter, several statistical methods to quantify the relationship between genotype and phenotype are described. For some clinical trials, particularly when a single gene is associated with the disease, the analysis takes the form of comparing proportions among groups of patients. For the first example, cystic fibrosis is used because of the clear link to a specific gene locus. This example illustrates some of the issues involved in comparing genotypes and phenotypes.

Cystic fibrosis (CF) is associated with changes in a specific gene located on chromosome 7 (Kerem *et al* 1989; Riordan *et al* 1989; Rommens *et al* 1989). Genetic analysis studies found that approximately 70% of the mutations in CF patients correspond to a deletion of three base pairs that result in the loss of a phenylalanine residue at position 508 of the protein that is the product of this gene. The remainder of the mutant gene pool consisted of other mutations. Clinically, there is a group of CF patients that have pancreatic insufficiency requiring enzyme-replacement therapy for digestion and another group that has sufficient pancreatic function for normal digestion. The larger group of patients (85%) have pancreatic insufficiency and are genetically more homogenous than the group with sufficient pancreatic function. A study by Kerem *et al* (1990) examined the relationship between phenotype and genotype for 293 patients with CF, taking into account pancreatic function and several clinical measures, including pulmonary function, growth measurements and sweat chloride levels. Table 7.1, reproduced from the Kerem paper, shows the relationship between genotypes and several of the clinical measures. The three genotypes are $\Delta F_{508}/\Delta F_{508}$ (the patient inherited

Table 7.1. Relationship between phenotype and genotype for a cystic fibrosis study: pancreatic function status and age at diagnosis of 293 patients with cystic fibrosis according to genotype (with respect to the ΔF_{508} mutation)

Variable	$\Delta F_{508}/\Delta F_{508}$	ΔF_{508}/Other	Other
No. of patients (% of cohort)	151 (52)	117 (40)	25 (8)
No. with pancreatic sufficiency at present (% of genotype group)[1]	2 (1)	33 (28)	16 (64)
Age at diagnosis (years)[2]	1.8 ± 3.3	4.4 ± 5.9	8.4 ± 8.3

[1] $P < 0.001$ by chi-squared test (2 d.f.).
[2] $P < 0.001$ by analysis of variance of three means; values are means ±SD.

Table 7.2. Relationship between phenotype and genotype for a cystic fibrosis study: current pulmonary function in 241 patients old enough for testing, according to genotype (with respect to the ΔF_{508} mutation)[1]

Variable	$\Delta F_{508}/\Delta F_{508}$		ΔF_{508}/Other			Other/Other			P Value[2]
	PI	PS	PI		PS	PI		PS	
No. of patients	121	1	66		32	6		15	—
FEV_1 (% predicted)	61	68	68	(0.01)	81	60	(0.10)	80	0.001
FVC (% predicted)	72	72	76	(0.02)	88	66	(0.13)	83	0.017
FEF_{25-75} (% predicted)	42	53	50	(0.02)	65	40	(0.05)	68	0.001

[1] Values are means adjusted to the overall mean age (20 years) computed from regression equations with a common slope for each pulmonary variable. Values in parentheses are P values for the comparison of means within genotype groups. PI, pancreatic insufficiency; PS, sufficient pancreatic function.
[2] By F test for differences among mean values in analysis of covariance.

the ΔF_{508} mutation from both parents: homozygous), ΔF_{508}/other (one ΔF_{508} mutation and one other mutation: heterozygous) or other/other (no ΔF_{508} mutation). The patients were also grouped according to pancreatic function status. From Table 7.1 it is seen that the diagnosis is made at an earlier age for the homozygous patients and that pancreatic function status appears to be related to the presence of the ΔF_{508} mutation. Table 7.2 extends this analysis to pulmonary function values, showing that there are significant differences in pulmonary function between genotype groups for this sample of 241 patients. Because pulmonary function decreases with age, the comparisons had to be adjusted for this effect using analysis of covariance. The analysis of covariance for this problem was represented in a concise form using a set of six scatterplots as shown in Fig. 7.4. An important conclusion of the study is that the variability in severity of the disease is due in part to multiple disease genotypes. For this study, classic statistical methods (F-tests, chi-squared test and analysis of variance) are sufficient to examine the relationship between genotype and disease phenotype. In the CF

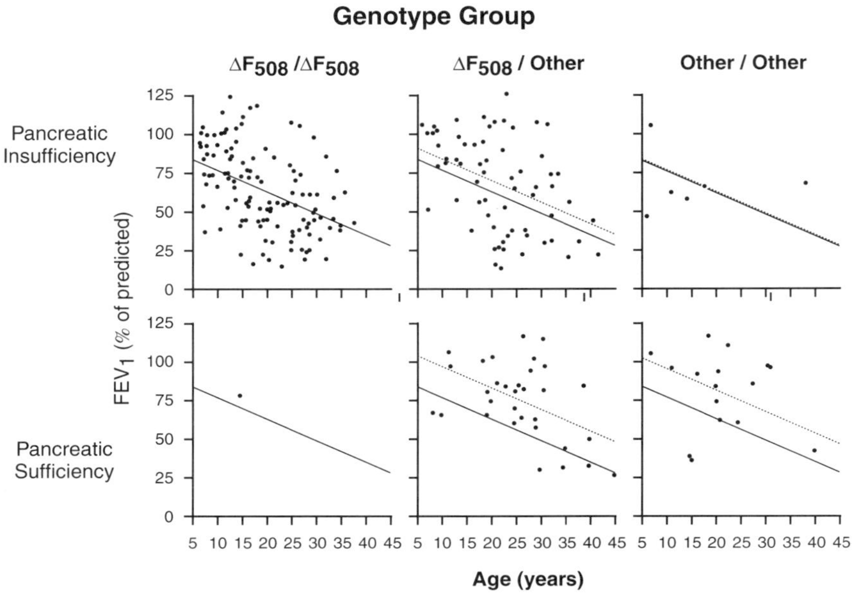

Fig. 7.4. Analysis of covariance for a type I diabetes study relating phenotype to genotype. In some genetic studies, a covariate is associated with change in response, possibly confounding the results of the analysis of the treatment variables. In this case, the study compares pulmonary function to genotype. However, because pulmonary function decreases with age, the comparisons had to be adjusted using analysis of covariance. The analysis of covariance for this problem was represented in a concise form using a set of six scatter-plots. Abscissa: age (years). Ordinate: fractional expiratory volume (FEV, %). (Reprinted with permission from Kerem E *et al* (1990) The relation between genotype and phenotype in cystic fibrosis–analysis of the most common mutation (ΔF_{508}). *N Engl J Med* 323: 1517–1522. Copyright © 1990 Massachusetts Medical Society. All rights reserved

study, three genotypes for a single gene were analyzed. For other diseases, the link between genotype and phenotype is more complex and requires different analysis. In the case of Alzheimer's disease, conflicting study results questioned the assumption that the disease is a single homogenous disorder (St George-Hyslop *et al* 1990).

Penetrance must also be considered in genetic models. If the assumption is made that all individuals having a specific genotype will express the corresponding phenotype, the penetrance is assumed to be unity. Individuals that do not have the specific genotype, in turn, do not express the phenotype and have a penetrance of zero. In some diseases, penetrances less than unity are observed and a given genotype can be associated with more than one phenotype, where other factors are involved in determining which phenotype is expressed. In these cases, a penetrance matrix is constructed to show the relationship between the genotypes and different phenotypes.

7.3 LINKAGE ANALYSIS

The oldest approach for assessing a relationship between genes and disease is linkage analysis: a method that requires knowledge of inheritance linkages in families. Linkage analysis also requires families that exhibit an inherited susceptibility to a disease. For each individual in the family, the observed phenotype or presence and severity of the disease are recorded in a pedigree. Finally, DNA samples are obtained from the family members and genetic markers are established on their entire genome.

There are several approaches to linkage analysis, many involving pairs of siblings. For example, if pairs of siblings both have the disease, alleles (different forms of a gene) at specific markers are examined to determine if the same genotype occurs frequently. Many methods involve the determination of how often recombination occurs, making understanding of recombination and mechanisms of inheritance key factors for understanding linkage analysis. Different alleles may be associated with different physical characteristics or the presence of a disease. If there is linkage between marker and disease loci (marker loci are close to the disease gene), then the markers will group individuals in the same family with the disease together. Markers that are far from the disease gene do not group individuals by disease status. Once markers that show strong linkage to the disease are located, positional cloning (Fig. 7.5) is done to identify the mutated gene or genes associated with the disease. In linkage analysis, no prior assumptions are made about the linkage between a gene and the disease. Both parametric and non-parametric statistical methods are used in linkage analysis, each having different sets of assumptions about the genetic model of inheritance.

In genetic terms, loci are linked if they are on the same chromosome and are physically located close enough so that their alleles do not separate independently at the time when gametes are formed. To search for disease linkage, investigators determine if genetic markers associated with a phenotype are linked to a gene located within 10–20 centimorgans that has allelic variation. The success of the method is due, in part, to a finely spaced set of markers, increasing the probability that the disease gene will be located near a marker.

Different forms of linkage analysis are used that either make assumptions about a model of genetic inheritance or are termed model free. In order to construct a statistical model for linkage, five components must be specified (Elston 1995): (i) the inheritance mechanism underlying the bivariate phenotype (marker and disease), which includes the number of loci and alleles in the model and any deviations from random mating; (ii) the population distributions of the genotypes involved (joint genotypic frequencies for marker and disease loci), which are usually estimated from the population; (iii) probability distributions of the phenotypes conditional on each genotype or the penetrance matrix; this component models the relationship

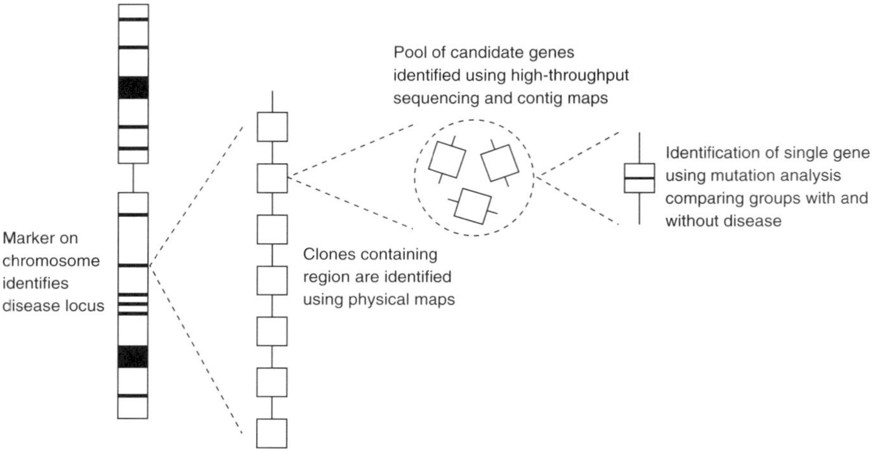

Fig. 7.5. Illustration of positional cloning. Blood or tissue samples from families having the disease under study are collected. Linkage studies are done to determine which, if any, markers show strong linkage to the disease. The markers are positioned at specific regions on the chromosomes to define disease locations (or loci). The marker data is analyzed to determine if the marker is associated with a candidate susceptibility gene or if the marker is polymorphic with no functional relationship to the disease. Physical maps of the chromosome are used to construct clones containing regions near the disease loci. The clones are used to construct a pool of candidate genes by sequencing and using contig maps an assembly of contiguous regions of the chromosome. With the candidate disease genes identified, further studies collect more data from people with and without the disease. Support or lack of support for a specific gene being a susceptibility gene or disease gene is given by analysis of the mutations in these groups of people

between phenotype and genotype, to account for cases where genotype does not correlate perfectly with phenotype; (iv) distribution of offspring genotypes conditional on their parents genotypes; and (v) sampling strategy in which the familial data were sampled from the population.

Results of linkage studies are often reported using a lod score: the common logarithm of the ratio of the likelihood of linkage with the likelihood of free recombination (no linkage). It is helpful to think about the lod score in the framework discussed in Chapter 6. An analogy can be drawn to pharmacological studies. In these studies, hypotheses are proposed to characterize the differences between control and treatment groups and sample observations are used to measure the strength of the hypothesis. The null hypothesis is usually no treatment effect and the alternative hypothesis is a treatment effect. In linkage analysis, the null hypothesis is no linkage (free recombination) and the alternative hypothesis is linkage. If θ is used to designate the recombination fraction, then $\theta = 0.5$ corresponds to free recombination and $\theta < 0.5$ indicates linkage. The lod score is defined as:

$$Z(\theta) = \log_{10}\left(\frac{L(\theta)}{L\left(\frac{1}{2}\right)}\right) \qquad \ldots(7.11)$$

where L indicates the likelihood function. Large, positive lod scores support the hypothesis of linkage; negative lod scores indicate lack of linkage. Historically, lod scores of >3 (corresponding to a likelihood ratio of $1:1000$) are used as the threshold for linkage and a limit of less than -2 (corresponding to a likelihood ratio of $1:100$) for absence of linkage. Computer programs based on maximum likelihood methods are used to estimate $Z(\theta)$ in practice.

Linkage analysis has been used with good results to identify susceptibility genes for diseases. Some examples include Duchenne muscular dystrophy (Davies *et al* 1983), Huntington's disease (Gusella *et al* 1983), cystic fibrosis (Tsui *et al* 1985), familial Alzheimer's disease (St George-Hyslop *et al* 1987) and adult polycystic kidney disease (Reeders *et al* 1985). Difficulties with interpretation of linkage studies include uncertainty about the genetic mode of inheritance and the paradox that as more familial data are collected, the strength or weakness of linkage may change, refuting earlier studies.

As an example of a linkage study, consider the two-locus model for insulin-dependent diabetes (Hodge *et al* 1983). Done during the early 1980s this study analyzed the results for 100 families and 27 genetic markers. The study is a good one to illustrate linkage analysis because it deals with many of the key issues involved in planning studies and interpreting linkage results: heterogeneity within the data set (admixture), uncertainty about the mode of inheritance and differential penetrance. Subsequent work has shown that the genetics of human type I diabetes is complex and polygenetic (Davies *et al* 1994). As is the case with some linkage studies, ensuing studies did not support all of the findings of the original study (Barbosa *et al* 1982). The results of the work were an important starting point for additional linkage and association analyses on different populations.

Because the mode of inheritance of type 1 diabetes was unclear, several genetic models were used: recessive with 20% penetrance, recessive with 50% penetrance; and dominant with 20% penetrance. As is the case with many diseases, the low penetrance values were supported by family and twin studies. Table 7.3, reproduced from Hodge *et al* (1983), is typical of the results reported for linkage studies. By inspection of the lod scores, it is seen that the maximal lod scores are seen for four markers: HLA, BF, GLO and Jk. Considering the number of comparisons made, statistical significance was determined to be a lod score of >3.

Non-parametric or distribution-free methods of linkage analysis do not require assumptions about the mechanism of inheritance. Sib-pair methods are an example of a non-parametric analysis. Sib-pair methods of linkage analysis are based on using data from pairs of siblings. The original description of the method was published in 1935 by Penrose and presented a study using sib-pairs that did not have a specified parentage (Penrose 1935). Analysis is based on a contingency

Table 7.3. Results of a linkage analysis for type I diabetes: summary of linkage analysis for recessive and dominant models with 20% penetrance[1]

Chromosome	Marker system		Recessive (Model 1)		Dominant (Model 3)		No. of informative families
			Lod	θ	Lod	θ	
1	PGD		0.71	0.01,0.01	—[2]	—	16
	Rh		−7.56	0.01,0.01	−4.53	0.01,0.01	82
	PGM1:	min	−3.89	0.01,0.01	−1.61	0.01,0.40	71
		max	0.55	0.30,0.20	1.02	0.30,0.05	
	AMY2		1.24	0.01,0.05	0.87	0.01,0.01	21
	Fy:	min	4.70	0.01,0.01	−1.76	0.01,0.01	84
		max	—	—	0.52	0.10,0.40	
2	Jk:	min	—	—	−1.48	0.05,0.40	78
		max	1.63	0.10,0.10	1.23	0.40,0.01	78
	ACP1		−4.90	0.01,0.01	−3.81	0.01,0.10	
4	GC		−5.50	0.01,0.01	−2.05	0.01,0.01	73
	MNSs		−10.11	0.01,0.01	−5.54	0.01,0.01	80
6	HLA		4.77	0.10,0.10	6.17	0.10,0.01	100
	BF		2.90	0.01,0.10	3.74	0.01,0.01	61
	GLO:	min	−1.41	0.01,0.01	−1.62	0.01,0.01	94
		max	1.16	0.10,0.20	0.88	0.40,0.01	
9	GALT		−0.85	0.01,0.01	—[2]	—	40
	ABO:	min	−3.25	0.01,0.01	−1.57	0.40,0.01	73
		max	0.55	0.02,0.40	—	—	24
	AK1		−1.27	0.01,0.01	—[2]	—	
13	ESD		−2.23	0.01,0.01	−3.04	0.01,0.01	38
16	HPA		−9.32	0.01,0.01	−4.56	0.01,0.01	76
	PGP		−2.48	0.01,0.01	−2.35	0.01,0.01	35
Unassigned	K		−2.21	0.01,0.01	−1.56	0.01,0.01	32
	P:	min	−1.87	0.01,0.01	−0.52	0.01,0.40	81
		max	0.84	0.40,0.01	0.53	0.40,0.01	
	GPT		−6.88	0.01,0.01	−4.24	0.01,0.01	80

[1] Note: minimum lod scores are shown only in <0.5; maximum lod scores are shown only in >0.5. First value of θ is male; second is female.
[2] In this model, all lod scores fell between −0.5 and 0.5.
Reprinted with permission from Hodge *et al* (1983). The search for heterogeneity in insulin-dependent diabetes mellitus (IDDM): linkage studies, two-locus models and genetic heterogeneity. *Am J Hum Genet* **35**: 1139–1155, published by The University of Chicago Press. Copyright 1983, The University of Chicago Press.

table of the observed phenotypes. For a pair of loci, if there is linkage, then the expected frequencies of pairs of siblings having both phenotypes or neither phenotype expressed is higher than the frequencies of sib-pairs where only one phenotype is expressed.

With the advent of genetic markers and genotyping, sib-pair methods have changed to be based on examination of pairs of affected siblings, e.g. cases where both siblings have the disease or characteristic being studied. The siblings are genotyped and the presence of specific marker alleles is recorded. If the siblings both have the same allele, the inheritance of the allele is termed 'identical by descent' (IBD). On the average, full siblings have half of their alleles shared identical by descent for a given locus. The relevant linkage therefore is between the marker locus (with the observable allele) and the disease locus associated with the disease allele. An interesting problem is how to handle more than a pair of siblings. Simple combinatorial math shows that $n(n-1)/2$ pairs of siblings can be formed from a set of n. However, these pairs cannot be considered independent samples. Hodge (1984) developed a method to measure the amount of information for larger collections of siblings based on Shannon's selective information theorem. For three siblings (where there are three possible sib-pairs), Hodge showed that after two independent sib-pairs were formed, the third pair provided no additional information three-quarters of the time and added additional information one-quarter of the time. The third pair did not add as much information as a third independent pair. Hodge (1984) proposed a weighting scheme for analysis using multiple sib-pairs to allow analysis to be performed that accounted for the difference in the number of siblings.

The earliest linkage studies were limited by the existence of a relatively small number of genetic markers. As a finer map is made of the human genome, more data are available for linkage studies. These studies involve both the genomic technologies to construct high-resolution genetic maps and linkage analysis to use the maps to identify candidate disease genes.

Figure 7.6 shows part of a genetic map of a human genome wide scan for susceptibility genes for type I diabetes (Davies *et al* 1994). This specific map was constructed from 290 different markers based on data from 96 affected sib-pair families. Each chromosome and associated marker locus form an entry on the map. The results of the linkage analyses are displayed in is the table featured in Fig. 7.6. The columns represent two different ways for dividing the data. In the column labeled total data, the number of pairs sharing either one or zero alleles from all of the sib-pairs, summed over paternal and maternal meioses, are displayed. In the second column, labeled HLA split, the families were split by HLA status: a marker associated with type I diabetes. This split is made by grouping sib-pairs that share two HLA alleles differently from sib-pairs sharing one or zero HLA alleles. After the split was made, the counts of the number of sib-pairs sharing either one or zero alleles at the marker locus were tabulated. The column labeled MLS is the maximum lod score. The linkage analysis included data from families that were not fully informative by using a maximum lod test proposed by Risch (1990a,b). Higher values of the MLS score are indicative of linkage. Values greater than unity were recorded in the table because they would be significant at the $P < 0.05$ level. The vertical bar indicates interlocus distances. These are expressed as gender-averaged, percentage recombination fractions. Distances with an asterisk indicate that the distance was derived from another map. Taken as a whole, the map presents a concise summary of the linkage analysis. Having multiple forms of the data (MLS scores plus counts

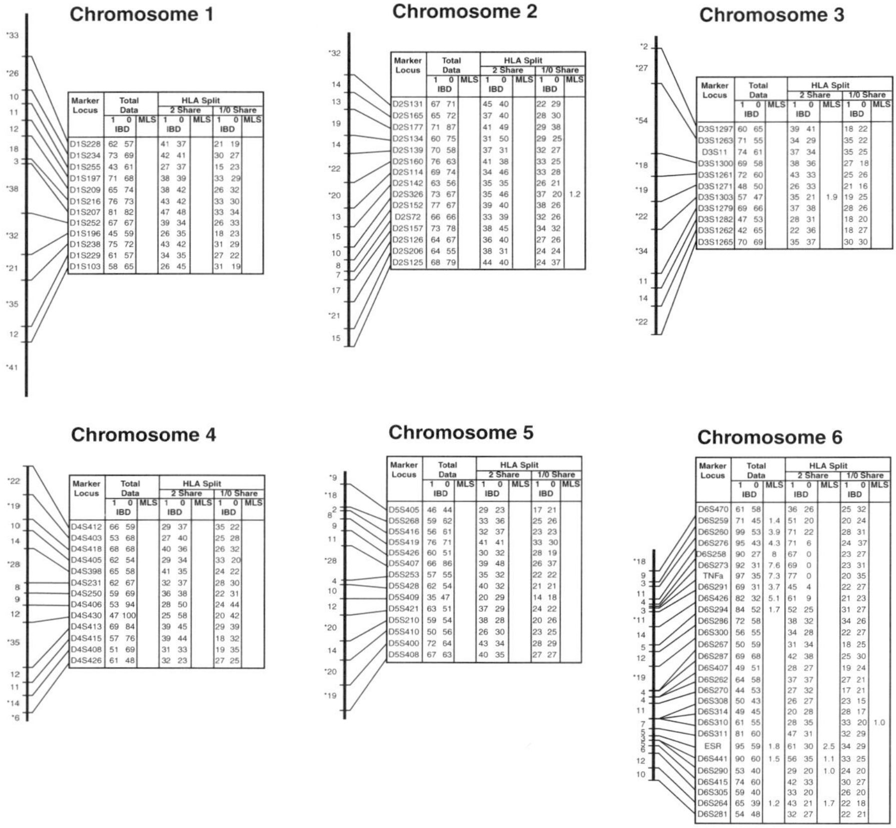

Chromosome 1

Marker Locus	Total Data 1 IBD	Total Data 0 IBD	Total Data MLS	HLA Split 2 Share 1 IBD	HLA Split 2 Share 0 IBD	HLA Split 2 Share MLS	HLA Split 1/0 Share 1 IBD	HLA Split 1/0 Share 0 IBD	HLA Split 1/0 Share MLS
D1S228	62	57		41	37		21	19	
D1S234	73	69		42	41		30	27	
D1S255	43	61		27	37		15	23	
D1S197	71	68		38	39		33	29	
D1S209	65	74		38	42		26	32	
D1S216	76	73		43	42		33	30	
D1S207	81	82		47	48		33	34	
D1S252	67	67		39	34		26	33	
D1S196	45	59		26	35		18	23	
D1S238	75	72		43	42		31	29	
D1S229	61	57		34	35		27	22	
D1S103	58	65		26	45		31	19	

Chromosome 2

Marker Locus	Total Data 1 IBD	Total Data 0 IBD	Total Data MLS	HLA Split 2 Share 1 IBD	HLA Split 2 Share 0 IBD	HLA Split 2 Share MLS	HLA Split 1/0 Share 1 IBD	HLA Split 1/0 Share 0 IBD	HLA Split 1/0 Share MLS
D2S131	67	71		45	40		22	29	
D2S165	65	72		37	40		28	30	
D2S177	71	87		41	49		29	38	
D2S134	60	75		31	50		29	25	
D2S139	70	58		37	31		32	27	
D2S160	76	63		41	38		33	25	
D2S114	69	74		34	46		33	28	
D2S142	63	56		35	35		26	21	
D2S326	73	67		35	46		37	20	1.2
D2S152	77	67		39	40		38	26	
D2S72	66	66		33	39		32	26	
D2S157	73	78		38	45		34	32	
D2S126	64	67		36	40		27	26	
D2S206	64	55		38	31		24	24	
D2S125	68	79		44	40		24	37	

Chromosome 3

Marker Locus	Total Data 1 IBD	Total Data 0 IBD	Total Data MLS	HLA Split 2 Share 1 IBD	HLA Split 2 Share 0 IBD	HLA Split 2 Share MLS	HLA Split 1/0 Share 1 IBD	HLA Split 1/0 Share 0 IBD	HLA Split 1/0 Share MLS
D3S1297	60	65		39	41		18	22	
D3S1263	71	55		34	29		35	22	
D3S11	74	61		37	34		35	25	
D3S1300	69	58		38	36		27	18	
D3S1261	72	60		43	33		25	26	
D3S1271	48	50		26	33		21	16	
D3S1303	57	47		35	21	1.9	19	25	
D3S1279	69	66		37	38		28	26	
D3S1282	47	53		28	31		18	20	
D3S1262	42	65		22	36		18	27	
D3S1265	70	69		35	37		30	30	

Chromosome 4

Marker Locus	Total Data 1 IBD	Total Data 0 IBD	Total Data MLS	HLA Split 2 Share 1 IBD	HLA Split 2 Share 0 IBD	HLA Split 2 Share MLS	HLA Split 1/0 Share 1 IBD	HLA Split 1/0 Share 0 IBD	HLA Split 1/0 Share MLS
D4S412	66	59		29	37		35	22	
D4S403	53	68		27	40		25	28	
D4S418	68	68		40	36		26	32	
D4S405	62	54		29	34		33	20	
D4S398	65	58		41	35		24	22	
D4S231	62	67		32	37		28	30	
D4S250	59	69		36	38		22	31	
D4S406	53	94		28	50		24	44	
D4S430	47	100		25	58		20	42	
D4S413	69	84		39	45		29	39	
D4S415	57	76		39	44		18	32	
D4S408	51	69		31	33		19	35	
D4S426	61	48		32	23		27	25	

Chromosome 5

Marker Locus	Total Data 1 IBD	Total Data 0 IBD	Total Data MLS	HLA Split 2 Share 1 IBD	HLA Split 2 Share 0 IBD	HLA Split 2 Share MLS	HLA Split 1/0 Share 1 IBD	HLA Split 1/0 Share 0 IBD	HLA Split 1/0 Share MLS
D5S405	46	44		29	23		17	21	
D5S268	59	62		33	36		25	26	
D5S416	56	61		32	37		23	23	
D5S419	76	71		41	41		33	30	
D5S426	60	51		30	32		28	19	
D5S407	66	86		39	48		26	37	
D5S253	57	55		35	32		22	22	
D5S428	62	54		40	32		21	21	
D5S409	35	47		20	29		14	18	
D5S421	63	51		37	29		24	22	
D5S210	59	54		38	28		20	26	
D5S410	50	56		26	30		23	25	
D5S400	72	64		43	34		28	29	
D5S408	67	63		40	35		27	27	

Chromosome 6

Marker Locus	Total Data 1 IBD	Total Data 0 IBD	Total Data MLS	HLA Split 2 Share 1 IBD	HLA Split 2 Share 0 IBD	HLA Split 2 Share MLS	HLA Split 1/0 Share 1 IBD	HLA Split 1/0 Share 0 IBD	HLA Split 1/0 Share MLS
D6S470	61	58		36	26		25	32	
D6S259	71	45	1.4	51	20		20	24	
D6S260	99	53	3.9	71	22		28	31	
D6S276	95	43	4.3	71	6		24	37	
D6S258	90	27	8	67	0		23	27	
D6S273	92	31	7.6	69	0		23	31	
TNFa	97	35	7.3	77	0		20	35	
D6S291	69	31	3.7	45	4		22	27	
D6S426	82	32	5.1	61	9		21	23	
D6S294	84	52	1.7	52	25		31	27	
D6S286	72	58		38	32		34	26	
D6S300	56	55		34	28		22	27	
D6S267	50	59		31	34		18	25	
D6S287	69	68		42	38		25	30	
D6S407	49	51		28	27		19	24	
D6S262	64	58		37	37		27	21	
D6S270	44	53		27	32		17	21	
D6S308	50	43		26	27		23	15	
D6S314	49	45		20	28		28	17	
D6S310	61	55		28	35		33	20	1.0
D6S311	81	60		47	31		32	29	
ESR	95	59	1.8	61	30	2.5	34	29	
D6S441	90	60	1.5	56	35	1.1	33	25	
D6S290	53	40		29	20	1.0	24	20	
D6S415	74	60		42	33		30	27	
D6S305	59	40		33	20		26	20	
D6S264	65	39	1.2	43	21	1.7	22	18	
D6S281	54	48		32	27		22	21	

Fig. 7.6. Genetic map of a human genome wide scan for susceptibility genes for type I diabetes. The map was created using genotyping data from 96 affected sib-pair families and fluorescence-based chromosome-specific microsatellite sets. Interlocus distances are shown as gender-averaged, percentage recombination fractions (except for fractions <0.01) when the evidence for linkage, by two-point analysis using the computer program LINKAGE, exceeded a lod score of 3. Data for sharing 1 or 0 alleles, identical by descent (IBD), summed over maternal and paternal meioses are shown at each microsatellite locus. Families were subdivided according to their IBD status at the HLA marker (sib-pairs sharing two alleles at HLA IBD: HLA split 2 share; and sib-pairs sharing 1 or 0 alleles at HLA IBD:HLA split 1/0 share). Maximum lod score (MLS) values were based on the 2,1,0 sharing of alleles IBD at each marker and are shown for values >1.0 ($P < 0.05$). Families in which the IBD status could not be determined unambiguously also contributed to the MLS values. (Reprinted with permission from Davies JL *et al* (1994) A genome-wide search for human type 1 diabetes susceptibility genes. *Nature* **371**: 130–136. Copyright 1994 American Association for the Advancement of Science.)

of number of alleles shared) is useful to run crosschecks on the results. For example, some of the largest MLS scores are on chromosome 6, marker loci D6S273 and TNF-α, with MLS scores of 7.6 and 7.3. At these loci, all of the sib-pairs that shared two HLA alleles have one allele shared IBD. For the dataset taken as a whole, roughly three times the number of sib-pairs that did not share an allele IBD shared one allele. Contrast these results with other markers on the same chromosome (D6S270 or D6S308, for example) or on other chromosomes where the counts are roughly equal. Having all the data arranged on the map allows evaluation of the possibility of chance significant MLS scores. If an MLS score exceeded the threshold of 1.0, then the number of sib-pairs sharing alleles was checked. If the number of sib-pairs sharing zero alleles exceeded the number sharing any number of alleles, then the MLS score was viewed as a random event.

Planning a linkage analysis involves all of the steps in the statistical framework: stating the research questions, analyzing the variables that are part of the study and estimating variances, making sample size estimates and planning the analysis. In stating the research question, it is important to understand what is known about the disease: population and case control studies, genetic mode of inheritance and biochemistry. Information on both phenotype and genotype are needed to obtain estimates of variability and to characterize the variables used in the study. For example, does severity of the disease or a laboratory result vary depending on genotype? If so, what are the differences in mean and standard deviation? Sample size estimates are calculated for the study using software or tables: for example, Table 7.4 estimates the number of recombination events required to detect linkage when given the recombination fraction and the desired statistical power, assuming that the phase is known. Many factors contribute to the estimate of sample size. If phase is unknown, then sample size must be increased. If the markers are not fully informative for linkage, then the number of families needed for the study is increased by multiplying by the inverse of the marker's heterozygosity. Finally, incomplete penetrance requires adjustment of sample size. If a sample of familial data has already been obtained and analyzed, computer software can be used to estimate sample size from this initial sample (Ott 1991).

Table 7.4. Sample size estimates for linkage analysis studies: number of known recombination events required to find $Z_{max} > Z_0$ with power Ø, given the true recombination fraction r

Z_0	Ø	True recombination fraction r					
		0.01	0.05	0.10	0.20	0.30	0.40
2	0.80	11	14	20	38	91	373
	0.90	11	16	23	46	111	461
3	0.80	16	21	28	54	126	516
	0.90	16	22	31	62	150	618

Given a candidate gene, one specific research question is how large must samples be in order to detect or disprove linkage. Candidate genes can result from a number of different hypotheses: for example, genomic maps may suggest a possible disease locus. The question then becomes how many non-recombinants must be observed to support linkage or how many samples must be examined in order to find a recombinant that rejects the hypothesis of linkage. If you assume that the true recombination fraction is 0.5, then Table 7.4 gives estimates of the number of potential recombination events (meioses) that must be observed to detect linkage. This number depends on the recombination fraction: larger samples sizes are needed for small recombination fractions. To support the hypothesis of a candidate disease gene requires a confidence interval for the recombination fraction (θ). If only non-recombinants are observed, the estimate of the recombination fraction is zero. Ott (1991) gives an example of calculating a confidence interval for the recombination fraction. The lower limit of the interval is zero, corresponding to the condition of no recombination. The upper limit of the interval corresponds to:

$$\theta = 1 - \alpha^{(1/n)} \qquad \ldots (7.12)$$

where n is the number of non-recombinants required for a significance level of α and a known recombination fraction. This equation can be solved for n:

$$n = \frac{\log(\alpha)}{\log(1 - \theta)} \qquad \ldots (7.13)$$

If the recombination fraction is small, then the number of recombination events that must be examined can be large, e.g. $n = 298$ for a significance level of 95% ($\alpha = 0.05$) and a true recombination of $\theta = 0.01$.

The inverse problem is finding a recombination event or, more likely, several recombination events to disprove linkage. If a lod score of less than -2 is assumed to be the threshold for absence of linkage, the number of recombination events that must be examined to support this level is (Ott 1991):

$$n \geq \frac{-4}{\log[4\theta(1 - \theta)]} \qquad \ldots (7.14)$$

This equation shows that, in general, few recombination events must be examined to reject a candidate gene in the absence of linkage. For this problem, larger values of θ will require a larger sample size. For the case of $\theta = 0.01$, a sample size of three is sufficient to give an expected lod score of -2, whereas a value of $\theta = 0.1$ requires a sample size of nine.

Because linkage analysis is based on concepts involving distances between loci (markers and disease genes), the search for a region containing a disease gene is narrowed if candidate genes are rejected. Support for a disease gene being on a specific chromosome can be reduced, depending on the spacing of the markers and the number of genes excluded. This concept is termed exclusion mapping.

There are examples of the use of exclusion mapping to find the location of disease genes, Marfan's syndrome (Kainulainen *et al* 1990) being one example. Although the candidate gene approach and exclusion mapping are sound methods theoretically and have demonstrated utility in several diseases, the complexity of many diseases (polygenetic basis, interaction of environmental factors and reduced penetrance) often makes their use in clinical genetics difficult.

7.4 ASSOCIATION ANALYSIS

In addition to linkage analysis, an important statistical tool for genetics studies is association analysis (Fig. 7.7). Association analysis does not require familial data. Instead, DNA samples and phenotype information are collected from large (thousands) samples of the population at random. Polymorphism in specific genes is correlated with the presence of disease. Association genetics builds on the capability of rapid genotyping of patients. If a subset of the population is identified as being susceptible to the disease, other members of the subset population can be tested to give further support to the hypothesis of an association between the gene or genes and the disease. Unlike linkage analysis, association analysis does assume a linkage between the disease and a gene or genes at the start of the analysis. Polymorphism of the gene is essential for association analysis: the

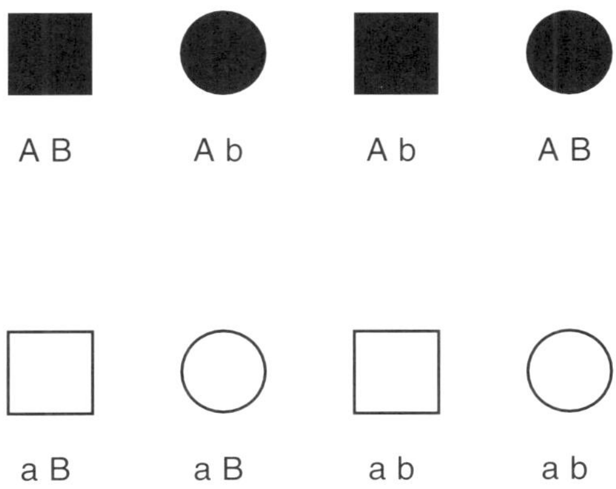

Fig. 7.7. Illustration of the concept of association analysis. In association analysis, correlations between phenotype and genotype are obtained from samples of a population. Polymorphism in specific genes is correlated with presence of the disease. Unlike linkage analysis, association analysis does assume a linkage between the disease and a gene or genes at the start of the analysis. In this figure, squares indicate males and circles indicate females. Phenotypes associated with the disease are indicated by a shaded symbol. The genotypes are recorded under each symbol. By testing the relationships between genotype and phenotype, support is given to the hypothesis that a candidate gene is in fact associated with increased susceptibility to a disease

gene must be polymorphic in order to associate the different marker alleles with expression of the disease. Association analysis is also appropriate for studies involving proteomics. As the different proteins coded by the genome are identified, association analysis is used to assess the strength of association between the proteins and disease.

Association analysis has several different forms: some are based on genetic models of inheritance and others not assume any genetic model. These differences correspond to differences between population-based and family-based tests. The model-free (population) methods are similar to other case-control designs and are analyzed using categorical data analysis. Genotypes are determined from random samples of patients having a disease and another random sample of control patients that do not have the disease. The data are tabulated in a familiar contingency table. An association between the genotype (marker allele) and disease is estimated using a chi-squared test. Relative risk in the population can be determined by drawing the control sample from the population at large instead of selecting patients without the disease.

As with any case-control study, associations can be due to any factor that segregates the samples and all of the caveats mentioned earlier apply to these studies as well. A variation on the method, termed the haplotype relative risk method (Elston 1995), examines the number of marker alleles passed from parents to offspring with and without the disease. The key to understanding this method is that it tests whether the marker allele and disease are transmitted from heterozygous parents to their offspring. If a sample of n people with a disease and their parents ($2n$) are typed for a marker, then data from the parents can be organized into a contingency table (Table 7.5). In this table, A indicates that the marker allele is transmitted to the children having the disease and A' indicates that the marker allele is not transmitted. Entries are made for both parents into the table. If a child and both parents have the genotype AA', it is not possible to identify which parent transmitted which allele so one parent is assigned to each of the AA' cells. The data are then reorganized into a table of the form of Table 7.6. Because each parent is entered twice into the reorganized table, the chi-squared

Table 7.5. Contingency table for association analysis based on the haplotype relative risk method: distribution of $2n$ parents according to the marker allele transmitted, and the marker allele not transmitted, to an affected offspring

Transmitted allele	Non-transmited allele		
	A	A'	Total
A	a	b	a + b
A'	c	d	c + d
	a + c	b + d	$2n$ = a + b + c + d

Reprinted with permission from Elston RC (1995) Linkage and association to genetic markers. *Exp Clin Immunogenet* **12**: 129–140. Copyright 1995 S. Karger AG.

Table 7.6. Reorganization of the contingency table (Table 7.5) for association analysis, to delineate transmission of the marker alleles to the offspring

Marker Allele	Transmitted	Non-transmitted	Total
A	a + b	a + c	2a + b + c
A'	c + d	b + d	b + c + 2d
Total	$2n$	$2n$	$4n$

statistic cannot be used. Instead, McNemar's test is used, based on the results from Table 7.5 and calculating the fraction:

$$Z = \frac{(b-c)^2}{(b+c)} \qquad \ldots(7.15)$$

This test does not assume independence of the transmitted and non-transmitted alleles. The test also only utilizes information from the heterozygous groups (AA') and does not involve a control group (Elston 1995). The odds ratio is estimated by:

$$\text{Odds ratio} = \frac{(a+b)(b+d)}{(a+c)(c+d)} \qquad \ldots(7.16)$$

If a random sample from the population is used to obtain the counts of non-transmitted alleles, then the odds ratio is termed the haplotype relative risk (Elston 1995). A potential difficulty exists with the AA' heterozygous parents. If these parents transmit more A alleles than A' alleles, regardless of whether their offspring have the disease, the haplotype relative risk ratio will be >1. Elston (1995) suggests that the marker should be tested for Mendelian segregation, or another table like Table 7.5 can be derived for a random sample of children and used to make a four-way table and tested for association.

The advantage to the association analyses described in the previous sections is that they do not assume any genetic mode of inheritance. The disadvantage is that associations can be due to many non-genetic factors: risk factors, environmental stratification, chance or non-Mendelian marker segregation. Association can also be the result of a tight linkage relationship between the marker and disease loci. Under the conditions of Hardy–Weinberg equilibrium, the alleles of the loci are independently distributed, assuming a recombination fraction of greater than zero. However, if the population has not reached equilibrium, the alleles of the linked loci can be correlated and they show an association in the population at large, analogous to the discussion about admixture in the linkage analysis discussion.

Genetic models can be constructed for association analysis. One simple model is a single susceptibility gene with full penetrance. The disease allele must be in linkage disequilibrium with the marker allele. Gene frequencies are tabulated for all of the alleles and the linkage disequilibrium equation (eq. 7.10) is used to determine whether the frequency of the haplotype associated with disease is greater than would be expected under equilibrium. The relevant statistical test is whether the disequilibrium parameter is significantly different from zero. The model assumes that the transmitted and non-transmitted alleles are independent if there is no linkage disequilibrium. McNemar's test is adapted for this model to:

$$Z = \delta(1 - 2\theta) \qquad \ldots(7.17)$$

where δ is the disequilibrium parameter and θ is the recombination fraction. This test is compared to zero and is considered a test of association (null hypothesis is $\delta = 0$) if θ is unequal to 0.5 or a test of linkage (null hypothesis is $\theta = 0.5$) if δ is unequal to zero (Elston 1995).

Elston (1995) points out that if there is any doubt about the presence of confounding risk factors then the model-based test should not be used unless the data suggest that there is no association between transmitted and non-transmitted alleles. The principal disadvantage of this model is the limitation of a single susceptibility gene.

Understanding the sampling conditions and assumptions about the data is critical to choosing model-free or model-based methods for either association or linkage analysis. Some investigators prefer to use the model-free methods for exploratory data analysis when little is known about the underlying genetics or the biochemical basis of a disease. Model-based methods, i.e. most parametric methods (Chapter 6), are used when the investigator wants maximal power to detect linkage or association and there is sufficient information about the genetic mechanism underlying the disease.

Both linkage analysis and association analysis are used in human genetics to formulate and test hypotheses about the relationship between genes and disease. However, identification of susceptibility genes is complicated by the fact that many diseases are polygenetic or associated with changes in multiple genes or environmental factors in addition to genetic changes. Some diseases have identical clinical characteristics but are caused by different combinations of susceptibility genes and environmental factors. In particular, some diseases are characterized by weak to moderate associations with mutations in several genes. Detecting weak associations with some degree of confidence requires a larger sample size.

There is considerable debate in the literature about whether linkage or association studies are better suited for identification of candidate disease genes from the human genome project. One major consideration in the debate is the statistical power of the two methods. Risch and Merikangas (1996) contend that association analysis has far greater power to detect genes with weak disease association and that, in the future, genetic studies of complex disease are likely to require large-scale testing by association analysis. Their analysis is especially important to consider in that the availability of the human genome will allow large-scale screening for candidate genes. Table 7.7 shows the number of families required to identify a disease gene for different types of analyses. The table compares a linkage analysis based on sibling pairs with an association analysis that compares

Table 7.7. Number of families required for identification of a disease gene for linkage and association studies

		Linkage			Association			
					Singletons		Sib-pairs	
Genotypic risk ratio (γ)	Frequency of disease allele A (p)	Probability of allele sharing (Y)	No. of families required (N)	Probability of transmitting disease allele A $P(\text{tr} - \text{A})$	Proportion of heterozygous parents (Het)	N	Het	N
4.0	0.01	0.520	4260	0.800	0.048	1098	0.112	235
	0.10	0.597	185	0.800	0.346	150	0.537	48
	0.50	0.576	297	0.800	0.500	103	0.424	61
	0.80	0.529	2013	0.800	0.235	222	0.163	161
2.0	0.01	0.502	296 710	0.667	0.029	5823	0.043	1970
	0.10	0.518	5382	0.667	0.245	695	0.323	264
	0.50	0.526	2498	0.667	0.500	340	0.474	180
	0.80	0.512	11 917	0.667	0.267	640	0.217	394
1.5	0.01	0.501	4 620 807	0.600	0.025	19 320	0.031	7776
	0.10	0.505	67 816	0.600	0.197	2218	0.253	941
	0.50	0.510	17 997	0.600	0.500	949	0.490	484
	0.80	0.505	67 816	0.600	0.286	1663	0.253	941

Reprinted with permission from Risch N and Merikangas K (1996) The future of genetic studies of complex human disease. *Science* **273**: 1516–1517. Copyright 1996 American Association for the Advancement of Science.

the genotypes for affected offspring and their parents. The table also allows comparison of these methods for a variety of genetic study problems characterized by different risks and disease frequency. In Table 7.7, genotypic risk ratio is the increased chance that an individual with a specific genotype has the disease. The table is derived using the assumption that the disease locus has two alleles, A and a. The disease allele is designated A. The possible genotypes are then AA, Aa and aa. The genotypic risk ratio (γ) is the increased risk that a heterozygous individual (Aa) has for developing the disease over the homozygous individual with a genotype aa (genotype of no disease alleles). The alleles interact in a multiplicative sense, so that the genotypic risk ratio for a homozygous individual with a genotype AA is γ^2. In other words, as the genotypic risk ratio decreases, there is less of a chance that an individual is at risk for the disease and there is a correspondingly weaker association between the disease and the gene. The frequency of the disease in the population is recorded in Table 7.7 over a wide range, from very infrequent (0.01) to frequent (0.80). All of the sample size estimates are derived assuming 80% power.

To make the sample size estimate for linkage analysis, a sib-pair method is used that calculates the probability that alleles are shared between siblings, assuming a closely linked marker locus (recombination fraction = 0) and a fully informative marker locus (heterozygosity = 1). If there were no linkage between disease and marker locus, random chance sharing of alleles would yield a probability of allele sharing of 0.5. From inspection of Table 7.7, it is seen that there are very small deviations from 0.5 for most of the combinations of genotypic risk ratio and disease frequency, particularly for the low-risk, low-frequency cases. When the number of families required to identify linkage is estimated, it is seen that the sample size is prohibitively large (greater than 60 000) for many cases.

For the association analysis, the transmission/disequilibrium test of Spielman *et al* (1993) is used in the calculation. This test looks at the frequency at which a specific allele is passed from heterozygous parents to affected offspring and therefore requires both affected children and parents. This frequency is compared to the frequency of alleles not transmitted to affected children. As with the linkage analysis, if this is simply by random chance, one would expect a 50% chance that any allele is passed from parents to offspring. If the allele is associated with the risk of a disease, there will be an increase of >50%. Risch and Merikangas tabulate the proportion of heterozygous parents and the probability that they will transmit a disease allele to offspring that will be affected by the disease. If data is available for affected and unaffected siblings, a more powerful test can be constructed to compare the allele frequencies between affected and unaffected siblings (Spielman, 1998). Again, the relevant test is whether the distribution of alleles is the same between these groups. Comparing the probability of transmitting the disease allele (association) with the probability of sharing of the allele between siblings (linkage) shows a large difference between the methods. The difference between random events is much larger for the association methods, particularly for those cases where there is a weaker signal (low genotypic risk). Examination of the sample size column shows that the association studies have samples sizes that are two orders of magnitude smaller: generally hundreds to several patients. The sib-pair method requires the smallest samples of all, particularly for cases where the genotypic risk is 4.0 or greater.

Although Table 7.7 shows the relative power of the association studies, it is important to account for multiple comparisons as with any statistical study. With the large number of genes that can be tested from the human genome project, the significance level for declaring a candidate gene can be very small ($<5 \times 10^{-8}$ for 1 million independent association tests). It is also important to recognize that one type of analysis does not preclude the other. If phenotype and genotype are recorded for patients, then association analyses can be run. If the familial relationships are known, then the full array of linkage and association analyses can be completed.

The underlying genetics and biochemical pathophysiology are investigated after identification of a susceptibility gene. For model-based methods, a hypothesis about the genetic mechanism has already been tested in order to complete the analysis. For model-free association analysis, the genetic mechanism is not specified. Although the statistical power of association analysis for testing many different candidate genes is clear, these analyses cannot distinguish between different genetic mechanisms (Hodge 1994). Implicit in many uses of genetic markers in disease association studies is the idea that the marker is located in close proximity and linked to the disease gene but is not the disease gene itself. The disease and marker loci are separate; the marker alleles are genotyped and associated with presence or absence of the disease. The disease allele may or may not be necessary for expression of the disease. Hodge (1994) describes this association model as the major disease gene or linkage disequilibrium model. Further work to genotype the distance between the marker and disease gene theoretically will allow investigators to identify the disease gene itself.

An alternative model is the susceptibility gene model (Hodge 1994). In this model, it is possible for a single locus to be associated with both the marker and disease allele. To build this model, Greenberg (1993) develops the concept of a necessary disease locus and a susceptibility locus. The necessary disease locus is one where the disease allele is necessary for expression of the disease. All people having the disease must have the disease allele in this model. Susceptibility, on the other hand, is associated with an increased risk for the disease. Other environmental and genetic factors determine whether a patient develops the disease. Environmental factors include other risk factors, viral or bacterial infection and toxins. Genetic factors include changing the penetrance for the necessary disease locus (Greenberg 1993). For an association test, which looks at correlations between specific alleles and presence or absence of the disease, similar results are likely to be obtained. However, for linkage analysis, the model of a susceptibility locus can make it difficult to detect linkage.

By simulating family data, Greenberg (1993) examined different interpretations of data from linkage and association analyses for the model of a susceptibility locus. If the estimated risk of a disease given an associated allele is designated α and the risk of a disease given an associated allele that is not present is designated β, Greenberg's results suggest that if α/β (relative risk) is 10 or less then it would be difficult to detect linkage and the lod scores would be negative for many values of the recombination fraction (Greenberg 1993). The framework for interpreting the results is based on the definitions of association and linkage. Association indicates that an allele, without regard to whether it is from a necessary or susceptibility locus, occurs more frequently in subjects having a disease than in the general population. Linkage analysis is an indication of whether the locus under study is close to a disease locus. Greenberg (1993) points out that linkage analysis is insensitive to the effects of individual alleles, but puts weight on the movement of alleles from generation to generation. Going back to the definition of relative risk, if α is low and β is zero (in other words, low penetrance), association analysis should still be able to detect association between the allele and the disease but linkage analysis will not indicate linkage. Many investigators believe that association analysis is a better statistical tool for cases where penetrance is low. However, association analysis may not be able to show the patterns of inheritance operating with the specific alleles.

The genetic susceptibility model is derived for cases where there are weak associations between genes and diseases. It is important to remember that the basis for this model is increased susceptibility to the disease, and not all patients having the susceptibility allele develop the disease. If every individual who had a specific allele has the disease, then this model would be inappropriate. When environmental factors are involved, this model is very useful: susceptibility is often a function of age, for example. One interpretation of this model is that the allele of the susceptibility gene has a weak effect on the pathophysiology of disease development. This is in contrast to the major disease model where the weakness of the association is explained by linkage disequilibrium or separation of the disease and marker loci. Within a family, if the major disease model is in operation, individuals having both the disease and marker allele for the disease will co-segregate or group together. If the susceptibility model was appropriate, the individuals would not necessarily co-segregate (Hodge 1994). The two models can be viewed as a continuum and in fact as submodels of a more general two-locus model (Hodge 1981; Hodge and Spence 1981). From this perspective, association analysis is a powerful filter for identifying interesting candidate genes, which are then investigated by additional studies, both genetic (further linkage and association studies) and genomic (finer map around the loci).

7.5 GENE EXPRESSION ANALYSIS

Human genetic studies provide information about candidate disease and susceptibility genes and the relationships between factors that increase the risk for disease. In order to develop drugs, it is necessary to understand the biochemical processes under control of genes in order to identify targets for small molecules. Knowing the sequence of a gene is only the first step in understanding disease etiology. Interactions between the proteins encoded by genes, the interactions of these proteins with endogenous hormones and receptors and interactions with the environment are necessary to understand the development of complex diseases. Genetic regulation is a complex control process: groups of genes are involved both directly and indirectly in many different signaling processes. The tools of genomics are used to map genes to their products (usually proteins) and to sequence and analyze data about the distribution of gene products. Generally, messenger RNA (mRNA) is used to analyze gene expression. By analyzing mRNA from different organs, tissues and cell types, the distribution of the products of a set of genes can be quantified.

There are several different methods available for gene expression analysis. The biochemistry, technology and experimental protocols are different between the approaches. One objective of the methods is to allow comparison of gene expression profiles. These profiles are obtained from diseased and normal tissues or from a single patient over the timecourse of therapy. Methods include hybridization assays using arrays of cDNA clones (DeRisi *et al* 1996; Schena *et al* 1995, 1996) or arrays of oligonucleotides (Lockhart *et al* 1996), serial analysis of gene expression (Velculescu *et al* 1995) and differential display (Laing and Pardee 1992). This section describes two of the methods: chip-based microarrays of cDNA clones and serial analysis of gene expression (SAGE).

The principles behind glass chip-based gene expression assays are shown in Fig. 7.8 (Schena 1996). Total mRNA is isolated from a tissue or other sample. Often, comparisons are made between diseased and normal tissues. For patients undergoing chronic treatment with drugs, blood samples provide the RNA to differentiate response to the treatment. Reverse transcription in the presence of fluorescent nucleotides is done to label the mRNA for detection. The resulting mixture, also known as a probe, is then hybridized to cDNA, which is linked to a glass slide, washed at high stringency and scanned for the fluorescence signal that will be proportional to the original expression level. The power of this approach is in part due to the extremely high densities that are possible on microarrays. The microarrays are printed on a glass surface using robotic methods to link the cDNA to glass. Array densities of greater than 1000 cDNAs/cm^2 make it possible to test against large portions of a genome; a 10 cm^2 array would contain cDNA for 10 000 different genes. The source of the cDNA can be fully sequenced clones, randomly chosen cDNA or expressed sequence tags. The probe is either a complex mixture of cDNA fragments generated from the mRNA or genomic DNA. Biological samples can be compared in the same assay using two-color fluorescence detection (Schena 1996). Using two-color detection, diseased and normal tissue can be compared without separate hybridization steps and image subtraction. The cDNA on the chip can be arranged to support kinetic studies

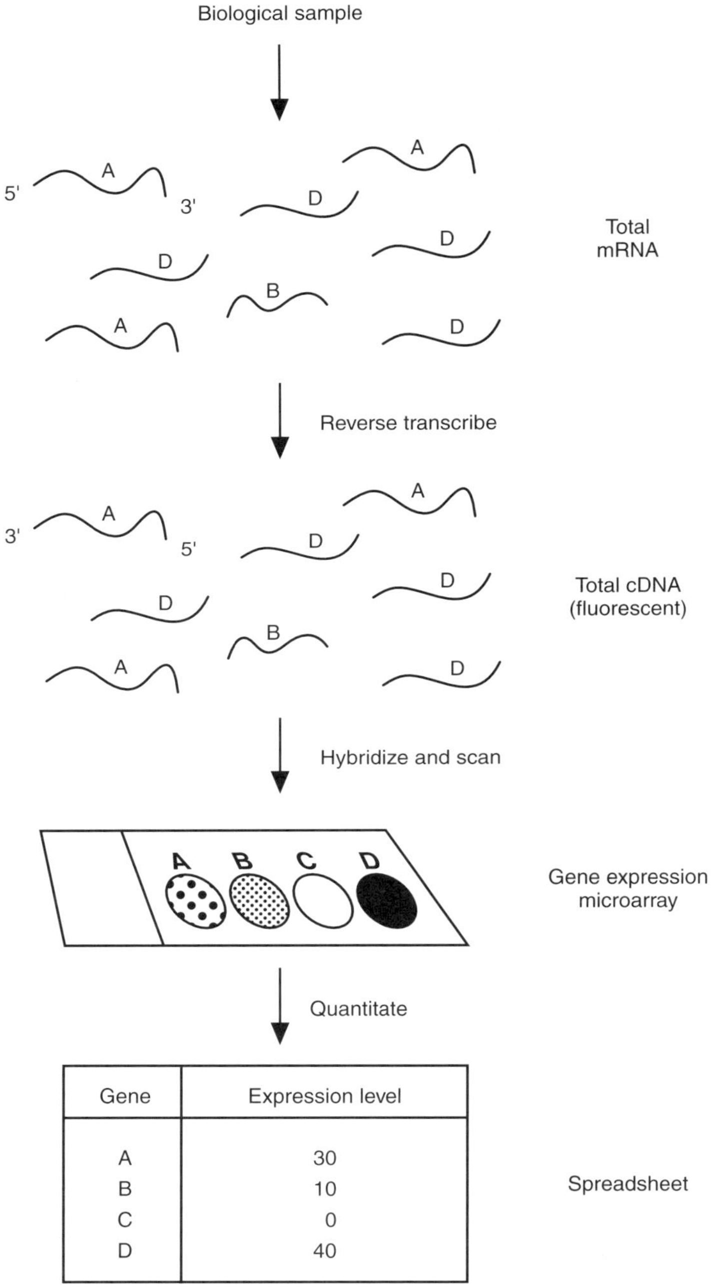
Biological sample
5'
3'
A
D
B
Total
mRNA
Reverse transcribe
3'
5'
Total cDNA
(fluorescent)
Hybridize and scan
A B C D
Gene expression
microarray
Quantitate
Gene
Expression level
A 30
B 10
C 0
D 40
Spreadsheet

to show the expression data over multiple time points. The volume of data from kinetic gene expression experiments is large: millions of observations are taken from tens of thousands of genes. Image analysis software is used to handle the image processing operations, quantify results and compare images. Interpretation of the data can be as simple as point-by-point comparisons of expression in normal and diseased tissues, or as complex as mathematical models for clustering and detection of interactions (Chapter 9).

Serial analysis of gene expression (SAGE) involves the use of short nucleotide sequence tags to identify gene transcripts (Velculescu *et al* 1995). The sequence tags are concatenated to allow analysis of the transcripts in a sequential manner. As with the chip-based methods, SAGE allows the analysis of expression patterns of thousands of genes and originates with mRNA. The short sequence tags allow identification of many different transcripts: 9–10 base pairs can distinguish over 260 000 different transcripts. In the SAGE method (Fig. 7.9), cDNA is synthesized from mRNA using a biotinylated oligo primer. The cDNA is cleaved using a restriction endonuclease with a four-base-pair recognition site. It is expected that each cDNA will be cleaved at least once. The cDNA is divided in half and ligated via the restriction site to linkers containing a restriction site for the tags. Cleavage of the ligated products produces linker with a short piece of the cDNA. The tags are easily amplified by the polymerase chain reaction (PCR). In an example by Velculescu (1995), short diagnostic sequence tags were isolated from mRNA from the human pancreas and concatenated. The tags were cloned into a plasmid vector and sequenced manually. Sequencing of the tags revealed a gene expression pattern that is characteristic of pancreatic function by comparison to GenBank. A pancreatic cDNA library was screened with probes for several enzymes. The abundance of the SAGE tags for the transcripts showed good agreement with results obtained from screening the library using hybridization analysis. The advantages cited for SAGE include the capability to measure accurately the abundance of the transcripts and to identify novel expressed genes. Serial analysis of gene expression does not require specialized robotic equipment, only the ability to run PCR and sequencing.

Gene expression analysis is the first step in understanding genetic control and regulatory networks. Examples include the lysis-lysogeny decision in the

---

Fig. 7.8. Chip-based gene expression assays. This diagram shows the steps to quantitate gene expression on a cDNA microarray. Total messenger RNA (mRNA) is isolated from a biological sample, primed with oligo-dT and fluorescently labeled using a single round of reverse transcription in the presence of fluorescein-12-dCTP. The fluorescent probe is hybridized to a cDNA microarray containing specific hybridization targets, washed at high stringency and scanned for fluorescence emission following laser excitation. Measurement of fluorescence intensity is used to quantitate gene expression. (Reprinted with permission from Schena M (1996) Genome analysis with gene expression microarrays. *BioEssays* **18**: 427–431. Copyright © 1996 Wiley–Liss, Inc., a division of John Wiley and Sons, Inc.)

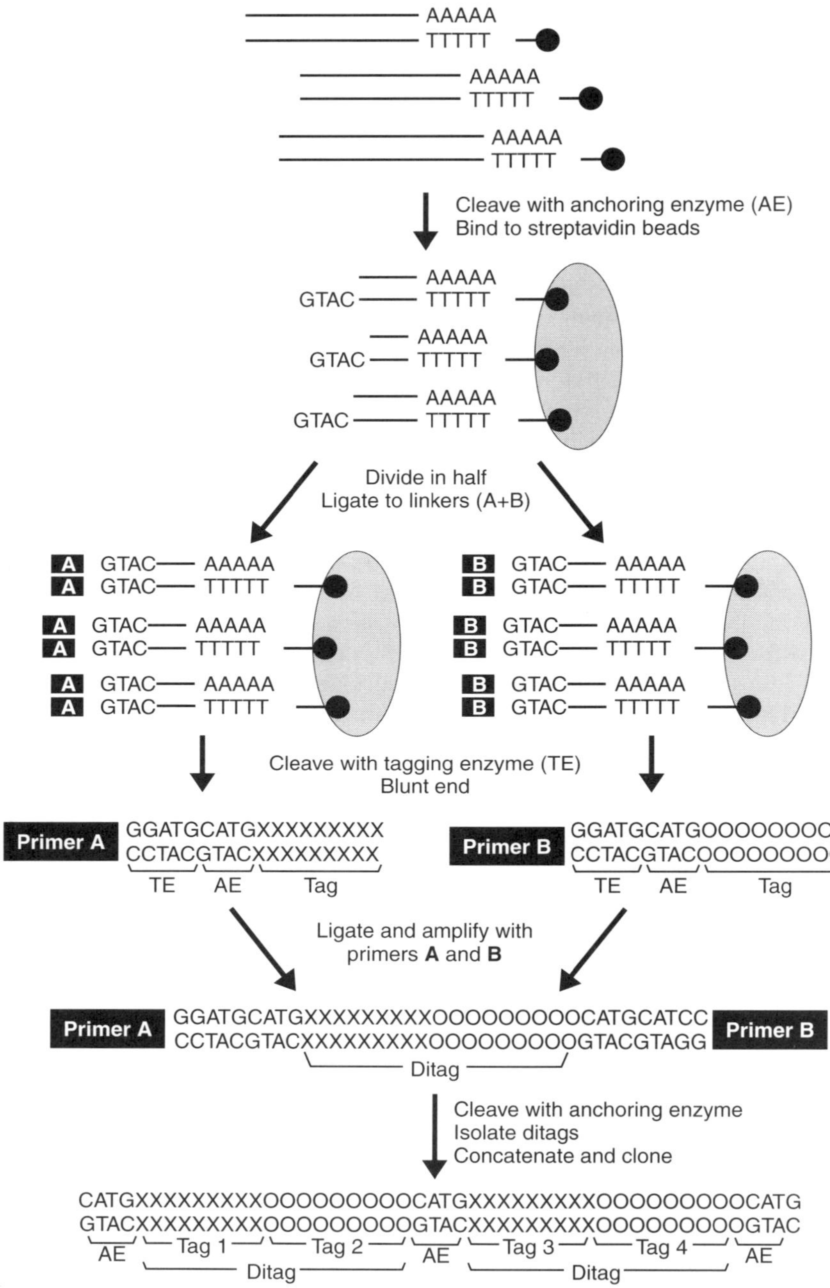
AAAAA
TTTTT
Cleave with anchoring enzyme (AE)
Bind to streptavidin beads
GTAC
Divide in half
Ligate to linkers (A+B)
A
B
Cleave with tagging enzyme (TE)
Blunt end
Primer A
GGATGCATGXXXXXXXXX
CCTACGTACXXXXXXXXX
TE
AE
Tag
Primer B
GGATGCATGOOOOOOOOO
CCTACGTACOOOOOOOOO
Ligate and amplify with
primers A and B
GGATGCATGXXXXXXXXXOOOOOOOOOCATGCATCC
CCTACGTACXXXXXXXXXOOOOOOOOOGTACGTAGG
Ditag
Cleave with anchoring enzyme
Isolate ditags
Concatenate and clone
CATGXXXXXXXXXOOOOOOOOOCATGXXXXXXXXXOOOOOOOOOCATG
GTACXXXXXXXXXOOOOOOOOOGTACXXXXXXXXXOOOOOOOOOGTAC
Tag 1
Tag 2
Tag 3
Tag 4

bacteriophage lambda (McAdams and Shapiro 1995) and the diauxic shift from anaerobic growth to aerobic respiration in yeast (DeRisi *et al* 1997). In the yeast study, monitoring was at the whole genome level. The diauxic shift is a consequence of the depletion of glucose. DeRisi *et al* (1997) developed an effective visual display for correlating changes in gene expression with changes in the metabolic cycle (Fig. 7.10). The yeast genes associated with each step are identified on a biochemical pathway diagram. By shading the diagram, it is easy to see which genes have higher expression levels because of the diauxic shift. There was significantly more expression for genes coding for aldehyde dehydrogenase, pyruvate carboxylase and acetyl-coenzyme A synthase. There is less expression of transcription of genes encoding pyruvate decarboxylase. Biochemically, these changes rechannel pyruvate away from acetaldehyde to oxalactetate, where it supplies the TCA cycle and gluconeogenesis. Other changes in gene expression were observed and correlated to other steps in the cycle. This work points to the power of combining information about gene expression with a known biochemical signaling pathway and is indicative of the type of control system modeling that will identify biochemical targets for new drugs. Using DNA microarrays to monitor differences in transcription provides an approach to analyze regulatory pathways and networks. Changes in genes can serve as markers for the activity of drugs and can be used to screen for drugs that are associated with a specific expression pattern.

7.6 THE HUMAN GENOME PROJECT

The human genome project brings a new perspective to drug discovery. At the most basic level, plans for the project are to finish the complete human genome sequence by the end of 2003. For drug discovery, the project provides information about the relationship between human genetics and disease. The availability of the entire human DNA sequence in addition to the sequences of other genomes (*Escherichia coli*, *Saccharomyas cerevisiae*, *Clostridium elegans*, Drosophila and the mouse) enables different types of studies and experiments both in clinical genetics and in biochemistry/molecular biology. Definition of the term 'sequence-based' biology is evolving as new methods for analysis of data are developed and work proceeds to understand the genetic component of disease and the biochemical function of gene products. This section describes the human genome project with an emphasis on the information that is available for drug discovery.

Fig. 7.9. Serial analysis of gene expression (SAGE). This figure is a schematic of the SAGE method to measure gene expression. See text for detailed explanation of the method. The X and O indicate nucleotides of different tags. (Reprinted with permission from Velculescu VE *et al* (1995) Serial analysis of gene expression. *Science* **270**: 484–487. Copyright 1995 American Association for the Advancement of Science.)

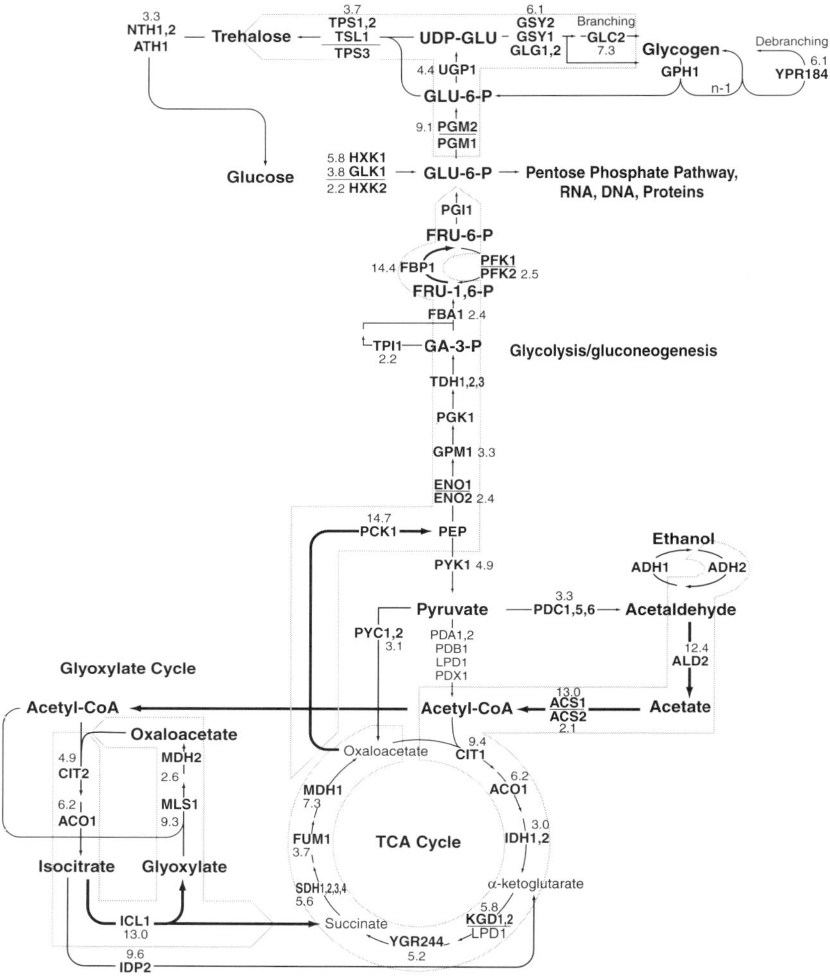

Fig. 7.10. Correlation of gene expression data with the metabolic cycle in yeast. This figure shows part of the yeast metabolic pathway associated with carbon and energy metabolism. Changes in gene expression for the mRNA encoding the enzymes are overlaid on the pathway. Yeast genes encoding the enzymes are identified by the different codes in the boxes. Shading is used to indicate genes where expression levels change in the diauxic shift, either increased or decreased expression. The magnitude of the change is shown using the small numerals adjacent to the boxes. For multimeric enzyme complexes, the number is an unweighted average of all of the genes in the box. (Reprinted with permission from DeRisi JL *et al* (1997) Exploring the metabolic and genetic control of gene expression on a genomic level. *Science* **278**: 680–686. Copyright 1997 American Association for the Advancement of Science.)

The US human genome project is a long-term project that is the joint effort of several government agencies (National Institutes of Health (NIH) and Department of Energy (DOE)) and an international network of collaborators. Planning is done on a five-year basis with open scientific debate about goals and methodology. Coordination of the effort is by the National Advisory Council for Human Genome Research at the NIH and the Biological and Environmental Research Advisory Council at the DOE (Collins *et al* 1998). The web addresses at the end of the chapter give details about the project from these agencies. In addition to issues related to the science and technology of molecular biology and sequencing, there are also substantial issues on integrating human genome data with clinical trials and biochemical research efforts. The genome project is viewed as a resource for the scientific community. Experimental results and databases are published on the Internet for the widest dissemination of information.

The most prominent goal of the human genome project is to finish the entire human genome sequence by the end of 2003. Quality of the data is a specific concern; the standard is an error rate of no more than one base in 10 000. The project plans to have full sequence data on 30% of the human genome (1 Gb) published by the end of 2001 and to have a working draft of 90% of the genome based on mapped clones completed at the same time. The concept of the working draft is a consequence of the sequencing process. In the first phase, random (shotgun) sequences are derived from mapped clones. Because the sequences are random, there are gaps and discrepancies in the data that must be resolved. The second or finishing phase resolves the gaps and discrepancies. The working draft sequence data will not be as complete, e.g. it will contain gaps, but will be useful for finding genes and exons through sequence searches.

For drug discovery, the human genome project offers an unprecedented chance to study human sequence variation and to compare human sequence data to sequences from other species. There are two main areas of sequence variation of interest to drug discovery projects: understanding variation associated with disease, and mapping differential responses to therapy. The capability to compare sequences from different species has ramifications for the use of animal models of disease and the development of high-throughput screening assays.

7.6.1 SINGLE NUCLEOTIDE POLYMORPHISMS

Sequence variation is the basis of human genetic studies, whether linkage or association analysis is used. With the advent of the human genome, one specific type of variation, single nucleotide polymorphism (SNP), is being utilized for genetic studies, particularly for complex polygenetic diseases. Single nucleotide polymorphisms are single base-pair differences that occur every 1000 bases (kb) on average between two haploid genomes. By looking at the polymorphism of these positions over a large number of people, it becomes possible to correlate the changes with the number of people affected by a disease. Single nucleotide polymorphisms are widely distributed across the genome and are well suited for automated analysis, e.g. using DNA chips (Schena 1996). The promise of SNPs is to allow genome-wide association studies to identify susceptibility genes from the population at large (e.g. large-scale association studies). These genes contribute to an increased susceptibility to disease and are likely to be interacting with each other. For complex diseases like cancer, atherosclerosis, schizophrenia and diabetes, this approach shows promise.

Maps of SNPs are also envisaged to predict individual responses to drugs and adverse reactions to drugs. For the human genome project, it is a goal to create a publicly available SNP map of at least 100 000 markers (Collins *et al* 1998). For target identification, SNPs in the coding sequences are of particular interest. Use of SNPs for target identification is not a panacea. They are subject to the same problems of interactions between genes, environmental factors, temporal relationships and complexity as other approaches (Pennisi 1998). Specifically, recombination can occur and decrease the strength of association between a set of SNPs and disease, particularly if it is viewed over time. Information is needed about the patterns of disease and history of the population in order to augment the information from the SNPs. Studies based on SNPs are anticipated for large-scale genome-wide association studies. The sample sizes for these studies will be large. Within a family, shared regions of the genome are often millions of bases long. This makes it possible to conduct linkage studies by working with shared chromosomal regions. A large linkage study requires about 500 000 genotypes probed with a few hundred markers (1000 individual types with 500 markers). Between unrelated individuals, only 10 kb or less may be common. A large-scale association study based on the human genome could require 100 million genotypes derived from 1000 individuals typed for 100 000 markers or SNPs (Kleyn and Vesell 1998).

Other goals and projects are part of the US Human Genome Project. Improvement of sequencing technology, design of linkage and association studies, a better understanding of what populations and diseases are best addressed by each method and addressing legal and ethical concerns are all part of this multidisciplinary effort. Further work to develop databases and bioinformatics software tools for analysis of data are critical components of the project. As more information about gene function is collected, methods are needed to visualize and develop models of interactions between genes and associated proteins. For target identification, the challenge is using information from the genetic studies and associations between DNA polymorphism and phenotype to understand biochemical function in order to elucidate control points that can be modulated by small molecules or gene therapy.

7.7 FROM GENETIC STUDIES TO TARGET IDENTIFICATION

The previous parts of this chapter give the background to understanding genetic studies and a framework for interpreting genetic data in terms of association and linkage. The result is identification of candidate or susceptibility genes and some idea of the significance of the evidence. Results should never be viewed in isolation. Further studies, different methods of analysis and different populations may either support or refute previous studies. Moving from human genetic studies to targets that can be tested against potential drugs is a process that involves many different sources of information and types of analyses (Spence 1998). There is clearly a filtering role in target identification that will become more important as the entire human genome is sequenced. From a collection of 100 000 genes in the human genome, the challenge is to identify the genes that are associated with tractable biochemical targets. The power of genetic studies is that clinical information is used to drive the drug discovery process. Specific disease studies are the first step in a process from candidate gene to understanding gene expression to identification of a biochemical target (enzyme, receptor, ion channel) that will become the focus of testing against potential drugs. Alternatively, the genetic study may be designed to understand differential responses to drugs: why individuals respond differently to drug treatment. The term pharmacogenetics is used to refer to studies of differential pharmacological responses to drugs (Kleyn and Vesell 1998).

Gene expression analysis allows a comparison of the genetic profile of diseased and normal tissues. Development of databases of expressed sequence tags (ESTs) are a further step towards comparison of genes between normal and diseased individuals and between species. These databases are limited to sequences obtained from exons of expressed genes and contain a short, unique DNA sequence that is easily recognized by a PCR assay. Alternatively, linkage maps are constructed to relate polymorphic loci to disease phenotypes (Donis-Keller *et al* 1987). These linkage maps can cover a substantial part of the human genome, some up to 95%. From either of these approaches, elucidation of function is needed before a biochemical target for screening can be identified. The yeast two-hybrid system (Fields and Song 1989; Chien *et al* 1991) is one approach for identifying protein–protein interactions and in turn deducing protein function (Fig. 7.11). In this assay, a bait protein already characterized in terms of function is fused to the DNA binding domain of a yeast protein and is used to screen against the gene products of thousands of genes hybridized to the activation domain of the same yeast protein. If the proteins interact, the association is used to 'back track' to a specific gene. This method is particularly useful for associating genes with proteins in a biochemical signal transduction pathway. These proteins often make good targets for small molecule inhibitors or activators.

A complete evaluation of the physiological and pharmacological profile of a target protein is the result of the target validation process. These evaluations are made in various models of human diseases to identify function at the cellular level and to establish a physiological rationale for a new drug (antagonist or agonist). Prior to this validation, analysis of the function of the target is done using full-length complementary DNA (cDNA) clones and biochemical assays to understand and characterize the target. Spence (1998) describes a 'gene triage'

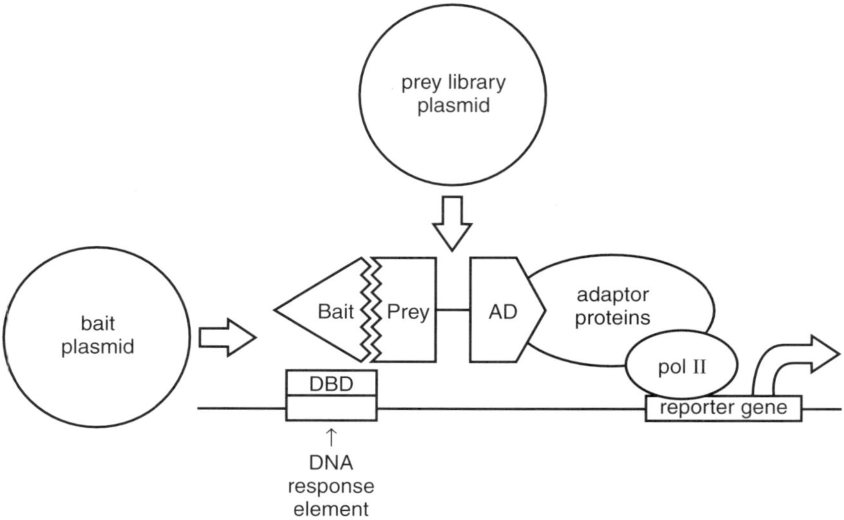

Fig. 7.11. Diagram of the yeast two-hybrid assay system. This system is designed to detect proteins that are capable of reacting with another protein and to identify the cloned gene for the interacting protein. The system is built based on the properties of some yeast proteins that contain separable domains for DNA binding and transcription. Plasmids are constructed to encode two hybrid proteins from yeast: the bait and the prey. The bait protein is a known protein and is fused to the DNA binding domain (DBD) in the DNA response element of the yeast transcriptional activator protein. The gene fusion of the bait protein and the DBD is the first of the two hybrids. This hybrid will not activate transcription of the reporter gene. A second sequence or plasmid (cDNA) library is fused to the activation domain (AD) of the same yeast protein, forming the second hybrid. This second hybrid is known as the prey. The activation domain is linked through a series of adaptor proteins (factors that mediate chromatin remodeling and assembly of the basal transcription machinery) and pol II to the reporter gene. Interaction between the bait and prey proteins will initiate transcriptional activation of the reporter gene, producing a measurable readout. (From Richard Buckholtz, Glaxo Wellcome.)

process based on how the candidate gene was identified (Fig. 7.12). For genes identified through homology cloning, database mining, expression cloning or yeast two-hybrid analysis, a biochemical function can usually be inferred from the sequence of proteins used in the process. Bioinformatics methods are used to compare nucleotide and protein sequences with databases from several species in order to gather more information about function and potential areas for cross-reactivity. The gene expression pattern in organs and tissues of interest and a comparison to the target sequence can give support to the therapeutic validity of the target. Analysis of gene expression patterns from cancerous tissues can be done because RNA can be obtained from biopsies and surgical samples.

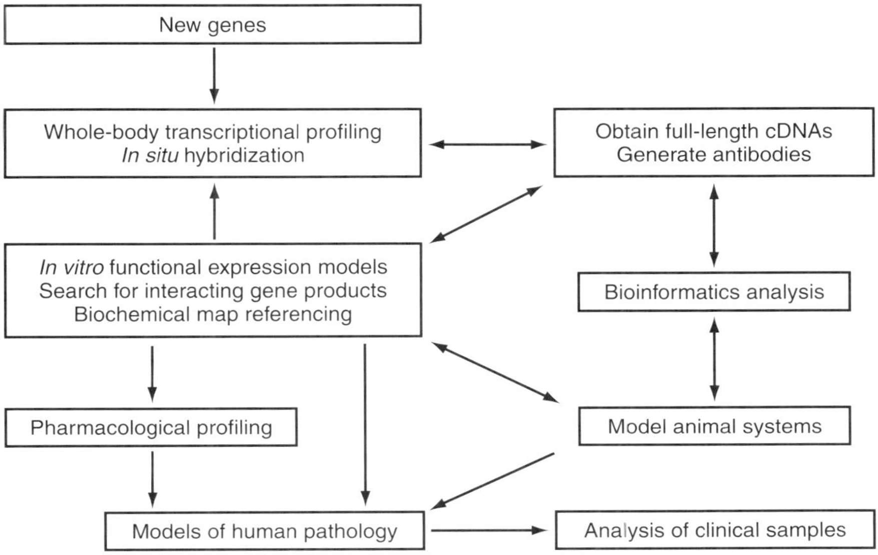

Fig. 7.12. The gene triage process. This diagram shows several places where genetic information is used in the target identification process. As new disease susceptibility genes are identified, a variety of models and analyses arise. *In situ* hybridization is used to profile the gene in different tissues. Bioinformatics analysis is carried and to compare the sequence against other genomes, including human. Pharmacological profiling is done to help construct models of human pathology. (Reprinted from *Drug Discovery Today*, Volume 3, Spence P (1998) Obtaining value from the human genome: a challenge for the pharmaceutical industry, pp 179–188. Copyright 1998, with permission from Elsevier Science.)

In situ hybridization can be performed for many targets of interest in order to determine expression levels. Candidate genes with expression in a tissue of interest would take priority over genes not expressed in the tissue or having more generalized expression. Chemical evaluation of a target is done by screening the target, expressed in a human cell line against compound collections and combinatorial libraries. These collections are generally biased for chemical series that were synthesized for prior drug discovery programs. Based on prior knowledge, compound selection or library design methods are used to identify compounds that are likely to bind to the target and provide more information about therapeutic validity or possibly be a starting point for synthetic design effort.

7.8 SUMMARY

Human genetic studies are an important part of the target identification process and are the basis for understanding differential responses to drugs. The statistical issues involved in designing and interpreting these studies are related to issues in other aspects of drug discovery: sample size estimates, measurement of variability, sampling theory, multiple comparisons, outliers and optimization. Understanding the strength of association between genes and disease, untangling cause and effect relationships, identifying interactions and being aware of the limitations of both genetic studies and the genomic methods used to generate data are critical steps in the target validation process. The principal methods for genetic studies, linkage and association analysis continue to evolve as studies for more complex, polygenetic diseases are published. The capability of correlating polymorphisms to data from unrelated individuals is of particular interest for drug discovery. These studies hold the promise of using information from the human genome project and high-throughput genotyping of specific populations to identify susceptibility genes for clinically significant diseases. With an understanding of the genetic basis of disease, further work in molecular biology and biochemistry identifies proteins, nucleic acids or other biochemical entities that are the functional control points or targets for new drugs.

7.9 REFERENCES

Barbosa J, Rich SS, Dunsworth TS and Swanson JJ (1982) *J Clin Endocrinol Metab* **55**: 93–195.

Botstein D, White RL, Skolnick MH and Davies RW (1980) Construction of a genetic linkage map in man using restriction fragment length polymorphisms. *Am J Hum Genet* **32**: 314–331.

Chien GT, Bartel P, Sternglanz R and Fields S (1991) The two-hybrid system: a method to identify and clone genes for proteins that interact with a protein of interest. *Proc Natl Acad Sci USA* **88**: 9578–9582.

Collins FS, Patrinos A, Jordan E, Chakravarti A, Gesteland R, Walters L and members of the DOE and NIH planning groups (1998) New goals for the US Human Genome Project: 1998–2003. *Science* **282**: 682–689.

Corder EH, Saunders AM, Strittmatter WJ, *et al* (1993) Gene doses of apolipoprotein E type 4 allele and the risk of Alzheimer's disease in late onset families. *Science* **43**: 1467–1472.

Davies KE, Pearson PL, Harper PS, *et al* (1983) Linkage analysis of two cloned sequences flanking the Duchenne muscular dystrophy locus on the short arm of the human X chromosome. *Nucleic Acids Res* **11**: 2303–2312.

Davies JL, Kawaguchi Y, Bennett ST, *et al* (1994) A genome-wide search for human type 1 diabetes susceptibility genes. *Nature* **371**: 130–136.

DeRisi J, Penland L, Brown PO, *et al.* (1996) Use of a cDNA microarray to analyze gene expression patterns in human cancer. *Nature Genet* **14**: 457–460.

DeRisi JL, Vishwanath RI and Brown PO (1997) Exploring the metabolic and genetic control of gene expression on a genomic level. *Science* **278**: 680–686.

Donis-Keller H, Green P, Helms C, Cartinhour S, Weiffenbach B, Stephens K, Keith TP, Bowden DW, Smith DR and Lander ES (1987) A genetic linkage map of the human genome. *Cell* **51**: 319–337.

Elston RC (1995) Linkage and association to genetic markers. *Exp Clin Immunogen* **12**: 129–140.

Fields S and Song O (1989) A novel genetic system to detect protein–protein interactions. *Nature* **340**: 245–246.
Greenberg DA (1993) Linkage analysis of 'necessary' disease loci vs. 'susceptibility' loci. *Am J Hum Genet* **52**: 135–143.
Gusella J, Wexler NS, Conneally PM, *et al* (1983) A polymorphic DNA marker genetically linked to Huntington's disease. *Nature* **306**: 234–38.
Hartl (1988) *A Primer of Population Genetics*. Sinauer Associates Sunderland MA.
Hellman NE, Grant EA and Groate AM (1998) Failure to replicate a protective effect of allele 2 of NACP/α-synuclein polymorphism in Alzheimer's disease: an association study. *Ann. Neurol* **44**: 278–281.
Hodge SE (1981) Some epistatic two-locus models of disease. I. Relative risks and identity-by-descent distributions in affected sib pairs. *Am J Hum Genet* **33**: 381–395.
Hodge SE (1984) The information contained in multiple sibling pairs. *Genet Epidemol* **1**: 109–122.
Hodge SE (1994) What association analysis can and cannot tell us about the genetics of complex disease. *Am J Med Genet* **54**: 318–323.
Hodge SE and Spence MA (1981) Some epistatic two-locus models of disease. II. The confounding of linkage and association.
Hodge SE, Anderson CE, Neiswanger K, Sparkes RS and Rimoin DL (1983) The search for heterogeneity in insulin-dependent diabetes mellitus (IDDM): Linkage studies, two-locus models and genetic heterogeneity. *Am J Hum Genet* **35**: 1139–1155.
Kainulainen K, Pulkkinen L, Savolainen A, Kaitila I and Peltonen L (1990) Location on chromosome 15 of the gene defect causing Marfan syndrome. *N Engl J Med* **323**: 935–939.
Kerem B, Rommens JM, Buchanan JA, *et al* (1989) Identification of the cystic fibrosis gene: genetic analysis. *Science* **245**: 1073–1080.
Kerem E, Corey M, Kerem B, *et al* (1990) The relation between genotype and phenotype in cystic fibrosis–analysis of the most common mutation (ΔF_{508}). *N Engl J Med* **323**: 1517–1522.
Kleyn PW and Vesell ES (1998) Genetic variation as a guide to drug development. *Science* **281**: 1820–1821.
Laing P and Pardee AB (1992) Differential display of eukaryotic messenger RNA by means of the polymerase chain reaction. *Science* **257**: 274–280.
Lockhart DJ, Dong H, Byrne MC, *et al* (1996) Expression monitoring by hybridization to high-density oligonucleotide arrays. *Nature Biotechnol* **14**: 1675–1680.
McAdams HH and Shapiro L (1995) Circuit simulation of genetic networks. *Science* **269**: 650–656.
Nei M and Roychoudhury AK (1973) Sampling variances of heterozygosity and genetic distance. *Genetics* **76**: 379–331.
Ott J (1991) *Analysis of Human Genetic Linkage*. Johns Hopkins University Press, Baltimore.
Pennisi E (1998) A closer look at SNPs suggests difficulties. *Science* **281**: 1787–1789.
Penrose LS (1935) The detection of autosomal linkage in data which consist of pairs of brothers and sisters of unspecified parentage. *Ann Eugen (London)* **6**: 133–138.
Reeders ST, Breuning MH, Davies KE, *et al* (1985) A highly polymorphic DNA marker linked to adult polycystic kidney disease on chromosome 16. *Nature* **317**: 542–544.
Riordan JR, Rommens JM, Kerem B, *et al* (1989) Identification of the cystic fibrosis gene: cloning and characterization of complementary DNA. *Science* **245**: 1066–1073.
Risch N (1990a) Linkage strategies for genetically complex traits. II. The power of affected relative pairs. *Am J Hum Genet* **46**: 242–253.
Risch N (1990b) Linkage strategies for genetically complex traits. III. The effect of marker polymorphism on analysis of affected relative pairs. *Am J Hun Genet* **46**: 242–253.

Risch N and Merikangas K (1996) The future of genetic studies of complex human disease. *Science* **273**: 1516–1517.

Rommens JM, Iannuzzi MC, Kerem B, *et al* (1989) Identification of the cystic fibrosis gene: chromosome walking and jumping. *Science* **245**: 1059–1065.

Roses A (1994) Apolipoprotein E affects the rate of Alzheimer's disease expression: β-amyloid burden is a secondary consequence dependent on APOE genotype and duration of disease. *J Neuropathol Exp Neurol* **53**: 429–437.

Schena M (1996) Genome analysis with gene expression microarrays. *BioEssays* **18**: 427–431.

Schena M, Shalon D, Davis RW and Brown PO (1995) Quantitative monitoring of gene expression patterns with a complementary DNA microarray. *Science* **270**: 467–470.

Schena M, Shalon D, Heller R, *et al.* (1996) Parallel human genome analysis: microarray-based expression monitoring of 1000 genes. *Proc Natl Acad Sci* **93**: 10614–10619.

Spence P (1998) Obtaining value from the human genome: a challenge for the pharmaceutical industry. *Drug Discovery Today* **3**: 179–188.

Spielman RS, McGinnis, RE and Ewens WJ (1993) Transmission test for linkage disequilibrium: the insulin gene region and insulin-dependent diabetes mellitus (IDDM) *Am J Hum Genet* **52**: 506–516.

Spielman RS and Ewens WJ (1998) A sibship test for linkage in the presence of association: the sib transmission/disequilbrium test. *Am J Hum Genet* **62**: 450–458.

St George-Hyslop PH, Tanzi RE, Polinsky RJ, *et al* (1987) The genetic defect causing familial Alzheimer's disease maps on chromosome 21. *Science* **235**: 885–890.

St George-Hyslop PH, Haines JL, Farrer LA, *et al* (1990) Genetic linkage studies suggest that Alzheimer's disease is not a single homogeneous disorder. *Nature* **347**: 194–197.

Tsui LC, Buchwald M, Barker D, *et al* (1985) Cystic fibrosis locus defined by a genetically-linked polymorphic DNA marker. *Science* **230**: 1054–1057.

Velculescu VE, Zhang L, Vogelstein B and Kinzler KW (1995) Serial analysis of gene expression. *Science* **270**: 484–487.

Xia Y, Rohan de Silva HA, Rosi BL, *et al* (1996) Genetic studies in Alzheimer's disease with an NACP/a-synuclein polymorphism. *Ann Neurol* **40**: 207–215.

7.10 FURTHER READING

Chevray PM and Nathans D (1992) Protein interaction cloning in yeast: identification of mammalian proteins that react with the leucine zipper of Jun. *Proc Natl Acad Sci USA* **89**: 5789–5793.

Clerget-Darpoux F (1982) Bias of the estimated recombination fraction and the lod score due to an association between a disease gene and a marker gene. *Ann Hum Genet* **46**: 363–372.

Corder EH, Saunders AM, Strittmatter WJ, *et al* (1993) Gene dose of apolipoprotein E type 4 allele and the risk of Alzheimer's disease in late onset families. *Science* **26**: 921–923.

Diatchenko L, Lau YF, Campbell AP, *et al.* (1996) Suppression subtractive hybridization: a method for generating differentially regulated or tissue-specific cDNA probes and libraries. *Proc Natl Acad Sci* **93**: 6025–6030.

Ehm MG and Wagner M (1998) A test statistic to detect errors in sib-pair relationships. *Am J Hum Genet* **62**: 181–188.

Gelbert WM (1998) Databases in genomic research. *Science* **282**: 659–661.

Greenberg DA and Hodge SE (1989) Linkage analysis under random and genetic reduced penetrance. *Genet Epidemol* **6**: 259–264.

Gress TM, Hoheisel JD, Lennon GG, Zehetner G and Lehrach H (1992) Hybridization fingerprinting of high-density cDNA-library arrays with cDNA pools derived from whole tissues. *Mammal Genome* **3**: 609–619.

Hodge SE (1981) Some epistatic two-locus models of disease. I. Relative risks and identity-by-descent distributions in affected sib pairs. *Am J Hum Genet* **33**: 381–395.

Hodge SE (1993) Linkage analysis versus association analysis: distinguishing between two models that explain disease-marker associations. *Am J Hum Genet* **53**: 367–384.

Hodge SE and Spence MA (1981) Some epistatic two-locus models of disease. II. The confounding of linkage and association. *Am J Hum Genet* **33**: 396–406.

Jones HB (1998) The relative power of linkage and association studies for the detection of genes involved in hypertension. *Kidney Int* **53**: 1446–1448.

Lander ES (1996) The new genomics: global views of biology. *Science* **274**: 536–539.

Maier E, Meier-Ewert S, Bancroft D and Lehrach H (1997) Automated array technologies for gene expression profiling. Drug Discovery Today **8**: 315–324.

Mayeux R, Saunders AM, Shea S, *et al* (1998) Utility of the apolipoprotein E genotype in the diagnosis of Alzheimer's disease. Alzheimer's Disease Centers Consortium on Apolipoprotein E and Alzheimer's Disease. *N Engl J Med* **338**: 506–511.

Monaco AP and Kunkel LM (1988) Cloning of the Duchenne/Becker muscular dystrophy locus. *Adv Hum Genet* **17**: 61–98.

Oliver S (1996) From DNA sequence to biological function. *Nature* **379**: 597–600.

Ott J (1989) Computer-simulation methods in human linkage analysis. *Proc Natl Acad Sci USA* **86**: 4175–4178.

Pennisi E (1997) Genomics: a catalog of cancer genes at the click of a mouse. *Science* **276**: 1023.

Pericak-Vance MA, Bass MP, Yamaoka LH, *et al* (1997) Complete genomic screen in late-onset familial Alzheimer disease. Evidence for a new locus on chromosome 12. *J Am Med Assoc* **278**: 1237–1241.

Risch N (1990) Linkage strategies for genetically complex traits. I. Multilocus models. *Am J Hum Genet* **46**: 222–228.

Roses AD (1995) Apolipoprotein E genotyping in the differential diagnosis, not prediction, of Alzheimer's disease. *Ann Neurol* **38**: 6–14.

Roses AD (1996) Apolipoprotein E and Alzheimer's disease. A rapidly expanding field with medical and epidemiological consequences. *Ann N Y Acad Sci* **802**: 50–57.

Roses AD (1997) Apolipoprotein E, a gene with complex biological interactions in the aging brain. *Neurobiol Dis* **4**: 170–185.

Roses AD (1998) Alzheimer diseases: a model of gene mutations and susceptibility polymorphisms for complex psychiatric diseases. *Am J Med Genet* **81**: 49–57.

Roses AD (1998) Genetic associations. *Lancet* **351**: 916.

Roses AD, Pericak-Vance MA, Saunders AM, *et al* (1994) Complex genetic disease: can genetic strategies in Alzheimer's disease and new genetic mechanisms be applied to epilepsy. *Epilepsia* **35** (Suppl. 1): S20–S28.

Serrano M, Hannon GJ and Beach D (1993) A new regulatory motif in cell-cycle control causing specific inhibition of cyclin D/CDK4. *Nature* **366**: 704–707.

Strachan T, Abitbol M, Davidson D and Beckman JS (1997) A new dimension for the human genome project: towards comprehensive expression maps. *Nature Genet* **16**: 126–131.

Strittmatter WJ and Roses AD (1996) Apolipoprotein E and Alzheimer's disease. *Annu Rev Neurosci* **19**: 53–77.

Wang H, Peters GA, Zeng X, *et al* (1995) Yeast two-hybrid system demonstrates that estrogen receptor dimerization is ligand dependent *in vivo*. *J Biol Chem* **270**: 23322–23329.

7.11 WEB ADDRESSES FOR THE HUMAN GENOME PROJECT

National Human Genome Research Institute at the National Institutes of Health
  www.nhgri.nih.gov
  www.ornl.gov
NCBI map of the human genome
  www.ncbi.nlm.nih.gov/SCIENCE96
Stanford human genome center
  http://shgc.stanford.edu
Whitehead Institute/MIT Center for Genome Research
  www.genome.wi.mit.edu

8 Experimental Design

8.1 INTRODUCTION

The basis of the framework described in Chapter 6 is that statistical methods are utilized throughout the process of drug discovery, not simply at the end when all of the data are collected. Statistical experimental design is the piece of the framework that is used during the earliest phases of chemical synthesis or biochemical assay design. Design is an integral part of the experimental planning process. Interpretation of data from designed experiments is facilitated by having a statistical model as the basis for the analysis. Pragmatically, experimental design saves time and resources during optimization of reactions and experiments by setting up a systematic plan for data collection that maximizes the information gained from the collected data. The larger numbers of compounds synthesized and tested in drug discovery laboratories and the introduction of robotic systems place a premium on effective design. The time necessary to identify optimal conditions for reactions used in library synthesis must be shortened for rapid production of novel chemistries. Utilization of scarce proteins and reagents often must be minimized in high-throughput screening laboratories that test libraries against an array of molecular targets. Assay variability must be minimized to use experimental results for structure–activity relationship analysis and to design subsequent combinatorial libraries.

This chapter illustrates the use of statistical experimental design for problems in drug discovery, both in biology and chemistry laboratories. The development and optimization of assays designed for high-throughput screening is the problem addressed for biology. Three examples illustrate the diversity of problems in assay development that experimental design can address in biochemistry. The first problem involves optimization of assay conditions for a scintillation proximity assay. The second example is concerned with optimization of a plate reader for a time-resolved fluorescence assay. This example illustrates how detection of interactions is critical for optimization of an assay. The third case study focuses on the choice of sampling and data analysis protocols for a radioligand binding experiment based on retrospective analysis of screening data. For chemistry, three problems are considered: optimization of a chemical reaction for yield, substituent selection for a series of compounds and selection of compounds to maximize coverage of chemical space.

8.2 PRINCIPLES OF EXPERIMENTAL DESIGN

The design of an experiment begins with a set of objectives and questions to be addressed. All experiments have some component of design, a plan to obtain data and methods for interpretation of the data. Statistical experimental design is a well-defined and versatile method for collecting data that provides the maximal amount of information in the fewest number of runs. Statistical experimental design uses the objectives of the experiments to formulate an analysis model. Information about the factors that can either be varied by the investigator or characterized as random are used in the modeling process. The statistical significance of varying a factor can be tested, either as a main effect or as an interaction with other factors. Some designs are constructed simply to identify important factors or to optimize conditions for an assay or chemical reaction. Other designs are used to gather data to model a relationship between the factors and a response or set of responses. Although formal experimental design is commonly performed in manufacturing and engineering environments, e.g. scale-up of chemicals and process optimization, it is rarely done in the context of biochemical experimentation. For the design of high-throughput screening assays, combinatorial libraries and optimization problems in drug discovery, statistical experimental design is a powerful method for collection and analysis of data. Designs can be so simple that they are constructed and analyzed on a piece of paper, or they can be complex and require computer software. Complexity depends on the number and type of factors involved and the underlying statistical model.

All designs involve a response or set of responses to be modeled. These are the dependent variables in the design. Often, the response is a variable to be optimized: yield in a chemical reaction or signal-to-noise ratio in a biological assay. As with all statistical analysis, an estimate of the variability of the response must be obtained. The design itself provides an estimate of variability by assuming replication at some or all of the design points. This calculation is made using the standard analysis of variance model described in Chapter 6 (Fig. 6.5). Many designs include several runs where the factors are set at their corresponding mean levels in order to estimate response variability. Finally, if historical data are available, they can be used to estimate response variability. In biochemistry, some designs are constructed to minimize variability. These designs require multiple tests at each combination of factor levels in order to calculate the variance that is used as the response.

In addition to an estimate of variability, it is important to understand the magnitude of change in the response that is of practical or functional importance. In terms of the statistical framework, knowledge of variability and magnitude of the difference that is of interest allows interpretation of the results of the experiment. This information is necessary to make sample size estimates for the experiment (see Fig. 6.14). If the estimate of variability is obtained during the design process, one result is often that a larger number of runs must be done in order to collect a sample that is large enough to test for statistical significance. The first set of experimental designs for a problem tend to have large numbers of factors and levels, leading to a complex model, particularly if interactions are considered. Because of the large numbers of factors and levels, it is important to be cautious about interpreting measures of statistical significance. Rather than setting arbitrary limits for significance, it is more important to examine plots of response versus the different factors, assess the strength of interaction between the factors and question the factor levels that were chosen in the design. This information and the estimate of variability of the response are used to plan a subsequent design with an adequate number of replicates to make the statistical tests.

The simplest designs to understand are full-factorial designs. All combinations of factor settings are tested in order to estimate main effects and interactions. Many factors for biochemical assays are aspects of the assay that are under the control of the experimenter and of interest. Choice of factors is balanced between a list of everything that could have an effect on the response and the complexity of the model. Factors include reagent concentrations, buffers, pH, temperature and time. For a robotic assay system, the list of factors includes parameters for individual dispensing steps: volumes, speeds and gaps. In chemistry, factors for designs that optimize reaction conditions are similar to those used in biochemistry: reagents and operating conditions. Designs that model the relationship between chemical structure and a response involve factors that are based on fragments of the structure, e.g. substitutions at multiple R groups or monomers for a combinatorial library.

After factors are identified, settings or levels for the factors must be determined. In many designs, two settings are used. Although two settings allow for testing of the main effects and interactions, they do not provide adequate data if the response is non-linear with respect to the factor. For biochemistry and chemistry, sequential design is an effective strategy: First, full- or fractional-factorial models of many factors with two settings to identify important factors, then response surface models of a subset of the factors tested at three or more settings.

Full-factorial designs are often depicted by geometric shapes that show the factors and settings on the vertices. For example, a cube can be used to show a three-factor, two-level design. After the experiment is run and data collected, the response is recorded at each vertex and inspection of the cube identifies factor settings for optimal response. A linear model is fit to the results and used to assess the statistical significance of the factors and interactions.

It is important to recognize that experimental design provides a structure to plan experiments rather than simply a means to find systematically the optimal settings for a group of factors. Effective analysis of data and, in turn, interpretation of the experimental results starts when the experiment is planned, and not after all of the data are collected. There are numerous pitfalls associated with interpreting data from a poorly planned experiment: data can be inconsistent between runs, effects can be confounded and nonsensical correlations can appear. Statisticians are often asked to analyze experimental data after months of laboratory work are completed and large amounts of data have been collected. Consulting with statisticians before the experimental work starts has the advantage of designing the experiments to provide the data necessary for the planned analysis. Design includes determining an appropriate sampling strategy, treatment structure and number of replications.

Designed experiments are used when the problem or experimental objective can be clearly stated. Objectives are either hypotheses to test or specific outcomes. For example, 'what is the minimum dose of a drug to cause at least a 20% increase in heart rate in dogs over a 1-h i.v. infusion' is a problem statement that has experimental design implications: how many dogs in each sample, randomization of the treatment and dose groups and statistical testing to interpret the results. An example of an outcome is 'design a high-throughput scintillation proximity assay (SPA) that will reliably (95% confidence) detect 50% inhibition for single tests of compounds'. Data analysis and design for the first problem involves determining the number of doses to test, the numbers of dogs to test in each sample, determining a treatment structure (randomized doses, controls) and an analysis model, perhaps repeated-measures analysis of variance because heart rate is measured over a period of 1 h. Clearly, there is a link between the experimental protocol and the data analysis planned for the experiment. The investigator and statistician need to address questions about carryover of drug between doses, equilibration time and treatment of the control dogs. Other research questions and variables often are considered later. Assessing the effect of the drug on other hemodynamic variables modifies the analysis to a repeated-measures multivariate analysis of variance model. Although it is not possible to include every possible analysis during design, adjustments to sampling can be made if multivariate analysis is planned. For the second problem, the design of the SPA assay, several phases of experimental design are used to optimize the assay: a screening design to identify critical factors and a multilevel factorial design to optimize the most important factors in order to minimize variance. After the design is complete and experimental results recorded for the designed experiment, probability calculations are performed to determine if the assay is sufficiently reliable to detect 50% inhibitors.

Interactions are an important consideration in experimental design. Part of the appeal of statistical experimental design is the capability of detecting interactions. However, as the number of factors and levels increases, the number of potential interactions increases even faster, leading to complex models that potentially require more runs than are feasible. Because of the straightforward interpretation of the main effects, it is tempting to consider a strategy of using a model that includes only the main effects and then following up with designs that include interactions for the significant or important main effects. At times, this is a good strategy. However, interactions can be between a factor that has a significant effect on a response and another factor that, in isolation, has either a modest or an insignificant effect on response. The presence of interactions can obscure interpretation of the main effects model. If interactions are missed during the screening phase of the experiment, it is possible that the factors that contribute to the interaction are also dropped at some point, making it impossible later to detect the interaction. The most common interactions to test for are two-factor interactions. Higher level interactions (e.g. among three or more factors) can be included in a model but are difficult to interpret. When setting up fractional factorial designs, it is common to allow higher level interactions to be confounded. Generally, unless there are valid theoretical or pragmatic reasons for ignoring two-factor interactions (e.g. number of runs is prohibitive), they should be included in the model.

The statistical framework for an experimental design involves:

(1) Stating the research problem (reaction to be optimized, maximize coverage of a defined chemical descriptor space).
(2) Calculating the variability of the response or responses to be optimized and using this information to set sample sizes.
(3) Enumerating the factors to be studied and choosing the factor ranges, settings or lists of structures (independent variables). This step identifies the sources of variability. Sources of variability that are random in nature must also be included.
(4) Characterization of the sample and underlying sampling assumptions, e.g. determining whether the observations are random and independent.

The best practices described in Chapter 6 (clear and concise presentation of results; choice of appropriate methods for visualization and interpretation; and linkage between experimental design and data analysis) are important when using statistical experimental design.

8.2.1 THEORETICAL BASIS

Experimental design is a classical statistical method that has found utility in many disciplines that require finding optimal conditions and modeling response surfaces. The theoretical basis for experimental design is described in detail in numerous statistical texts (Cochran and Cox 1957; Hicks, 1974; Box *et al* 1978; Montgomery 1984). Application case studies can be found in the literature associated with a given field. Examples include industrial process optimization (Daniel 1976; Murphy 1977) and chemistry (Deming and Morgan 1987; Austel 1983). There are some specific problems and methodologies associated with using experimental design for biological and chemical applications that are addressed in several review articles and books (Haaland 1989; Lutz *et al* 1996).

In many applications, the underlying question is how to optimize something in a systematic way in order to extract the most information in the fewest numbers of runs. In some cases, identification of the optimal conditions is sufficient. In other cases, the relationship between the factors and the response is modeled mathematically; this is termed response-surface modeling. Interactions are particularly important because they are more the rule than the exception in the natural sciences and can greatly increase the complexity of a problem or, in the worst case, make interpretation of the data impossible. A special subset of experimental designs is variance-reduction experimental designs. In these designs, the variance of a response is optimized rather than the response itself. These designs are important in biology because by minimizing variance an assay becomes more robust. In other applications, cost is an important factor and optimization is desired to minimize the use of scarce resources such as proteins or reagents. In many cases, multiple responses are to be considered, e.g. minimizing variance while keeping receptor concentration low in order to conserve protein. Finally, the advantage to having a model, even a simple one such as the linear model that forms the basis of classical experimental designs, is that you can make predictions. It is also possible to test whether the data does, in fact, correspond to the model.

In classical experimental design, investigators choose a number of controllable factors that they believe may have an effect on the response. Depending on the number of factors and their possible ranges, different types of designs are chosen to sample the space defined by these factors. All of the factors are varied simultaneously in a systematic way that later allows measuring the significance of a specific factor or interaction between factors. Typically the factors are sampled at the extremes of the operating range and a linear or polynomial interpolating function is used to model the region between the design points. In this chapter, space-filling designs — experimental designs that are not based on linear models — are discussed because of their potential for application to intrinsically non-linear problems in the optimization of complex processes such as biochemical assays.

8.2.2 CLASSICAL EXPERIMENTAL DESIGNS

8.2.2.1 Nomenclature

Throughout this discussion, factors refer to experimental variables that can be changed by the investigator, e.g. temperature, time in incubator, pH or choice of buffer. Levels are the specific settings of factors for a given run in an experiment. Interactions arise from combinations of factors, e.g. an interaction between pH and temperature. Response is the variable or quantity being optimized. Examples include yield in a chemical reaction and variability of a measurement in an assay. Effect is the magnitude of the change in response as factors are varied. A run is an experiment performed at specific settings of all the factors. Because linear statistical model is often used to interpret the results of an experimental design, some of the terminology commonly used with these models is also used in experimental design. Main effects refer to the parameters associated with single factors, an estimate of their standard deviation or significance in the model.

Interactions are effects that depend on two or more factors. If estimates of the coefficients associated with each term in the model (main effects and interactions) are independent, the design is said to be orthogonal. The mathematical form of the linear model, in matrix terms, is:

$$Y = X\beta + \varepsilon \qquad \ldots (8.1)$$

where X is the design matrix (independent variables or predictors), β is a vector of parameter estimates obtained by linear regression, Y is the vector of observed responses and ε is a vector of error terms, assumed to be normally distributed with a mean of zero and a standard deviation of σ.

8.2.2.2 Full-factorial Designs

Full-factorial designs are conceptually the easiest experimental designs to set up and interpret. In a full-factorial design data are obtained for all combinations of factors, allowing the effect of each factor and interaction between factors to be studied. Often, two levels are chosen for each factor, representing the upper and lower limits of the operating range, although more than two levels can be chosen to understand better the relationship between a factor and response. Graphically, simple full-factorial designs can be described by geometric shapes, e.g a square for a two-factor, two-level design (Fig. 8.1) and a cube for a three-factor, two-level design (Fig. 8.2). If each factor is sampled at two levels, you will need 2^k runs for k factors for a full-factorial design.

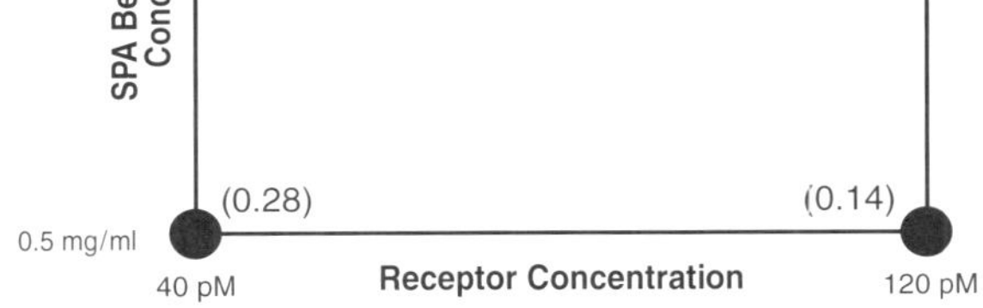

Fig. 8.1. Experimental design for a two-factor, two-level full-factorial design. The two factors and their levels are concentration of receptor (40 pM, 120 pM) and concentration of radioligand in a scintillation proximity assay (SPA) (0.1 nM, 0.4 nM). The response — variability of the pK_i estimate in standard deviation units — is listed under the factor level in parentheses. (Reprinted from *Drug Discovery Today*, Vol. 1, Lutz MW, *et al*, Experimental design for high-throughput screening. Copyright 1996, pp. 277–285, with permission from Elsevier Science.)

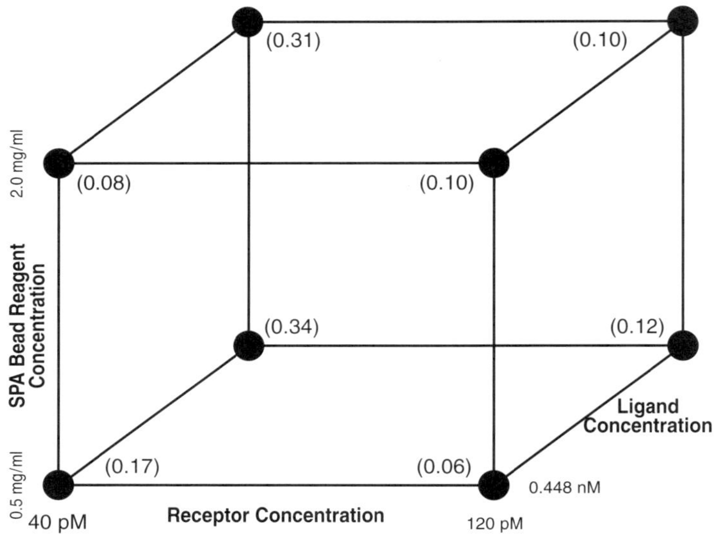

Fig. 8.2. Experimental design for a three-factor, two-level full-factorial design. The three factors and their levels are concentration of receptor (40 pM, 120 pM), scintillation proximity assay (SPA) reagent bead concentration (0.5 mg/ml, 2.0 mg/ml) and concentration of ligand (0.448 nM, 1.792 nM). The response — variability of the pK_i estimate in standard deviation units — is listed under the factor level in parentheses. (Reprinted from *Drug Discovery Today*, Vol. 1, Lutz MW, *et al*, Experimental design for high-throughput screening. Copyright 1996, pp. 277–285, with permission from Elsevier Science.)

Although is it clear that if there are many factors and/or levels these designs will require a large number of runs, they are very useful in practice, particularly for assay development. Interpretation of the design can usually be done graphically, often simply by recording the response at the vertex of each design point (Figs 8.1 and 8.2). For factor levels that are under software control in a robotic assay system, large numbers of runs are easily set up and an assay is optimized in a short time.

8.2.2.3 Fractional-factorial Designs

A fractional-factorial design is a compromise between the information on all interactions that can be obtained in a full-factorial design and the total number of runs. Fractional-factorial designs can be grouped by their resolution, the types of effects and the interactions that can be estimated by a class of designs. Higher resolution designs allow more interactions and effects to be studied than lower resolution designs but require more runs. The term 'confounding' is used to refer to effects (factors and/or interactions) that cannot be estimated independently. A resolution V design allows all main effects and two-factor interactions to be estimated, sacrificing only the ability to estimate higher order interactions. A resolution IV design does not allow any two-factor interaction to be confounded with a main effect but allows confounding among the two-factor interactions; it is a good design when four to eight variables are being investigated with anticipated interactions to detect. Resolution III designs allow main effects to be confounded with interactions but do not allow confounding among main effects; these designs are often chosen for screening designs when there are a large number (more than eight) of factors being tested.

A variation on the fractional-factorial designs is the Plackett–Burman (Plackett and Burman 1946) design. This is a low-resolution design that requires even fewer runs than fractional-factorial designs. It is most useful for screening large numbers of factors rather than testing for interactions.

For a fractional-factorial design, the easiest way to set up a design is to use experimental design software or consult a listing of designs. The design digest in Haaland's (1989) book provides a list of the most common designs encountered in practice.

8.2.2.4 Response Surface Modeling

Response surface modeling involves the use of a regression equation to describe mathematically the relationship between factors and response. Often graphs such as contour or surface plots are used to describe these relationships. Response surface modeling can be based on the full and fractional designs already described, but additional data need to be collected to define better the curvature of the surface. First a center point, located at the centroid of all of the design points, is added. Points are then added to cover regions away from the design points. These are commonly called star points (Fig. 8.3). These designs are known as central composite designs. In terms of the linear model, terms for quadratic effects are included for interpretation of results. Box–Behnken designs (Box and Behnken 1960) are another class of designs that are used for response surface modeling. These designs allow estimation of all main effects, two-factor interactions and quadratic effects. Box–Behnken designs are useful when a set of factors settings can be considered to be the standard operating conditions, perhaps as a follow-up design after some factorial screening designs. The Box–Behnken design allows you to do a more detailed analysis of the region, spanning out from the central design point.

8.2.2.5 D-optimal Designs

D-optimal designs (Kiefer 1959; Galil and Kieser 1980; Snee 1985) are an example of a computer-generated design. Many experimental design software programs such as JMP® (SAS Institute, Cary, NC) and DISCOVER (BBN Software Products, Boston, MA) include D-optimal designs as an option. In using a D-optimal design or any other similar computer-generated design, a user chooses

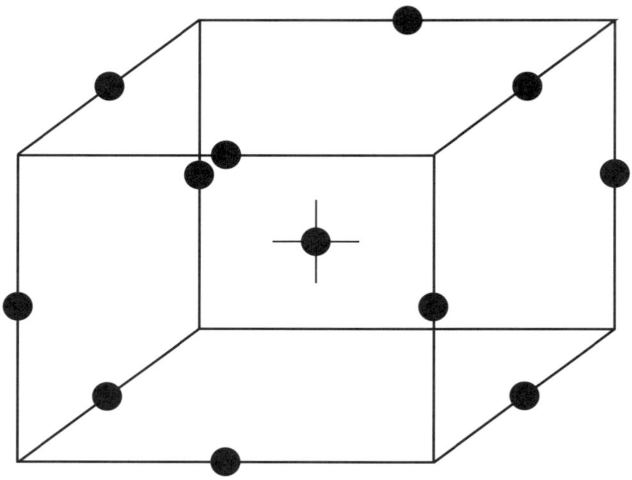

Fig. 8.3. Star points for a central composite design. These points augment the corner points of factorial designs, to give a better estimate of the curvature of response, and are used for response surface modeling. (Reprinted from *Drug Discovery Today*, Vol. 1, Lutz MW, *et al*, Experimental design for high-throughput screening. Copyright 1996, pp. 277–285, with permission from Elsevier Science.)

an acceptable number of runs, identifies the main effects and interactions of interest and allows the computer to generate the design. In a D-optimal design, this is done iteratively by maximizing the determinant $|(X'X)^{-1}|$. The determinant can always be maximized by increasing the number of points or by expanding the range of values for the predictors. These methods will define better a linear relationship between a set of predictors and a response. In a D-optimal design, however, the number of runs and ranges for the predictors is fixed. Computationally, the algorithm chooses design points that maximize the determinant and minimize the region that bound the parameters of the linear model used to interpret the design by iteratively adding and deleting points from a set of candidate design points. The main disadvantage to these designs is that they assume that the form of the mathematical model is known *a priori*. D-optimal designs are not necessarily orthogonal. If there are interactions present that are not included in the design, the other effects may not be estimated properly.

Computer-generated designs are useful in the biochemical sciences because the software often allows substitution for points that are inaccessible due to physical constraints. For example, it may not be possible to run an experiment at the factor settings for a particular design point in a full-factorial design. A computer-generated design, however, can substitute a different point and still choose a design that maximizes the D-optimal criteria, giving the best design for that series.

8.2.3 IDENTIFICATION OF FACTORS AND THEIR LEVELS

Identifying critical factors and associated levels is very important and is an area where many designs fail. Often, factor levels are chosen that do not cover a suitable range or cover such a broad range that they are not adequately modeled by a linear model. Factors and settings are usually determined from prior experience and knowledge of the assay system. Some practitioners argue that it is worth running a series of undesigned experiments to define suitable ranges for the factors; these experiments are often run at the limits of the ranges. Others prefer to start with a low-resolution screening design. Ideally, the settings are chosen so that the investigator expects to see a difference in response over the range of factor settings. For screening designs it is important to include all factors that are known to influence the response or are suspected to influence the response. Unfortunately, the designs get large as the number of factors increase. Design is a sequential process. For a poorly characterized process, the initial designs are large and involve many different factors. Many factors can be omitted from the later designs when it is clear that they do not influence the response either directly or as part of an interaction with another factor. Whether variables that do not show a significant linear effect can still be part of a significant interaction is the subject of some debate. In linear model application, it is a standard practice to require that if a factor is part of an interaction it must also be included as a main effect. Designs are run several times to determine if the results are reproducible.

Discrete factors must be treated as discrete variables in the design; they cannot be considered continuous variables. A continuous variable enters the model used to interpret the design as a term that has a single coefficient (e.g. similar to a term in a linear regression equation). Discrete variables have coefficients assigned to each level, similar to the ANOVA models discussed in Chapter 6. Many discrete factors do not have a defined response at the levels implied by a transformation to a continuous space. Often, selection of discrete factors must be made from a larger set, e.g. when selecting monomers or substituents for a chemistry design. Cluster analysis using descriptors of the different factors is one approach that works well to segment the space. Representative compounds are selected from the clusters to use in the design. The section on designs for chemistry (section 8.4) describes this approach in more detail.

Different experimental designs are chosen for specific objectives. When a robotic assay is first set up, many factors potentially can influence the robustness of the assay. Table 8.1 shows some of the mechanical factors in a robotic assay system that can have an effect on biological response. Reagent volumes, dispensing and aspiration gaps and speed can all have an effect on variability. To find the important factors, a low-resolution fractional-factorial design might be run. This is known as factor screening. In Table 8.1, coefficient of variation (%CV) is the response being optimized. It is shown that %CV can vary from 1% to 10% as a function of the mechanical factors. After the important factors are identified, a full-factorial design might be set up to find the optimal settings

Table 8.1. Possible factors to optimize in a robotic biochemical assay system and associated coefficient of variation (%CV)

Sample volume (ml)	100	100	100	100	40
Sample dispensing speed (ml/s)	600	50	50	600	600
Sample aspiration speed (ml/s)	400	50	400	400	300
Sample cutoff speed (ml/s)	600	50	50	600	500
Trail gap (mm)	30	0	0	0	20
Travel gap (mm)	15	0	0	0	15
Air gap aspiration speed (ml/s)	20	0	0	0	70
Air gap dispensing speed (ml/s)	100	0	0	0	800
Tip height (mm)	Z_{Bot}	Z_{Bot}	Z_{Bot}	Z_{Bot}	Z_{Disp}
Tip touch	No	No	No	No	No
Transfer in liquid	No	No	No	No	No
No. of a sample dispenses	6	1	1	1	1
No. of pre-dispenses	1	0	0	0	0
No. post-dispenses	1	0	0	0	0
Dispensing delivery (ml)	50	50	50	50	50
Aspiration delivery (ml)	50	50	50	50	50
Prime	0	0	0	0	0
Sample type	Aq/System	Aq/System	Aq/System	Aq/System	Aqueous
Tip type	Regular teflon outer coating	Regular Teflon outer coating	Regular teflon outer coating	Regular teflon outer coating	Regular teflon outer coating
Syringe size (ml)	1000	1000	1000	1000	1000
%CV	4.52	1.56	1.23	0.91	9.09

Data provided by Peter Leitner, Glaxo Wellcome.

for the factors. Response surface modeling could be done to explore the region around the optimal settings, to assess sensitivity of the factors and to test for higher order interactions.

8.2.4 ASSAY VARIABILITY AND REPLICATION

It is important to have a good estimate of assay variability, both to interpret the results of the experimental design and later to be able to identify significant differences between compounds tested in the assay. When assay variability itself is the response being optimized, each design point must be replicated. For other designs, a replicated center point is an efficient means to calculate an estimate of assay variability. The center point is obtained from a run where all factors are set in the middle of their range. This point is replicated several (typically 3–10) times in order to provide an estimate of the variance of the response. It is important to recognize that the replicates must be independent. For example, observations obtained from adjacent wells on a 96-well microtiter plate are not independent and will measure only the error in pipetting. Although the center point is a reasonable initial estimate of variability, it is also possible that the factors themselves may have an effect on variability, requiring replication at other design points.

As with most statistical analysis, the interpretation of an experimental design using a linear model assumes that the error in the response is random variation. All systematic effects are assumed part of the model. The investigator must be careful to guard against the introduction of inadvertent systematic effects by examining the experimental protocol for sources of systematic effects, such as time effects, lack of homogeneous reaction conditions and differences between reagents and biological materials. In many cases, if a systematic effect is suspected, it is possible to modify the design and add terms to the model to account for the effect; this a technique is known as blocking. Most standard experimental design software allows for blocking effects. In practice, they are terms that are allowed to be confounded with high-order interactions.

8.2.5 ANALYSIS AND INTERPRETATION OF EXPERIMENTAL DESIGNS

Analysis and interpretation of experimental designs can be as simple as sorting a table of results to find the conditions that give the lowest assay variability, or as complex as fitting a model to the data and using interactive graphics to study the effect of interactions.

For designs that have few factors, simple plots that resemble the shapes of the design (squares and cubes) are useful to identify general trends. These plots can be drawn easily with pencil and paper. The response value is simply filled in at the design points. Tabulating the data in a statistical analysis or spreadsheet program allows sorting by different criteria in order to find trends in the data. Sorting by size of effect is often helpful for two-level factors. The average response at each factor setting is calculated and subtracted to yield the size of the effect.

Because a statistical model is used to interpret an experimental design, all of the tools appropriate for the evaluation of the model apply, including standard deviations and significance tests on the parameters, residual plots, normal probability plots and analysis of variance tables (Snedecor and Cochran 1989). It is particularly important to compare the variance of estimated effects to assay variability.

Although the statistical analysis will calculate a significance test for each interaction term, it is often helpful to make interaction plots for two-factor interactions as well. Figure 8.4 is an example of an interaction plot. The ordinate is the response value and the abscissa gives the factor settings for one of the two factors. The second factor is used to group points on the plot; all points that share a factor setting are connected by a line. If there is no interaction, then the lines are parallel. If the lines are not parallel then the viewer can determine whether

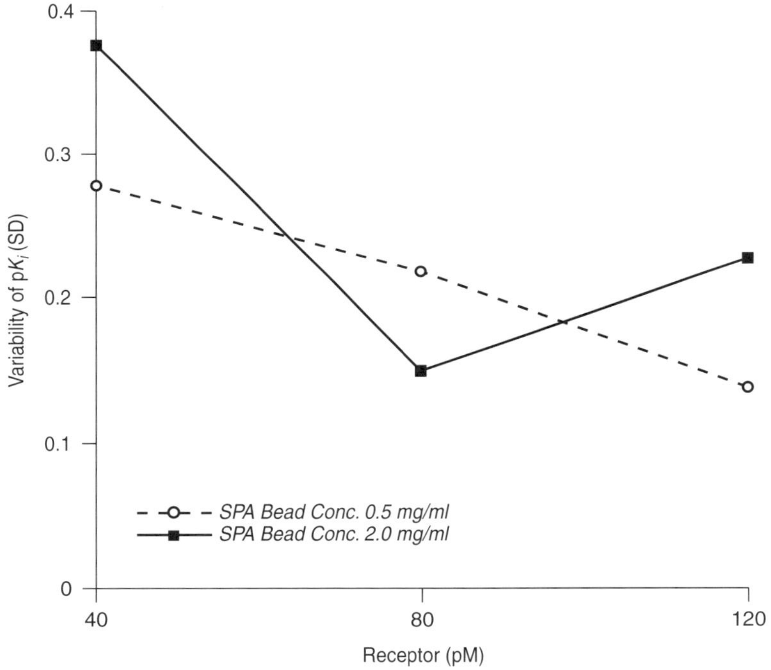

Fig. 8.4. Example of an interaction plot. Abscissa: receptor concentration. Ordinate: response value, i.e. variability of the pK_i in standard deviation units. SPA = scintillation proximity assay. (Reprinted from *Drug Discovery Today*, Vol. 1, Lutz MW, *et al*, Experimental design for high-throughput screening. Copyright 1996, pp. 277–285, with permission from Elsevier Science.)

the interaction is working to their benefit, e.g. lowering variability while keeping the first factor at a preferred level for example, or whether it is a detrimental interaction. Interactions are also clearly shown on contour plots. Figure 8.5 shows the interaction between two protein concentrations used in a biological assay on the signal-to-noise ratio of the assay. The abscissa is the concentration of one protein and the ordinate is the ratio of the two proteins. The jagged edges in the contour plot show that the relationship between signal-to-noise ratio and protein concentration is not a simple one.

The software package JMP® (SAS Institute, Cary, NC) has visualization capabilities that allow investigators to interpret an experimental design interactively. A graph pane is created for each effect in the model, where the abscissa reflects the factor settings and the ordinate is the response. For example, Fig. 8.6 shows

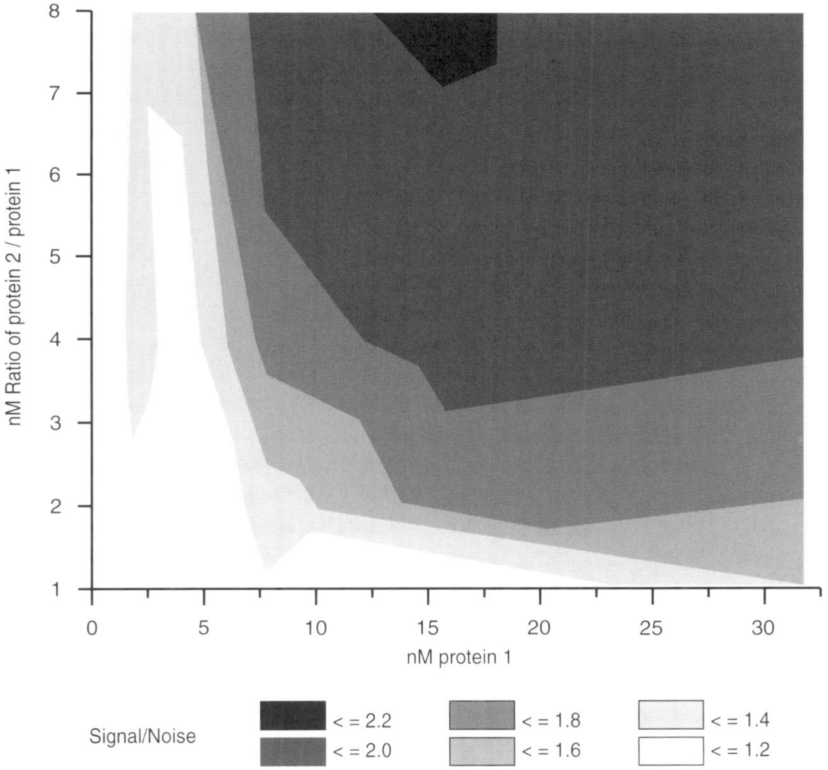

Fig. 8.5. Contour plot showing the interaction between two protein concentrations on the signal-to-noise ratio in a biochemical assay. Abscissa: concentration of the first protein (nM). Ordinate: ratio of the concentration of the second protein to the first. Each contour shows the signal-to-noise ratio. (Data provided by Peter Leitner, Glaxo Wellcome.)

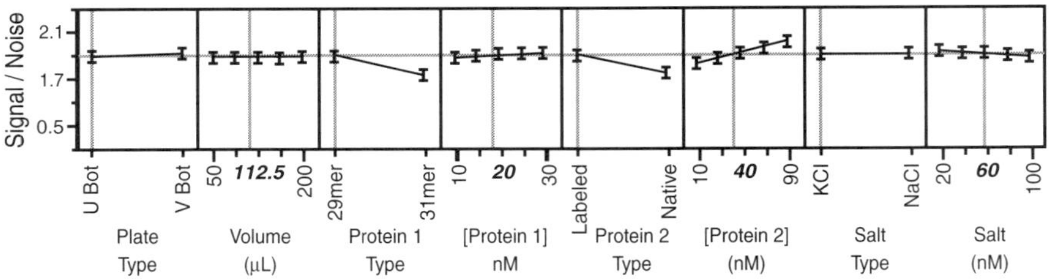

Fig. 8.6. Interactive plots for an experimental design examining the effects of protein concentrations, plate type, sample volume and buffer on signal-to-noise ratio in a biochemical assay. Abscissa: settings for each factor in the experimental design. Ordinate: signal-to-noise ratio. The vertical lines show the current setting for each of the factors, e.g. all the plots are made by assuming these settings. The horizontal line is the grand mean of all the response values. (Data provided by Peter Leitner, Glaxo Wellcome.)

the results of a fractional-factorial design to study the effect of various protein and salt types and concentrations on signal-to-noise ratio in a biochemical assay. The vertical slider, originally set at the mean for each factor, can be moved to any factor setting. As the slider is moved, all panes update to show the effect on response. It is important to realize that the effect on response is calculated from the full model of the data, including interactions. The error bars allow the viewer to identify significant differences between response at different factor levels and differences from the mean response (horizontal line). The software package JMP® also allows contour plots to be constructed easily from experimental design results. Depending on the number of factor levels used, contour plots can be helpful to identify optimal operating regions. However, their usefulness is limited by their bivariate nature.

Strategies for using the results from one experimental design to plan the next utilize the methods described in Chapter 6. First, the fit of the model should be tested using the analysis of variance model and by examining residual plots to ensure that there is only random variation. Determining which, if any, variables are significant is difficult if there are many variables in the model or if there are high correlations between the variables. One approach is to test the model after removing variables that appear to be insignificant. For the reduced model, the ANOVA is tested to see if it is still significant and the residual plots are examined to determine if the model fits the data. If the reduced model appears appropriate, another experimental design based on this model can provide confirmation of the importance of the factors and possibly test additional levels of the factors.

5.2.6 SPACE-FILLING EXPERIMENTAL DESIGNS

Space-filling designs extend experimental design theory and methods from linear models to non-linear models. When experimental designs fail, it is often because the response surface is irregular or contains spikes and valleys that are difficult to model with linear or polynomial functions. In a space-filling design the factor settings are chosen to distribute evenly the design points throughout the region or space of interest, rather than sample the extrema. Particularly for problems in biology and chemistry, these designs may be more appropriate as compared to the classical designs that work well for optimizing reaction yield or for an industrial process. The designs tend to require many data points. Unlike classical designs where factors often are sampled at two levels, each factor may be sampled at 10–20 levels. For high-throughput screening and robotic assay systems, it is possible to achieve the sampling protocols required by these designs. Although algorithms to create the designs have existed for some time (Kennard and Stone 1969), good software has only recently become available. Programs for space-filling designs include ALEX (Sacks *et al* 1989; Welch *et al* 1992) and SAS/QC (SAS/QC Software 1993). Cluster analysis (Zemrock 1986) can also be used to create the designs.

Interpretation of results from a space-filling design requires different methods from traditional designs. In general, a non-parametric regression model is necessary to interpret the data due to the possibility of curvature in the design space. Examples of methods used to analyze the results of space-filling designs include polynomial regression models, Gaussian stochastic processes, thin-plate splines and neural networks (Haaland *et al* 1994). The simplest models are the polynomial models that can be fitted using ordinary least squares regression. Neural network models can be fitted using a variety of commercial software packages. For Gaussian stochastic processes and thin-plate splines, the model has a component in addition to the linear term of eq. 8.1:

$$Y = X\beta + Z + \varepsilon \qquad \ldots(8.2)$$

The Z term has a distribution that is normal with a mean of zero and a standard deviation $\sigma^2 R$, where R is the correlation function. The essential point is that a flexible function is used rather than a strict linear model. In some cases, only the intercept term of the linear model is used and all of the flexibility is put into the Z term. The idea behind the correlation function is a simple concept: points located together in space should be more highly correlated, an assumption that is valid for many problems in biology and chemistry.

For space-filling designs, choice of an analysis method depends both on the experimental objective and the nature of the surface itself. For screening problems, simple graphical methods or sorting the design table by the response variable are often sufficient. When response surface modeling is done, choice of an analysis method is based on the plausibility and stability of the parameter estimates and the model fit.

Some recent examples of space-filling designs in biology are a design to determine optimal conditions for protein construct storage conditions (Menius *et al* 1994) and a study of buffer conditions on DNA strand displacement amplification (Haaland *et al* 1994).

8.3 EXPERIMENTAL DESIGN FOR BIOCHEMISTRY

8.3.1 OPTIMIZATION OF A SCINTILLATION PROXIMITY ASSAY

This example considers optimization of a scintillation proximity assay (SPA). These assays are often considered as an alternative to radioligand binding assays that are more amenable to high throughput and automation. It is not necessary to separate bound and free radioligand in an SPA assay. For this example, human α_{1B}-adrenoceptors were expressed in mammalian cells and the membranes were partially purified. Scintillation proximity assay beads with a wheatgerm agglutinin coating were used to bind the glycosylated proteins of the membranes containing the expressed receptors. A competition binding assay was performed where concentration–response curves were generated in response to 5-methyl-urapidil displacement of tritiated prazosin. For screening, compounds of different affinity for α_{1B}-adrenoceptors will be tested. 5-methylurapidil was chosen to use for the design in order to provide a full concentration–response curve.

The primary objective for this experimental design was to minimize variation in the K_i estimate. A secondary objective was to minimize the amount of receptor needed. Several responses were considered: variability of the K_i, signal-to-noise ratio, range of counts and adequacy of the non-linear regression fit to the concentration–response curve. For this example, the multiple response variables were considered separately.

Three factors were considered in the design: receptor concentration (40, 80 and 120 pM), SPA bead reagent concentration (0.5 and 2.0 mg/ml) and ligand concentration (0.448 and 1.792 nM). Because the number of factors and levels was small, it was possible to use a full-factorial design of 12 runs ($2 \times 2 \times 3$). Because variability in K_i was the primary response, the design was repeated three times to estimate the variance at each design point.

8.3.1.1 Variability

The results for variability are shown in Fig. 8.7. Each panel is obtained from the statistical analysis/visualization software package JMP®. The ordinate in each case is variability of the negative log of K_i (pK_i). The abscissa gives the factor levels used in the design. The vertical lines represent the values for particular factor settings that can be changed by the user. When the factors are changed, the plots update to show the effect on response predicted by the linear model, including interactions. The horizontal lines show the response obtained when all factors are set at their middle level. The error bars show the pooled standard deviation estimate for the model.

From the plots, it is seen that for the factor levels chosen there is a significant effect of receptor concentration (Fig. 8.7A). The highest receptor concentration

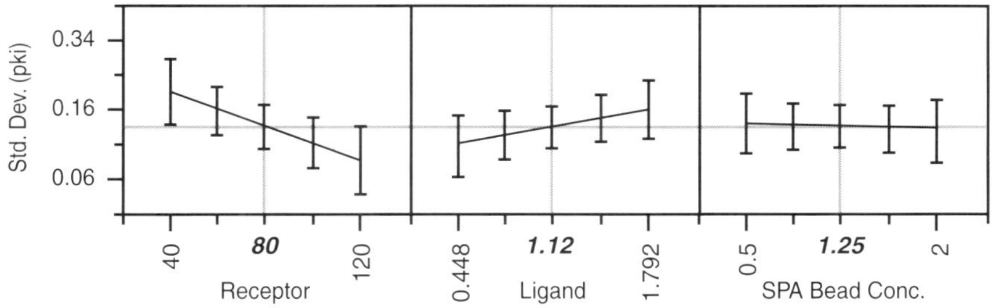

Fig. 8.7. Experimental design results for a full-factorial optimization of a scintillation proximity assay (SPA) to reduce variability. Factors are: receptor concentration (pM), ligand concentration (nM) and SPA reagent bead concentration (mg/ml). Abscissa: settings for each factor in the experimental design. Ordinate: variability of the pK_i (expressed in standard deviation units) calculated from three replicate measurements. (Reprinted from *Drug Discovery Today*, Vol. 1, Lutz MW, *et al*, Experimental design for high-throughput screening. Copyright 1996, pp. 277–285, with permission from Elsevier Science.)

minimizes variance. An inverse relationship is observed between ligand concentration and variability: the lowest ligand concentration has the smallest variance (Fig. 8.7B). The SPA bead reagent concentration does not have a significant effect on variability in this experiment (Fig. 8.7C). A contour plot (Fig. 8.8) is an alternative way of viewing these results. This plot is helpful in visualizing operating regions, particularly if many factor levels are used, but is limited by its bivariate nature.

8.3.1.2 Signal-to-noise Ratio

The optimal factor settings for signal-to-noise ratio (S/N) are slightly different from variability. The best S/N is obtained at the highest receptor concentration (Fig. 8.9A). Lowering ligand concentration (Fig. 8.9B) and SPA bead reagent concentration (Fig. 8.9C) also improves S/N.

8.3.1.3 pK_i Trend

Figure 8.10 and Table 8.2 show the trend of pK_i values for the different design points. It is seen that pK_i is not invariant with respect to receptor and ligand concentrations. Inspection of Table 8.2 shows that although most combinations of factor levels have a variance that is typical of this type of assay, the two rows with the lowest receptor concentration have unusually high variance.

8.3.1.4 Summary

The best conditions are low ligand and SPA bead reagent concentration and high receptor concentration. These conditions minimize variance and provide the best

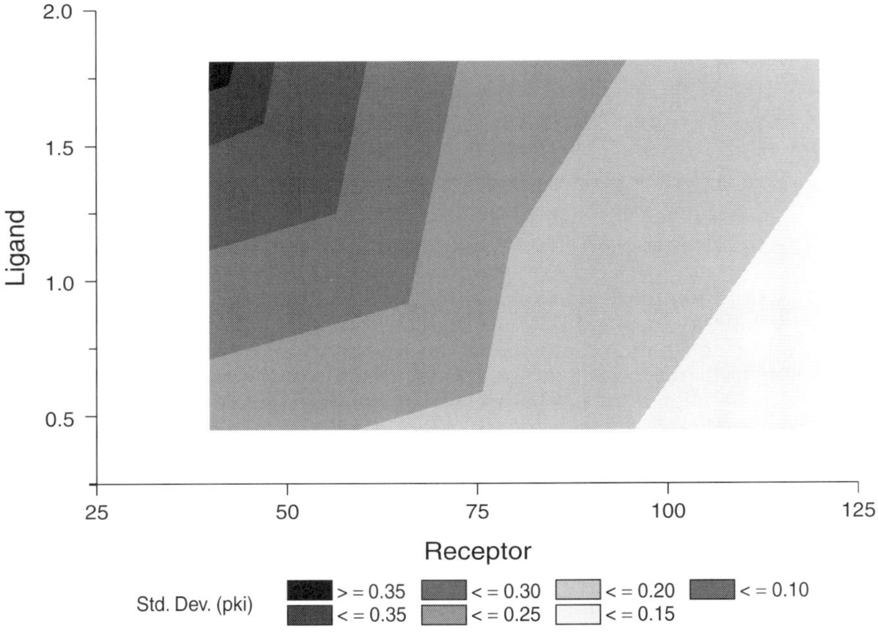

Fig. 8.8. Contour plot for experimental design results of a full-factorial optimizaton of a scintillation proximity assay to reduce variability. Abscissa: receptor concentration (pM). Ordinate: ligand concentration. Contours are variability of the pK_i (expressed in standard deviation units) calculated from three replicate measurements. (Reprinted from *Drug Discovery Today*, Vol. 1, Lutz MW, *et al*, Experimental design for high-throughput screening. Copyright 1996, pp. 277–285, with permission from Elsevier Science.)

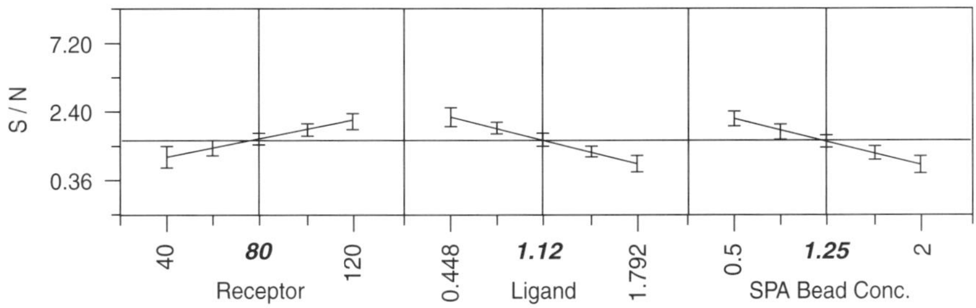

Fig. 8.9. Experimental design results for a full-factorial optimization of a scintillation proximity assay (SPA) to reduce variability. Factors are: receptor concentration (pM), ligand concentration (nM) and SPA reagent bead concentration (mg/ml). Abscissa: settings for each factor in the experimental design. Ordinate: signal-to-noise ratio. (Reprinted from *Drug Discovery Today*, Vol. 1, Lutz MW, *et al*, Experimental design for high-throughput screening. Copyright 1996, pp. 277–285, with permission from Elsevier Science.)

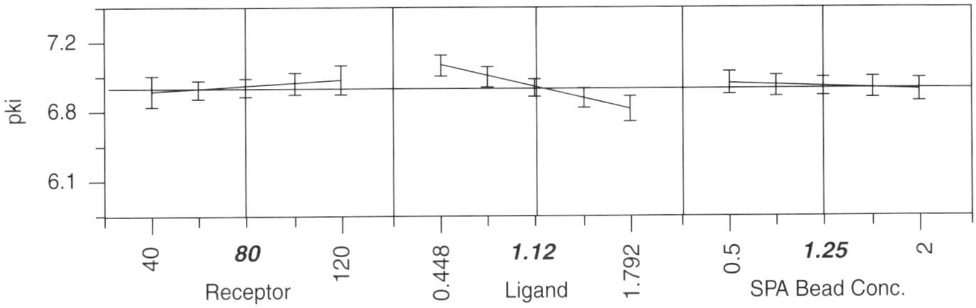

Fig. 8.10. Consistency of the pK_i estimates for a full-factorial optimization of a scintillation proximity assay (SPA) to reduce variability. Factors are: receptor concentration (pM), ligand concentration (nM) and SPA reagent bead concentration (mg/ml). Abscissa: settings for each factor in the experimental design. Ordinate: pK_i. (Reprinted from *Drug Discovery Today*, Vol. 1, Lutz MW, *et al*, Experimental design for high-throughput screening. Copyright 1996, pp. 277–285, with permission from Elsevier Science.)

Table 8.2. Trend of pK_i values for a full-factorial experimental design to optimize a scintillation proximity assay (SPA)

Receptor (pM)	Ligand (nM)	SPA concentration (mg/ml)	*n*	Mean (pK_i)	Standard deviation (pK_i)
40	0.448	0.5	3	6.99	0.17
40	0.448	2.0	3	6.96	0.08
40	1.792	0.5	3	6.72	0.34
40	1.792	2.0	3	6.38	0.31
80	0.448	0.5	3	7.03	0.13
80	0.448	2.0	3	6.89	0.19
80	1.792	0.5	3	6.71	0.17
80	1.792	2.0	3	6.81	0.12
120	0.448	0.5	3	6.93	0.06
120	0.448	2.0	3	7.07	0.10
120	1.792	0.5	3	6.72	0.12
120	1.792	2.0	3	6.68	0.10

Reprinted from *Drug Discovery Today*, Vol. 1, Lutz MW, *et al*, Experimental design for high-throughput screening. Copyright 1996, pp. 277–285, with permission from Elsevier Science.

S/N. If the amount of receptor was minimized, perhaps because quantities were limited, it would be desirable to use the lowest ligand concentration level. For the best S/N, the lowest concentration of SPA bead reagent should be used. Although data are not shown for a concentration–response curve fit, the higher SPA bead reagent concentrations appeared to give the best fits. These conclusions address the trade-off between the different factors. After completing the design and constructing a model, it is possible to be more quantitative about making decisions. For example, if lack of fit to the concentration–response curve data does not seem to be a problem, it might be possible to lower SPA bead reagent concentration in order to obtain a better signal.

8.3.2 OPTIMIZATION OF A PLATE READER FOR A HOMOGENEOUS TIME-RESOLVED FLUORESCENCE ASSAY

Biochemical assays generally involve a large number of potential factors to optimize. Three to thirty factors are the order of the number of factors optimized in non-automated assays. For automated robotic assays, if all systems are included there can be hundreds of candidate factors. Table 8.3 shows the factors used in a screening experimental design to optimize settings of a plate reader used for homogeneous time-resolved fluorescence assays. The response to be optimized is signal-to-background ratio (S/B): higher values of this response are preferred. The nine factors are all variables that control the time resolution of the reader. Optimization of the entire robotic system would involve additional variables for each stage (e.g. liquid handling, scheduling and plate handling).

The first design is a screening design involving all nine factors. A fractional factorial design is used that will estimate the main effects only. Six replicates are measured at each combination of factor settings. To obtain a general assessment of the significance of the factors, interpretation of the data starts with a screening model of the main effects in JMP®. A histogram of the response (Fig. 8.11)

Table 8.3. Factors to optimize in a screening experimental design to optimize time-resolution settings for a plate reader used in a homogeneous time-resolved fluorescence assay

Abbreviation	Interpretation
Capacitor Size	Discharge capacitor size for the flash lamp. Low = 10 nF, high = 2 × 10 nF
Flash Energy	Flash energy level: voltage setting of the flash lamp discharge capaitor
Max Exc Energy	Choice of two types of capacitors. This will define the maximum excitation energy
Light Integ Ref	Light integration reference level: excitation energy
Emm Aperture	Emission aperture: aperture size and shape (three levels)
Count Delay	Time between excitation pulse and start of emission pulse reading in microseconds
Count Window	Total counting time of the emission pulse reading in microseconds
Count Cycle	Time between excitation pulses in microseconds
Means Height	Height of the plate filled with the samples

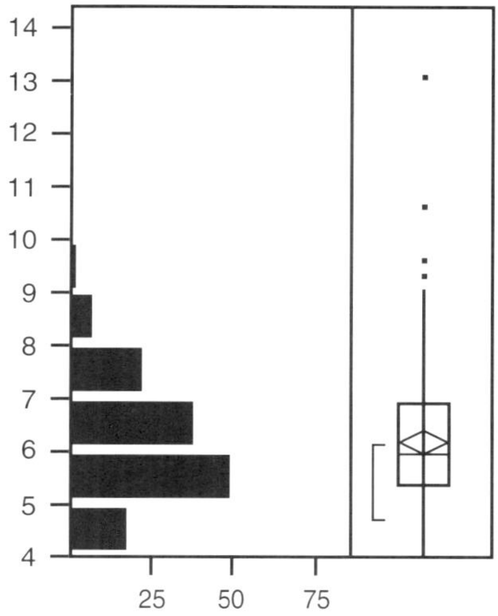

Quantiles

maximum	100.0%	13.20
	99.5%	13.20
	97.5%	9.51
	90.0%	7.85
quartile	75.0%	6.97
median	50.0%	6.00
quartile	25.0%	5.40
	10.0%	4.80
	2.5%	4.26
	0.5%	4.10
minimum	0.0%	4.10

Moments

Mean	6.26
Std. Dev.	1.31
Std. Err. Mean	0.11
Upper 95% Mean	6.47
Lower 95% Mean	6.04
N	144.00
Sum Weights	144.00

Fig. 8.11. Histogram of signal-to-background ratio for a screening experimental design to optimize time-resolution settings for a plate reader used in a homogeneous time-resolved fluorescence assay. Abscissa: signal-to-background ratio. Ordinate: count. (Data provided by Peter Leitner, Glaxo Wellcome.)

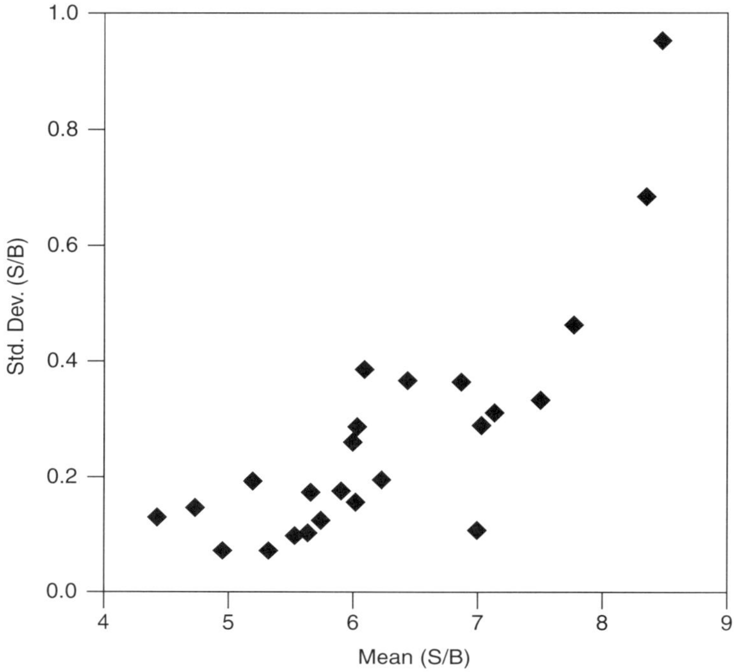

Fig. 8.12. Variability of signal-to-background ratio (S/B) as a function of the magnitude of S/B for the screening design. Abscissa: mean S/B. Ordinate: standard deviation of S/B

shows that S/B varied from approximately 4 to 13 in the different runs of the experimental design. Analysis of the six replicates indicates some caveats about interpreting the response. First, variability of S/B increases as S/B increases (Fig. 8.12). This will present difficulties for interpreting the design and residual plots. A histogram of the mean of the replicate values (Fig. 8.13) for the response is representative of the range of S/B that is covered by the settings of the factors in the screening design, e.g. 4–8. Analysis of the screening design is shown in Table 8.4 and Fig. 8.14. From Table 8.4 it is seen that the root-mean-square (RMS) error is approximately 0.91. This is an estimate of the overall variability, although the number varies as a function of the magnitude of S/B (Fig. 8.12). The ANOVA is significant, showing that the experimental design model explains a significant portion of the variation but that the fit is poor, as evidenced by the low R^2 (0.6) and the significant lack of fit ($P = 0.003$). Significant lack of fit can result from a variety of sources. If important factors or interactions are not included in the model then there can be lack of fit. The standardized residual plots for this model show a strong dependency of the residual on the magnitude

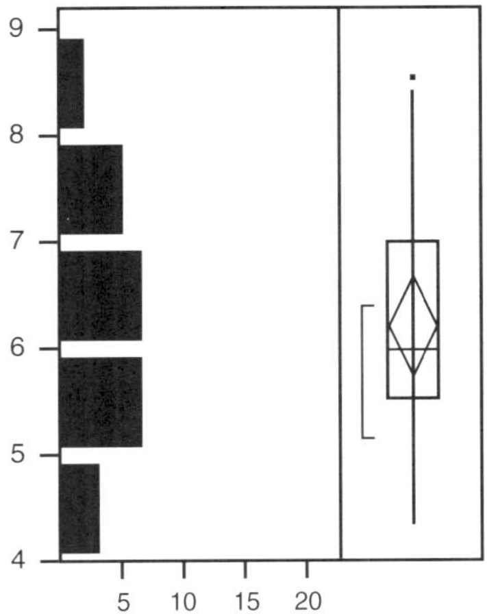

Quantiles		
maximum	100.0%	8.48
	99.5%	8.48
	97.5%	8.48
	90.0%	8.07
quartile	75.0%	7.04
median	50.0%	6.03
quartile	25.0%	5.57
	10.0%	4.85
	2.5%	4.42
	0.5%	4.42
minimum	0.0%	4.42

Moments	
Mean	6.26
Std. Dev.	1.08
Std. Err. Mean	0.22
Upper 95% Mean	6.71
Lower 95% Mean	5.80
N	24.00
Sum Weights	24.00

Fig. 8.13. Histogram of mean ($n = 6$ replicates) signal-to-background ratio measurements for a screening experimental design to optimize time-resolution settings for a plate reader used in a homogeneous time-resolved fluorescence assay. Abscissa: signal-to-background ratio. Ordinate: count

of S/B (Fig. 8.15A) and stratification of the residual values with respect to some of the factors in the model (Fig. 8.15B). There are several interpretations of these residuals plots. First, the relationship between variability of S/B and the magnitude of S/B was recognized, explaining part of the linear relationship between the variables, particularly in Fig. 8.15A. However, the plots also indicate that the

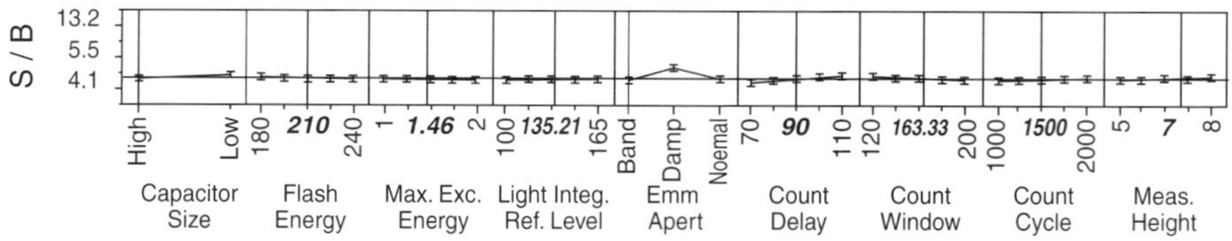

Fig. 8.14. Experimental design results for a screening experimental design to optimize time-resolution settings for a plate reader used in a homogeneous time-resolved fluorescence assay. Factors are listed in Table 8.3. Abscissa: settings for each factor in the experimental design. Ordinate: signal-to-background ratio. (Data provided by Peter Leitner, Glaxo Wellcome.)

predictive ability of the model decreases as S/B increases, suggesting that other factors or interactions must be considered.

Inspection of the table of effect tests (Table 8.4) shows that four factors are highly significant: one is borderline and four are not significant using an arbitrary threshold of 0.05. However, the lack of interactions in the model and caveats about the variability of the response require caution in the interpretation of these results. The pictorial representation of the effects (Fig. 8.14) reinforces the fact that changes in S/B for this design over the range of all of the factors is small. Emission aperture is identified as a significant factor because the error bars do not overlap and a setting of damped improves S/B. Count delay and count window are both significant factors by inspection of Table 8.4 and, from the plots, appear to have optimal settings in opposite directions of magnitude. For the best S/B, count delay should be set at larger values. Smaller values of count window are associated with a better S/B.

Interpretation of the screening design suggests that it is possible to reduce the number of factors in subsequent designs and to propose some optimal settings. Because of the large number of factors and levels, the initial design choice precludes investigation of interactions. The analysis used must match the design, so it is not possible to reanalyze the data with a different model to test for interactions. However, combinations of factors can be examined to determine if the data suggest factors for the next design. Because interactions are suspected, possibilities for the next design include a smaller factorial screening design or perhaps several designs to investigate interactions between the factors. The relationship between count delay and count window is important, both from an understanding of the physics of the system and from the observation in the screening design that they have a significant effect on S/B. Therefore, a small factorial design to investigate these factors is chosen as the next step.

Because the objective of the second design is to evaluate the interaction between two factors, a full-factorial design is constructed. Three levels are chosen for count window and six levels for count delay, in order to allow the response surfaces to be visualized on the contour plots and to correspond to the resolution

Table 8.4. Experimental design results for a screening experimental design to optimize time-resolution settings for a plate reader used in a homogeneous time-resolved fluorescence assay

Summary					
R^2	0.55				
R^2 (adjusted)	0.52				
Root mean square (RMS)	0.91				
Mean of response	6.26				
Number of observations	144.00				
Analysis of variance	**Degrees of freedom**	**Sum of squares**	**Mean square**		
Model	10	136	13.60		
Error	133	111	0.84		
Corrected total	143	247			
F ratio	16.28				
Prob > F	<0.0001				
Lack-of-fit test	**Degrees of freedom**	**Sum of square**	**Mean square**		
Lack of Fit	13	24.57	1.89		
Pure error	120	86.54	0.72		
Total error	133	111.11			
F ratio	2.620				
Prob > F	0.003				
Max. R^2	0.650				
Effect tests	**Degrees of freedom**	**Sum of square**	**Mean square**	**F ratio**	**Prob > F**
Capacitor size	1	11.32	11.32	13.56	0
Flash energy	1	1.68	1.68	2.01	0.16
Max. excitation energy	1	1.68	1.68	2.01	0.16
Light integration reference level	1	1.08	1.08	1.29	0.26
Emission aperture	2	57.57	28.79	33.86	0
Count delay	1	23.50	23.50	30.35	0
Count window	1	6.60	6.60	7.90	0.01
Count cycle	1	1.51	1.51	1.80	0.18
Measurement height	1	2.87	2.87	3.43	0.07

Data provided by Peter Leitner, Glaxo Wellcome.

of these variables. The full-factorial design for this experiment will have 18 runs. As with the screening design, it is replicated six times. Replication of the design is relatively low cost because settings on the plate reader itself are being optimized. Replication involves reading a plate under different conditions. If other elements of the robotic system were included in the design, a more judicious choice of replication would be needed (e.g. center points rather than the entire design).

The full design and resulting means and standard deviations for S/B are shown in Table 8.5. The table is sorted by S/B with optimal values at the top of the

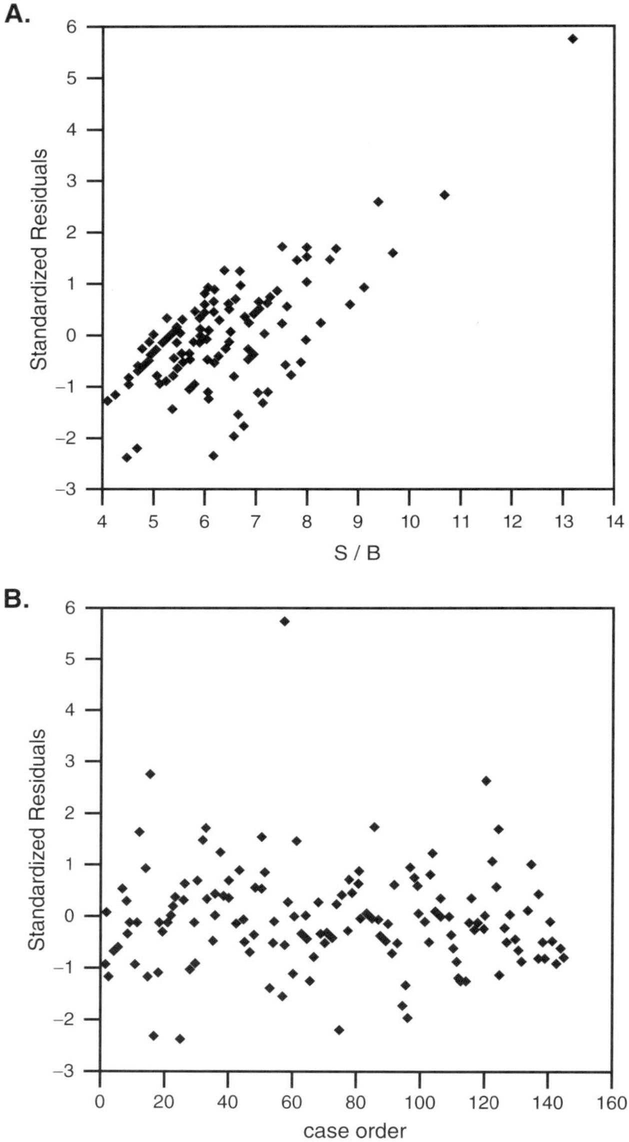

Fig. 8.15. Residual plots for a screening experimental design to optimize time-resolution settings for a plate reader used in a homogeneous time-resolved fluorescence assay. (A) Abscissa: signal-to-background ratio. Ordinate: standardized residual value. (B) Abscissa: case order. Ordinate: standardized residual value

Table 8.5. Experimental design results for a full-factorial experimental design to optimize time-resolution settings for a plate reader used in a homogeneous time-resolved fluorescence assay

Count window	Count delay	Mean (S/B) ($n = 6$)	Standard deviation ($n = 6$)
800	75	31.65	3.37
800	50	30.88	1.45
800	100	30.73	4.68
400	125	30.47	5.98
400	50	28.87	2.83
400	100	28.62	7.73
800	125	27.05	6.14
400	175	26.48	5.42
400	150	25.13	4.46
600	50	24.65	4.50
800	175	24.62	3.80
400	75	24.50	3.77
600	125	24.20	4.78
600	75	23.30	2.40
600	150	22.53	3.81
800	150	22.50	2.58
600	100	22.08	3.45
600	175	20.60	1.57

Summary					
R^2	0.13				
R^2 adjusted	0.10				
Root mean square	4.93				
Mean of response	26.05				
Number of observations	108.00				
Analysis of variance	**Degrees of freedom**	**Sum of squares**	**Mean square**		
Model	3	368.98	122.99		
Error	104	2528.35	24.31		
Corrected total	107	2897.33			
F ratio	5.06				
Prob > F	<0.0026				
Lack-of-fit test	**Degrees of freedom**	**Sum of squares**	**Mean square**		
Lack of fit	14	834.14	58.58		
Pure error	90	1694.21	18.82		
Total error	104	2528.35			
F ratio	3.17				
Prob > F	0.00				
Max. R^2	0.42				
Effect tests	**Degrees of freedom**	**Sum of squares**	**Mean square**	**F ratio**	**Prob > F**
Count window	1	128.92	128.92	5.3	0.02
Count delay	1	46.30	46.30	1.9	0.17
Count window × count delay	1	126.36	126.36	5.2	0.02

S/B, signal-to-background ratio.

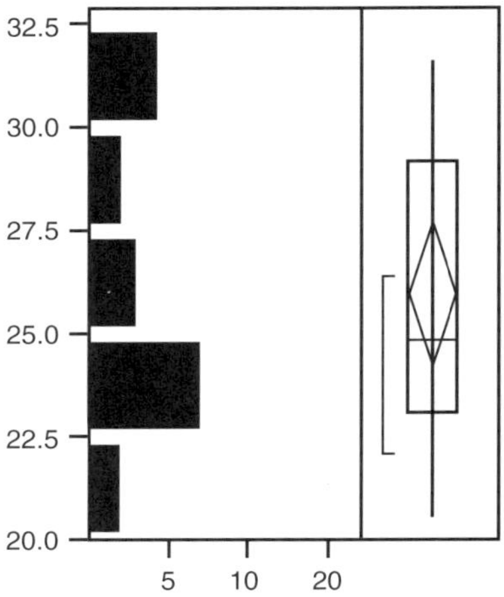

Quantiles

maximum	100.0%	31.65
	99.5%	31.65
	97.5%	31.65
	90.0%	30.96
quartile	75.0%	29.27
median	50.0%	24.89
quartile	25.0%	23.11
	10.0%	21.94
	2.5%	20.60
	0.5%	20.60
minimum	0.0%	20.60

Moments

Mean	26.05
Std. Dev.	3.43
Std. Err. Mean	0.81
Upper 95% Mean	27.76
Lower 95% Mean	24.34
N	18.00
Sum Weights	18.00

Fig. 8.16. Histogram of signal-to-background ratio for a full-factorial experimental design to optimize time-resolution settings for a plate reader used in a homogeneous time-resolved fluorescence assay. Abscissa: signal-to-background ratio. Ordinate: count

table. The histogram of S/B for this experiment is shown in Fig. 8.16. The S/B values range from approximately 20 to 32 in this experiment, with RMS = 4.93. As with the screening design, the variation in S/B appears to be a function of the magnitude of S/B (Fig. 8.17), although there are some higher values of S/B that have lower variation than would be predicted. These are points that are

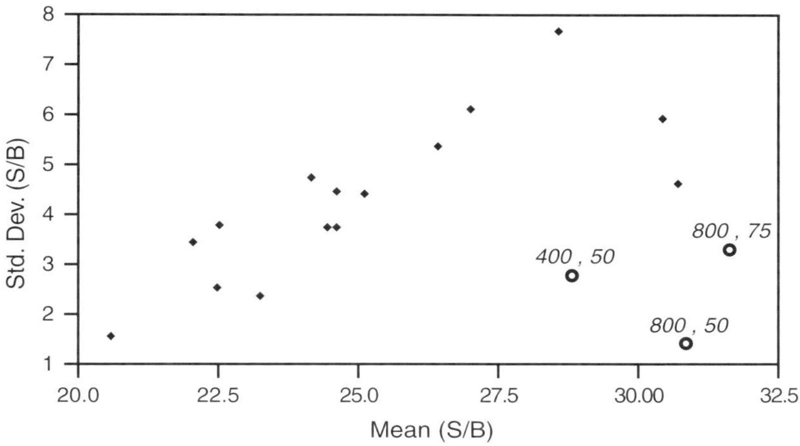

Fig. 8.17. Variability of signal-to-background ratio (S/B) as a function of the magnitude of the mean S/B for the full-factorial design. Abscissa: mean S/B. Ordinate: standard deviation of S/B. The labeled points show count window and count delay values, respectively

particularly interesting in that they have both optimal S/B and low variation: they are labeled on the plot with corresponding values of count delay and count window.

Analyses of the results of the design are shown in Fig. 8.18 and Table 8.5. There is significant lack of fit in this model. From the effect tests it is seen that count window and the interaction of count window and count delay are both significant (Fig. 8.18). Count delay as a main effect is not significant at the 0.05 level. The most important results are seen in the effect and interaction plots. With count window set at the highest level (800), lower values of count window are associated with higher S/B. Inspection of the interaction plots shows the same result: if count window is set high (700, 800), the count delay should be set low for a higher S/B, this result is demonstrated by the family of intersecting lines. If the count window is set low (400) then the setting of count window is unimportant or slightly favors a higher setting (Fig. 8.19). A contour plot of these factors is very useful for choosing optimal settings (Fig. 8.20). From the plot it is seen that at the higher settings of count window, there is an ideal region for optimal S/B associated with low settings of count delay. There is a very small region at low settings of count window with an intermediate setting (125) of count delay that has a high value of S/B.

Inspection of the residual plots (Fig. 8.21) shows some of the same problems encountered in the screening design. Although the case order plot (Fig. 8.21B) appears to be more random, there are still patterns in the standardized residuals,

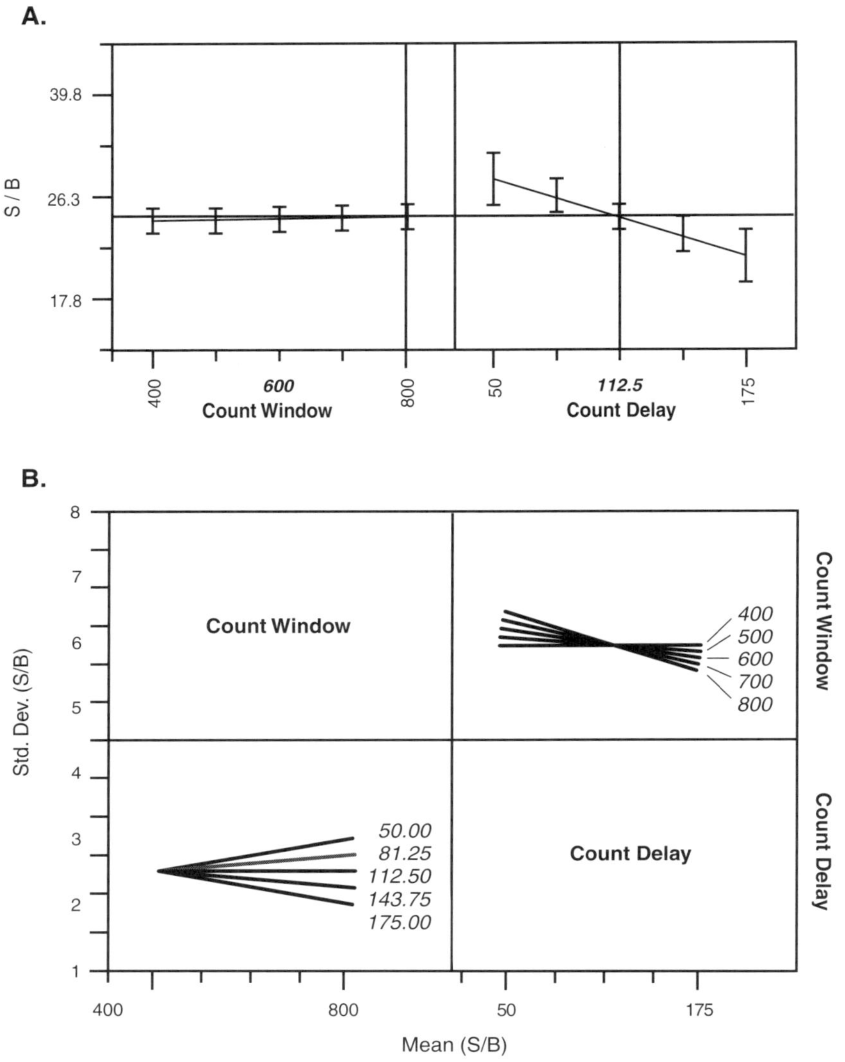

Fig. 8.18. Experimental design results for a full-factorial experimental design to optimize time-resolution settings for a plate reader used in a homogeneous time-resolved fluorescence assay. Factors are count window, count delay and their interaction. (A) Plots showing the effects of the two factors. Abscissa: settings for each factor in the experimental design. Ordinate: signal-to-background ratio. (B) Interaction plots. Abscissa: settings for the two factors. Ordinate: signal-to-background ratio. The curves are identified by the associated levels of the second (interacting) factor. (Data provided by Peter Leitner, Glaxo Wellcome.)

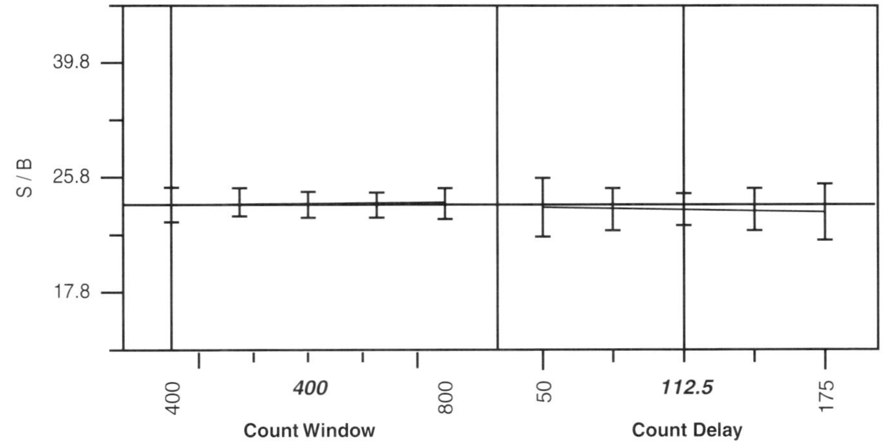

Fig. 8.19. Experimental design results for a full-factorial experimental design to optimize time-resolution settings for a plate reader used in a homogeneous time-resolved fluorescence assay. In this case, count window is set at the lowest level and count delay is at the mean. These plots should be contrasted to the plots of Fig. 8.18A. Abscissa: settings for each factor in the experimental design. Ordinate: signal-to-background ratio

and the plot of standardized residuals versus S/B (Fig. 8.21A) shows a strong linear relationship. The significant lack of fit is also problematic.

This example of optimization of the plate reader is typical of complex experimental designs that involve large numbers of factors and levels. One design rarely suffices to optimize the process and often the designs reveal other factors and variables that must be considered. Both the screening design and the second response surface design provided valuable information. From the screening design, several factors could be set on the plate reader, e.g. setting the emission aperture at 'damped' is reasonable. The second design points toward optimal settings of count delay at 50 and count window at 800. This design provided the important result that variability of S/B is low at these settings, as well as optimal. This is an important point, considering that these points are exceptions to the general observation that variability of S/B increases as a function of its magnitude. The next steps for this design work require balancing between refinement of the model and the amount of work required to generate the data needed for the model. The S/B is only one response to optimize, and if there are others (e.g. variability of S/B or magnitude of signal) then they must be considered. More interactions can be examined in pairs or a design involving several interactions can be constructed. Most importantly, the lack of fit of the model and the patterns in the residual plots require further investigation. The response variable S/B has some potential problems associated with it. Because readings from several areas on the plate

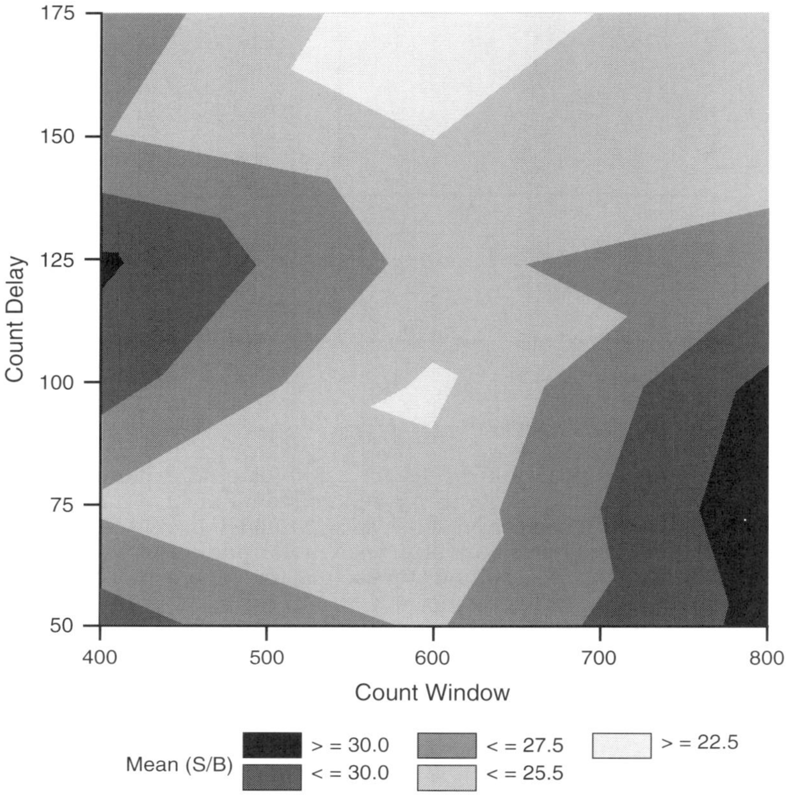

Fig. 8.20. Contour plot for experimental design results of the full-factorial experimental design to optimize time-resolution settings for a plate reader used in a homogeneous time-resolved fluorescence assay. Abscissa: count window (μs). Ordinate: count delay (μs). Contours are signal-to-background ratio; larger numbers are optimal

are associated with two variables — signal (S) and background (B) — systematic plate effects can cause problems such as those observed in the residual plots, or the model may have to be refined to include a covariate based on either signal or background.

8.3.3 OPTIMAL SAMPLING AND DATA ANALYSIS PROTOCOL FOR A RADIOLIGAND BINDING ASSAY

The choice of a sampling protocol for assays that involve non-linear curve fitting, such as radioligand binding experiments, is important to ensure the quality and reproducibility of the resulting K_i estimates or the precision of interpolated

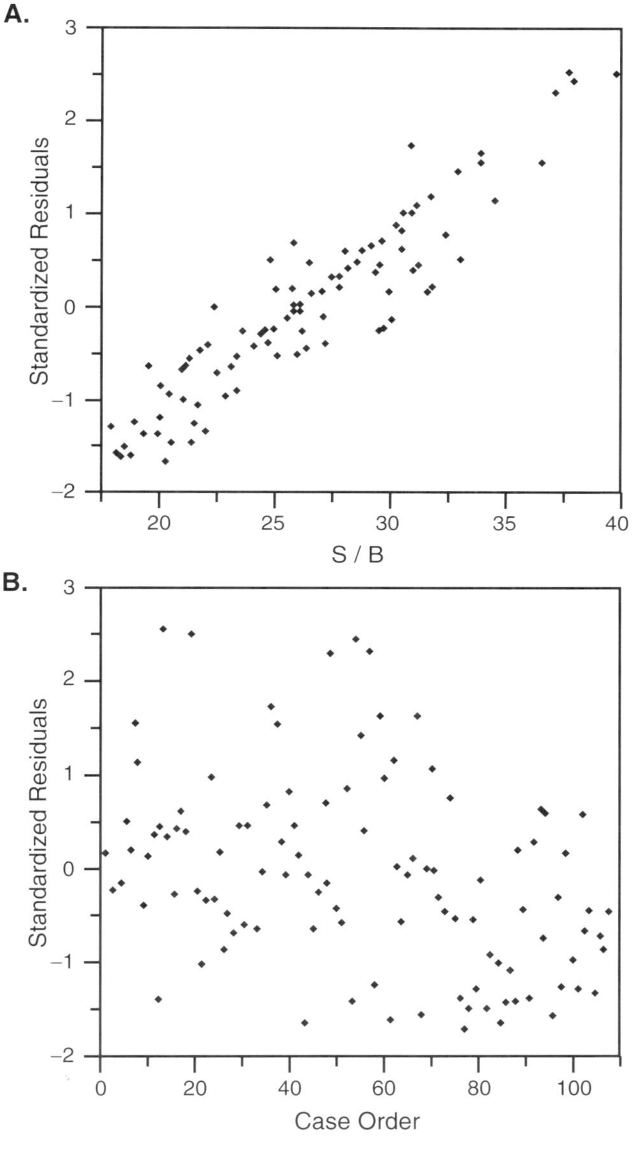

Fig. 8.21. Residual plots for a full-factorial experimental design to optimize time-resolution settings for a plate reader used in a homogeneous time-resolved fluorescence assay. (A) Abscissa: signal-to-background ratio. Ordinate: standardized residual value. (B) Abscissa: case order. Ordinate: standardized residual value

concentrations if the experiment is an a radioimmunoassay or SPA. The objective is to sample a sufficient number of concentrations over a suitable range to define the location and scaling parameters of the curve (IC_{50}; exponent if used; upper and lower asymptotes). Each point can also be replicated. Munson and Robard (1980) suggest that if the within-sample variance is smaller than that between samples (response at different concentrations), then the mean value of replicates should be used, otherwise all replicates should be used.

Figure 8.22 shows the experimental design used to address the question of optimal sampling and fitting of a concentration–response curve. The analysis is different from traditional experimental designs where experimental runs are performed to collect data after the design points are selected. Because we had a large collection of concentration–response curves for a variety of different compounds, we were able to do a retrospective analysis by sampling these curves in a Monte-Carlo simulation experiment.

The data were obtained from competition binding curves in a radioligand binding assay for three subtypes of α_1-adrenoceptors (Goetz *et al* 1994). The

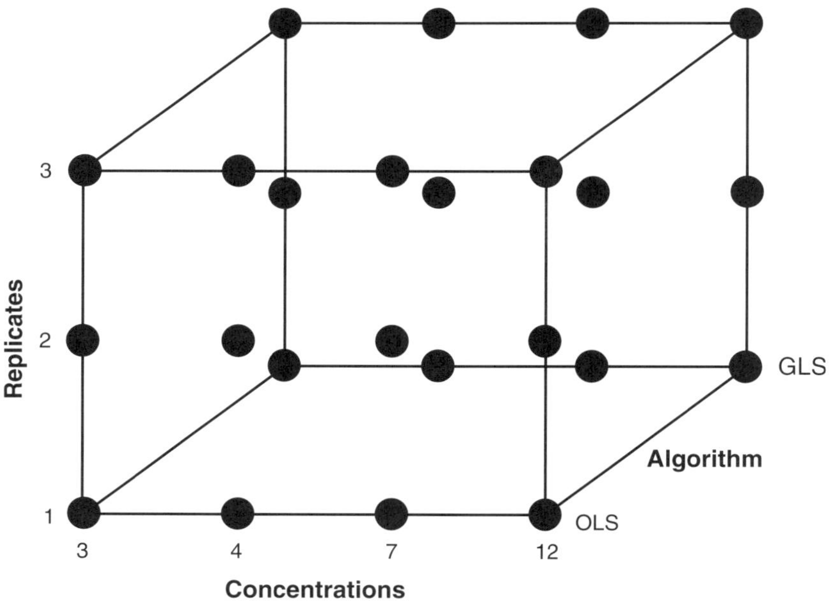

Fig. 8.22. Schematic of the experimental design for choice of a sampling and data analysis protocol for a radioligand binding assay. The three factors form the faces of the cube and list the factor levels. OLS = ordinary least-squares algorithm; GLS = generalized least; squares algorithm. (Reprinted from *Drug Discovery Today*, Vol. 1, Lutz MW, *et al*, Experimental design for high-throughput screening. Copyright 1996, pp. 277–285, with permission from Elsevier Science.)

receptors were human clones expressed in mammalian cells. In all cases, the radioligand was ^{125}I-HEAT, i.e. (±)-b-([^{125}I]iodo-4-hydroxyphenyl)ethylamino-methyltetralone. Roughly 200 inhibitors were chosen at random, covering a potency range of six log units. Each concentration–response curve was constructed by resampling the data for the number of concentrations and replicates at the design point (Efron and Tibshirani 1991, 1993; Lutz *et al* 1995). This approach allowed us to pool results across different drugs.

The objective of this experiment was to simulate calculations from a calibration curve. Three factors were considered: number of concentrations, number of replicates and choice of non-linear regression algorithm. The non-linear regression algorithms were either unweighted ordinary least-squares or a generalized least-squares approach that modeled the variance of the response by a power function (Carroll and Ruppert 1982; Giltinan and Ruppert 1989). Ordinate values were chosen at random within the range of values associated with a particular curve. These values were used to calculate concentrations and were compared to the true values (known from making the same calculation using all of the data available for the curve). The response used in the design is the standard deviation of the difference between the true and observed values for at least 100 simulations.

The results are presented in a cross-classification table (Table 8.6). The values are in standard deviation (SD) units. Differences of 5 SD are not statistically significant ($P > 0.05$); differences of 10 SD are marginally significant. In a large portion of the table there is statistically not much difference. In other words, 12 single doses give results that are as consistent as 12 doses in triplicate. Replication has less effect on reproducibility than the number of concentrations tested in this set of simulations. The results suggest that 12 single concentrations or 7 concentrations in duplicate appear to be optimal. The generalized least-squares algorithm appears to perform better than the ordinary least-squares algorithm for all of the combinations of replicates and number of doses, except in the cases where the curve is severely undersampled (lower right corner of table).

Table 8.6. Results of an experimental design to choose a sampling and data analysis protocol for a radioligand binding assay

No. of concentration doses	OLS replicates[1]			GLS replicates[2]		
	3	2	1	3	2	1
12	44	45	45	28	30	31
7	49	55	59	29	33	37
4	67	67	—	37	45	—
3	273	275	—	383	379	—

[1] OLS, ordinary least-squares algorithm.
[2] GLS, generalized least-squares algorithm.
Reprinted from *Drug Discovery Today*, Vol. 1, Lutz, MW, *et al*, Experimental design for high-throughput screening. Copyright 1996, pp. 277–285, with permission from Elsevier Science.

These results make intuitive sense. In curve fitting, points are necessary to define the asymptotes and the region around the IC_{50}. Replication is less important if there are neighboring points to influence the fit. The superiority of the variance function model also is reasonable: the variance of the binding data is noticeably a function of signal. The combination of a simple experimental design with a Monte-Carlo simulation in this case illustrates an effective method for determining an optimal sampling scheme for the most consistent results that minimizes the use of resources.

8.3.4 AUTOMATED EXPERIMENTAL DESIGN

Although robotic assay systems greatly increase the complexity of experimental designs, the execution of the design can be faster if factor settings can be changed under software control. Experimental designs are constructed using statistical software and the resulting design is converted into a series of instructions for the robot. The levels of the design are set at either discrete (high/low) points or are chosen by mapping continuous variables (such as gaps and arm travel speed) into bins. The design is a map for a number of different runs of the experiment, therefore it must provide a map both for the scheduling software and for the individual components: the liquid-handling system, the robot arm if used, the reader (counter or other device), water baths, incubators, shakers and other instruments. Automating the experimental design is facilitated if a scripting language is used to control the different components, e.g. map from a high-level scripting instruction (move plate, incubate for 30 min) to the low-level instructions associated with the different components. The scripting language handles the translation into many possible different communication protocols used by the different components. The liquid-handling system is a good first component to start with because many of the factors and variability of the assay are associated with reagent dispensing. Automated experimental design is well suited for the sequential design approach. A screening design is run and the results are used to create the script for a more detailed design on the most important factors.

8.4 EXPERIMENTAL DESIGN FOR CHEMISTRY

In the biochemistry applications of experimental design the objectives of the design are to optimize an assay or process (reduce variability, increase signal, minimize use of reagents) or to model the process with the intent of choosing operating conditions (understand main effects and interactions in order to design a robust assay). The concepts of experimental design that are used in biochemistry also apply in chemistry. Challenges for using experimental design in chemistry include the realization that for many optimization problems the factor levels are choices of substituents or building blocks for molecules. These substituents must be characterized, which introduces the complexity of many different methods to characterize chemical structures: size, shape, topology, other descriptors and physical properties. Spaces defined by these descriptors are complex in general, but may be more tractable to use for homologous series of compounds. Proposed experimental designs may not be synthetically feasible. Combinations of factor settings may give reaction conditions that do not work. However, experimental design is useful in chemistry and some simple designs that can be crafted without computer software can maximize the amount of information obtained from a series of molecules.

Experimental design is a natural tool to use for the design of combinatorial libraries (Gallop *et al* 1994; Gordon *et al* 1994). For these applications, the factors are either substituents or collections of monomers attached at a particular site on a molecule. The design process selects a set of compounds for synthesis that will allow interpretation of the chemical properties' space using a linear model. From this model, specific questions can be answered about the effect of a set of substituents at a specific site and interactions between sites. Young and Hawkins (1995) published an analysis of a full-factorial combinatorial peptide library designed to study the stereochemical binding properties of NK1 receptors (Wang *et al* 1993). Their work illustrates the use of analytical and visualization tools to analyze data from combinatorial libraries.

8.4.1 EXPERIMENTAL DESIGN FOR REACTION OPTIMIZATION

Optimization of chemical reactions is an ideal application for experimental design. Chemical reaction descriptions identify the factors that are varied, e.g. reactants and associated concentrations, temperature, catalysts and concentrations. Often, information is available about operating ranges for these factors from previous work that facilitates setting the levels for the factors in the experimental design. The scale of application of experimental designs used for reaction optimization is diverse. Some designs are used to optimize library chemistry in order to maximize the probability that the reaction will work for a variety of monomers. Experimental designs are used in scale-up chemistry, production and chemical engineering to ensure the maximum yield of these large-scale reactions. Finally, designs can be useful for determining reaction conditions for a single molecule that has proven to be difficult to make from literature protocols. Carlson (1992) provides specific guidelines for experimental design in chemistry and tables of data that are useful for reaction optimization designs.

The responses used in reaction optimization usually are reaction yield, minimization of side-products, minimization of reactant quantity or simply a binary indicator of whether a reaction worked. Factors include reactant concentrations and reaction conditions such as temperature, pressure and time. Response surface designs are very useful for reaction optimization because they characterize response as a function of the different factors. If the reaction is well characterized, the response surface design is often the first design, bypassing the

Table 8.7. Factors to optimize in a response surface experimental design to optimize yield in a chemical reaction

Abbreviation	Interpretation
*R*2/*R*1	Ratio of reactant 2 concentration to reactant 1 concentration
*Catalyst/R*1	Ratio of catalyst concentration to reactant 1 concentration
Temperature	Temperature (in °C)

large screening designs necessary to identify the factors to use in response surface modeling.

Table 8.7 lists the factors used in a simple experimental design for reaction optimization. Although this example is an oversimplification of the actual synthesis, it illustrates the principles of using experimental design for reaction optimization. The reaction involves two reactants and a catalyst. The experimental objective is to maximize yield. To provide a more meaningful mapping between the design and the synthesis, the ratio of the reactant concentrations (*R*2/*R*1) and the ratio of catalyst concentration to reactant *R*1 are used as factors rather than the individual reactant concentrations. Combinations of factors can be used in experimental designs if they make the implementation easier. Temperature is included as the final factor in the design. Because each run is a separate, time-consuming synthesis, it is desirable to keep the number of runs to less than 30. However, curvature in response as a function of the factors is possible and an estimate of response variability is needed. A design is needed that supports response surface modeling for relatively few factors (three), few interactions (three two-factor and one three-factor) and possible curvature. For this example, a central-composite design is chosen. In addition to the high/low/mean levels used in factorial designs, this design includes two axial or star points (additional levels) for each factor to define better the curvature. Nine center points are included to provide an estimate of variability.

The results of the experiment are shown in Table 8.8 and Fig. 8.23. From the nine center points, estimates of mean yield (55%) and standard deviation (2.8) are obtained. Prior experiments and practical considerations suggest that differences in yield greater than 10% are of significance in terms of changing the synthesis. The prediction profiles or interactive plots (Fig. 8.23) give a summary of the effect of the factors on response. Generally, yield increases as *R*2/*R*1 increases and as *Catalyst/R*1 increases, although there is some curvature in the *Catalyst/R*1 plot. Curvature is very marked in the temperature plot: the lower and upper temperatures are associated with higher yields. The fact that the error bars are smaller for the mean levels is somewhat of an artifact because there were nine observations at the mean of all factor levels. The very large error bars on the extreme factor settings show that yield varies over a wide range at these levels, perhaps as the result of an interaction. Because the observations made at

Table 8.8. Experimental design results for a response surface experimental design to optimize yield in a chemical reaction

Summary					
R^2	0.48				
R^2 (adjusted)	0.13				
Root mean square (RMS)	9.00				
Mean of response	60.83				
Number of observations	23.00				
Analysis of variance	**Degrees of freedom**	**Sum of squares**	**Mean square**		
Model	9	984.65	109.41		
Error	13	1052.65	80.97		
Corrected total	22	2037.30			
F ratio	1.35				
Prob > F	0.30				
Effect tests	**Degrees of freedom**	**Sum of squares**	**Mean square**	**F ratio**	**Prob > F**
*R2/R*1	1	0.34	0.00	0.00	0.95
Catalyst/R1	1	6.10	6.10	0.08	0.79
Temperature	1	26.15	26.15	0.32	0.58
*R2/R*1 × *R2/R*1	1	0.43	0.43	0.01	0.94
Catalyst × *R2/R*1 × *R2/R*1	1	312.50	312.50	3.86	0.07
Catalyst/R_1 × *Catalyst/R1*	1	11.16	11.16	0.14	0.72
Temperature × *R2/R*1	1	8.00	8.00	0.10	0.76
Temperature × *Catalyst/R1*	1	8.00	8.00	0.10	0.76
Temperature × Temperature	1	24.3	24.3	0.30	0.59

these levels include various settings of the other factors, it is important to look at the individual observations.

From the effect table (Table 8.8), interpretation of the data becomes clearer. Using standard threshold P values (e.g. $P < 0.05$) implies that none of the effects are significant. However, the interaction between *R2/R*1 and *Catalyst/R*1 is the most significant (at 0.07) and the F ratio of 4 suggests that it is an effect that should be examined in more detail. Examination of the interaction plots supports the interpretation of a significant interaction. The interaction plots of either *R2/R*1 or *Catalyst/R*1 with temperature are a series of parallel curves. The interaction plot of *R2/R*1 and *Catalyst/R*1 shows curves that intersect and several different profiles of yield, depending on the settings for these factors. A contour plot of these variables (Fig. 8.24) gives the clearest picture of the interaction. There is a small region where both *R2/R*1 and *Catalyst/R*1 are at their highest settings, which gives the largest yield. However, over most of the region defined by the settings of the factors for this experiment, yield is consistently under 60%. The model does not provide a particularly good fit to the data (Table 8.8). The R^2

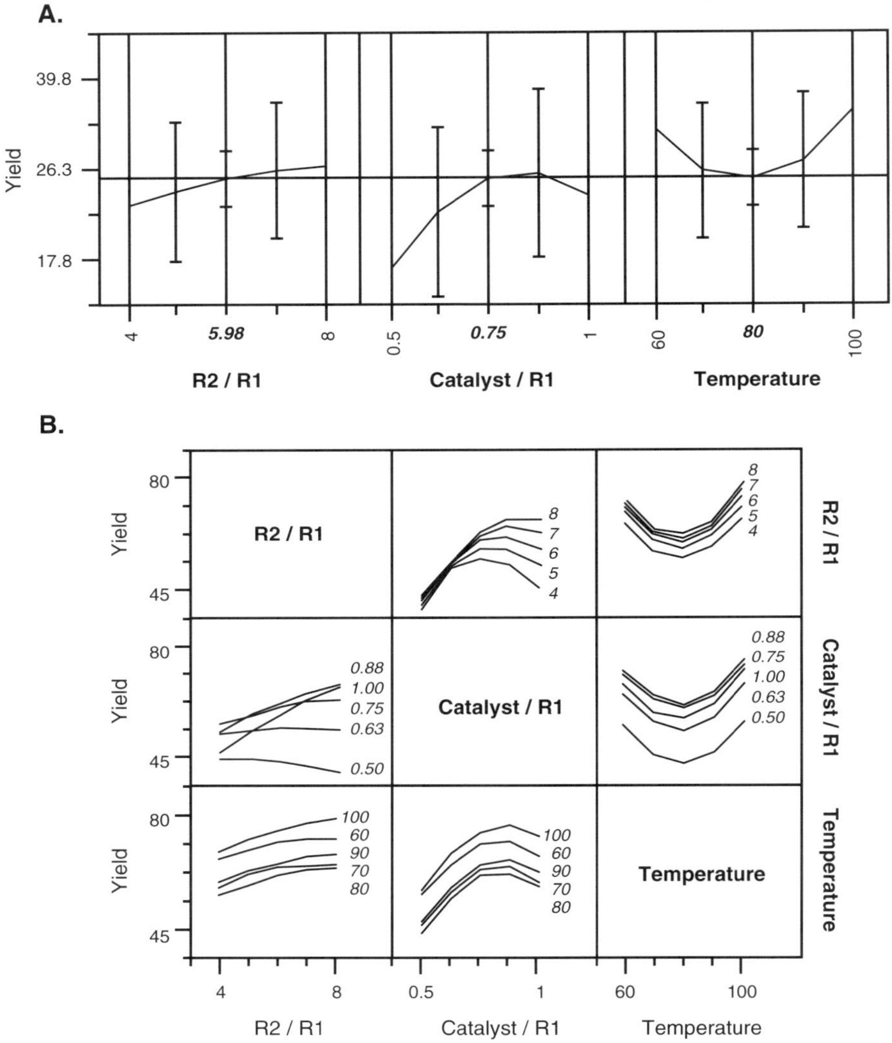

Fig. 8.23. Experimental design results for a response surface experimental design to optimize yield in a chemical reaction. Factors are listed in Table 8.7. (A) Plots showing the effects of the three factors. Abscissa: settings for each factor in the experimental design. Ordinate: yield (%). (B) Interaction plots. Each cell in the matrix of plots shows the interaction between two factors as listed at the right and bottom of the plot. Abscissa: settings for each of the three factors. Ordinate: yield (%). The curves are identified by the associated levels of the second (interacting) factor

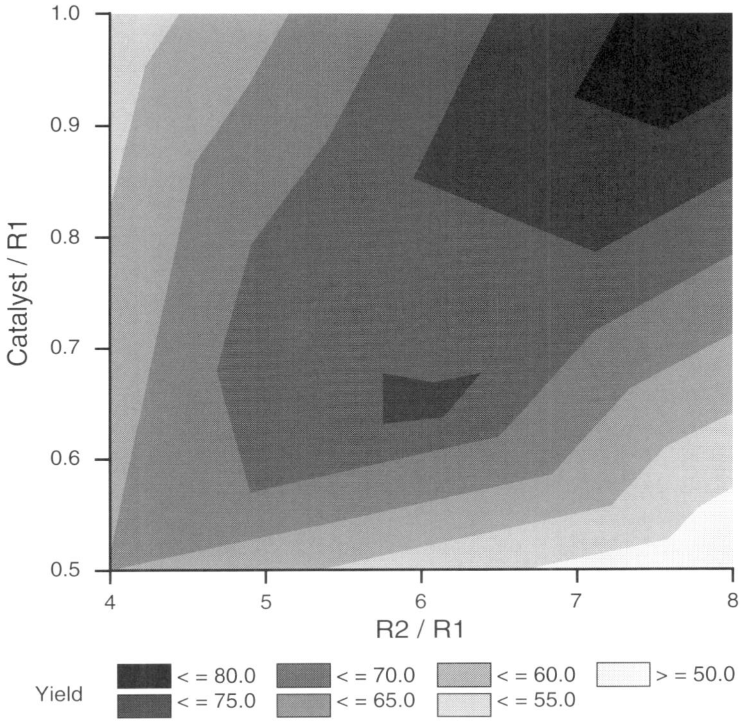

Fig. 8.24. Contour plot for a response surface experimental design to optimize yield in a chemical reaction. Abscissa: ratio of second reactant to first. Ordinate: ratio of catalyst to first reactant. Contours are yield (%); larger number are optimal

value is a low 0.48 and the RMS value of 9 is considerably larger than the standard deviation of the replicate measurements. The analysis of variance table (Table 8.5) shows that the error mean square is large relative to the model mean square.

For optimal yield, both *R*2/*R*1 and *Catalyst*/*R*1 should be set high. This may seem like an anticlimactic conclusion to the design, however the strength of the interaction between these factors is important. From interpretation of this design, temperature could be set at any level. The single observation at a temperature of 90°C (an axial point) was associated with a fairly high yield of 76. Some additional experiments in the region around this temperature could provide more data about an ideal setting of temperature. The optimized reaction yield of 90 coupled with a low estimate of variability may be sufficient for the objectives of the design. However, the poor fit of the model suggests that there are potentially other factors that influence the response; thus, if a second design is planned, investigation of other factors or response surface modeling around the new optimal conditions are reasonable experimental designs to construct.

8.4.2 EXPERIMENTAL DESIGN TO SELECT SUBSTITUENTS

For many problems in chemistry, optimization involves choosing substituents or building blocks rather than optimizing continuous variables such as temperature or concentration. For example, a designer of a combinatorial library is often interested in choosing a diverse set of monomers. The second example of statistical experimental design in chemistry involves the choice of phenyl ring substituents in order to provide maximal information for structure–activity relationship (SAR) studies.

In SAR studies, the goal is to correlate biological activity with a discernable aspect of chemical structure. If a homologous series is the focus of an SAR study, it is critical to choose substituents that define a range of biological activities. If a mathematical relationship (quantitative structure–activity relationship, or QSAR) is to be estimated for the series, quantitative descriptors must be associated with each compound.

Nielsen *et al* (1998) provide an example of using substituent-based experimental design for a homologous series of chalcones. Figure 8.25 shows the skeleton of the series. Substitutions are made at positions 2, 3 and 4 on both phenyl rings. Sixty-two substituents that could be used at these positions were considered. To characterize the substituents, four common parameters were used: molar refractivity (MR, steric effects), lipophilicity (π) and the Hammett constants

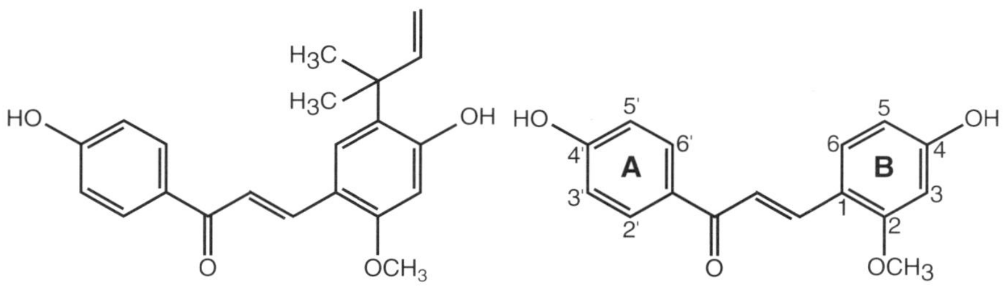

Licochalcone A (1) **Echinatin (2)**

Fig. 8.25. Chemical structures used in an experimental design to select substituents. (A) Licochalcone A: a compound isolated from Chinese licorice that inhibits proliferation of certain species of the protozoan parasite Leishmania. (B) Echinatin: the common skeleton for the structures in the experimental design. Substitutions were made at positions 2', 3', 4' and 2, 3 and 4 as indicated on this structure. (Reprinted with permission from Nielsen *et al*, Antileishmanial chalcones: statistical design, synthesis and three-dimensional quantitative structure–activity relationship analysis. *J Med Chem* **41**: 4819–4832. Copyright 1998 American Chemical Society.)

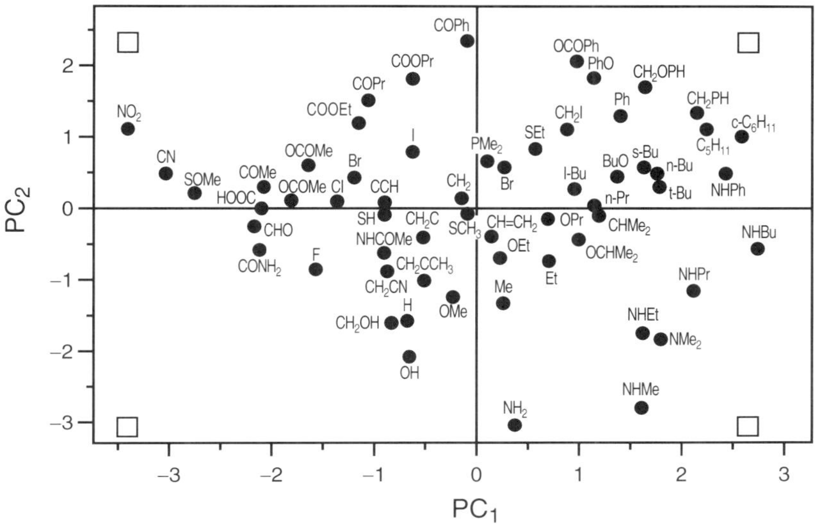

Fig. 8.26. Substituents used in the experimental design for compound selection. Each phenyl ring substituent used in the design is placed on this plot. The scales are principal components based on four descriptors: molar refractivity, lipophilicity (π) and the two Hammett constants (σ_p and σ_r). Abscissa: first principal component. Ordinate: second principal component. (Reprinted with permission from Nielsen *et al*, Antileishmanial chalcones: statistical design, synthesis and three-dimensional quantitative structure–activity relationship analysis. *J Med Chem* **41**: 4819–4832. Copyright 1998 American Chemical Society.)

(σ_p and σ_r, electronic effects). One interpretation of factorial experimental design would involve choosing substituents that covered the widest range of these parameters and constructing a factorial design. There are two problems with this approach: first, these parameters are highly correlated, which makes fitting a linear model to the data difficult, thus precluding analysis of the data; second, many of the compounds in the design are poorly soluble. Nielsen *et al* (1998) reported that a fractional-factorial design based on substitutions at the 4 and 4' positions provided data on only five out of eight compounds. To address the problem of correlations between the factors, Nielsen *et al* (1998) performed a principal component analysis on the factors, giving two principal components to explain 89% of the variation in the data. This allowed construction of a plot showing the various substituents arranged in the principal component space (Fig. 8.26). For a substituent, the contribution from each of the four original descriptors is obtained by examining the principal component loadings. If the original variables are scaled identically, larger values of the coefficients associated with the original variables indicate a stronger dependence on the variable for that principal component. Variables are centered (mean of zero) and scaled (standard deviation of unity) to ensure that the magnitude of the coefficients can be compared. To center and scale the variables, a new variable is constructed by subtracting the mean and dividing by the standard deviation (SD):

$$X' = \frac{(X - \overline{X})}{\mathrm{SD}\ (X)}.$$

A difficulty in using principal components is that it is more difficult to understand the relationships between the original variables (measurements or calculations) and the response. For the designs, substituents are chosen from the plot that are in different regions, e.g. high scores on both principal components, low scores on both or other combinations.

To address the problem of not being able to obtain data on many of the compounds, Nielsen *et al* (1998) fixed one of the ring substituents with a substituent known to give good activity in the biological assay. This was done for both rings in turn. This approach is a compromise. The design does not allow analysis of whether the substitutions at one ring interact with the other; each design must be considered separately. The resulting design is shown in Table 8.9. The biological activities are IC_{50} values for two assays: antileishmanial activity (potency against the leishmania parasite) and suppression of lymphocytes. The desired profile is maximal effect for antileishmanial activity and minimal suppression of lymphocytes. Results of the design were used to build a three-dimensional QSAR model of the series. Analysis of the data shows that large substituents in positions 4', 2, 3 and 4 are predicted to minimize suppression of lympocytes without sacrificing antileishmanial potency.

This study illustrates some of the difficulties in substituent-based experimental design. The design can suggest compounds that either cannot be synthesized or tested for activity. Bioisoteres or substituents that are considered to be roughly equivalent to a different substituent can alleviate the problems with synthesis, but in other cases entire parts of the design cannot be implemented. Characterization of substituents and compounds must be considered. The parameters chosen to characterize the structures are often highly correlated, necessitating a transformation or mathematical combination of the factors.

8.4.3 EXPERIMENTAL DESIGN FOR SELECTION OF COMPOUNDS AND COVERAGE OF SPACE

The last example of experimental design in chemistry to be discussed in this chapter involves the selection of structures from a larger set. Several scenarios involve this type of application. A virtual library of structures is constructed and is sampled to give a smaller number of representative structures that are synthesized and tested. Monomers for a combinatorial library are to be used in an experimental design to optimize the coverage of descriptors that are relevant to

Table 8.9. Results of an experimental design to select substituents for a series of chalcones designed for antileishmanial activity[1]

PC_1	PC_2	Ring A	Ring B	IC$_{50}$ ± SD (μM)	
				Leish	Lymph
+	+	2'-OC_4H_9	3.5-OCH_3	7.2 ± 1.2	13.0 ± 1.0
+	−	2'-$N(CH_3)_2$	3.5-OCH_3	7.3 ± 1.0	5.7 ± 0.3
−	+	2'-NO_2	3.5-OCH_3	3.4 ± 0.5	3.7 ± 0.2
−	−	2'-F	3.5-OCH_3	3.5 ± 0.8	6.0 ± 0.4
+	+	3'-OCH_4H_9	2.4-OCH_3	20.0 ± 3.0	65.0 ± 4.0
+	−	3'-$N(CH_3)_2$	3.5-OCH_3	5.8 ± 0.7	5.0 ± 0.2
−	+	3'-NO_2	2.4-OCH_3	8.6 ± 1.1	18.0 ± 2.0
−	−	3'-F	3.5-OCH_3	5.5 ± 0.3	9.2 ± 0.5
+	+	4'-C_6H_{11}	3.5-OCH_3	31.0 ± 2.0	39.0 ± 2.0
+	−	4'-$N(CH_3)_2$	3.5-OCH_3	NS	NS
−	+	4'-NO_2	3.5-OCH_3	16.0 ± 1.0	20.0 ± 1.0
−	−	4'-F	H	15.0 ± 1.0	23.0 ± 21.0
+	+	2',3',4'-OCH_3	2-OC_4H_9	26.1 ± 1.0	52.0 ± 3.0
+	−	2',3',4'-OCH_3	2-$N(CH_3)_2$		
−	+	2',3',4'-OCH_3	2-NO_2	5.9 ± 0.4	7.0 ± 0.6
−	−	2',3',4'-OCH_3	2-F	15.0 ± 1.0	11.0 ± 1.0
+	+	2',3',4'-OCH_3	3-OC_6H_5	8.7 ± 0.5	17.0 ± 1.0
+	−	2',3',4'-OCH_3	3-$N(CH_3)_2$	20.0 ± 3.0	26.0 ± 1.0
−	+	2',3',4'-OCH_3	3-NO_2	17.0 ± 2.0	36.0 ± 2.0
−	−	2',3',4'-OCH_3	3-F	9.7 ± 0.9	6.4 ± 0.4
+	+				
−	+	2',3',4'-OCH_3	4-$N(CH_3)_3$	21.1 ± 2.0	35.0 ± 2.0
−	+	2',3',4'-OCH_3	4-CN	18.6 ± 0.9	5.4 ± 0.3
−	+				

[1] PC, principal component. A plus sign represents a high value and a minus sign represents a low value. Reprinted with permission from Nielsen *et al*, Antileishmanial chalcones: statistical design, synthesis and three-dimensional quantitative structure-activity relationship analysis. J Med Chem 41: 4819–4832. Copyright 1998 American Chemical Society.

the pharmacokinetic profile of the library. A new biochemical target is available and the corporate collection of compounds is to be sampled for compounds to be tested in the high-throughput screen for the target.

In the example, a subset of a compound collection is to be chosen for testing. Figure 8.27 shows the distributions of four descriptors (number of hydrogen bond donors and acceptors, molecular weight and calculated log P) for approximately 1700 compounds constructed by sampling a database for any compounds with α_1-adrenergic activity. This set of compounds can be considered to define a region of chemical space where the dimensions are the four descriptors. A D-optimal design is performed to select compounds from this set that best define the space. The results of running a D-optimal design are shown in Fig. 8.28. Fifty-five compounds are chosen that maximize the D-optimal criteria. Comparison of the histograms of the original set of 1700 compounds and the selected set of 55 shows that both cover the same range for the descriptors. The relevant questions

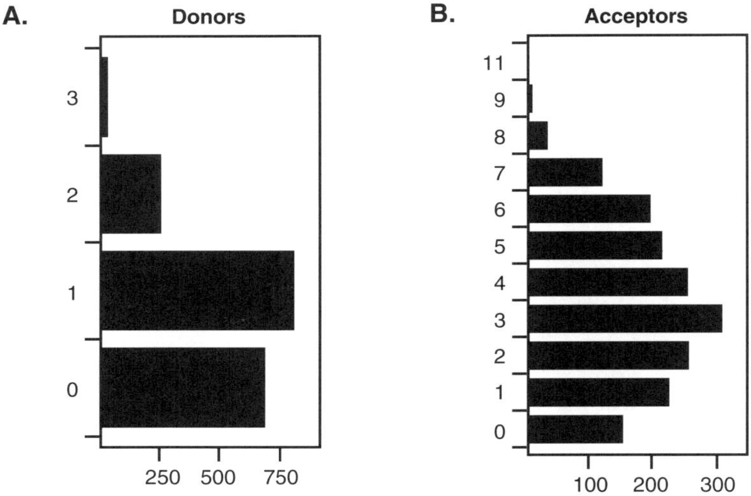

Frequencies

Level	Count	Probability	Cum. Prob.
0	700	0.39	0.39
1	800	0.45	0.84
2	241	0.14	0.98
3	42	0.02	1.00
Total	1783		

Frequencies

Level	Count	Probability	Cum. Prob.
0	156	0.09	0.09
1	228	0.13	0.22
2	258	0.14	0.36
3	304	0.17	0.53
4	259	0.15	0.68
5	214	0.12	0.80
6	199	0.11	0.91
7	118	0.07	0.97
8	36	0.02	0.99
9	10	0.01	1.00
11	1	0.00	1.00

Fig. 8.27. Distributions and descriptive statistics (counts, quantiles and moments) of chemical descriptors for a set of 1783 compounds tested for α_1-adrenergic receptor activity: (A) number of hydrogen bond donors; (B) number of hydrogen bond acceptors; (C) molecular weight; (D) calculated log *P*

are how well the 55 compounds cover the space and what aspects of the original set are not covered by the resulting design?

Inspection of the two sets of histograms shows that although the compounds selected using the D-optimal design cover the univariate space defined by each of the descriptors, the resulting distributions are not uniform. D-optimal designs sample the edges of a descriptor space rather than trying to fill optimally the space or fit to a specific distribution. This characteristic of D-optimal designs is observed for the example if a plot is made of molecular weight versus calculated log *P* for the original set of compounds and the selected compounds (Fig. 8.29).

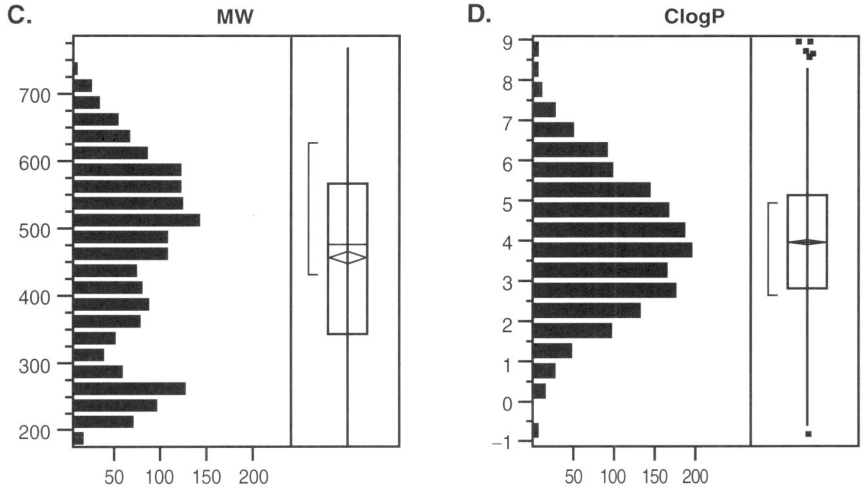

C. MW

Quantiles		
maximum	100.0%	766.84
	99.5%	726.02
	97.5%	697.72
	90.0%	629.62
quartile	75.0%	567.62
median	50.0%	477.59
quartile	25.0%	347.45
	10.0%	251.73
	2.5%	215.26
	0.5%	191.23
minimum	0.0%	175.23

Moments	
Mean	457.04
Std. Dev.	139.39
Std. Err. Mean	3.30
Upper 95% Mean	463.52
Lower 95% Mean	450.57
N	1783.00
Sum Weights	1783.00

D. ClogP

Quantiles		
maximum	100.0%	8.99
	99.5%	8.19
	97.5%	7.12
	90.0%	6.18
quartile	75.0%	5.06
median	50.0%	3.91
quartile	25.0%	2.75
	10.0%	1.85
	2.5%	0.95
	0.5%	0.22
minimum	0.0%	-0.88

Moments	
Mean	3.95
Std. Dev.	1.63
Std. Err. Mean	0.04
Upper 95% Mean	4.03
Lower 95% Mean	3.87
N	1695.00
Sum Weights	1695.00

Fig. 8.27. (*continued*)

In this plot the selected compounds have a marker symbol '+'. It is seen that the selected compounds sample the edges of the space but also provide some coverage of the center of the space. In a similar fashion, if principal component analysis is done using all four descriptors and a plot is made of the first two principal components (Fig. 8.30), it is seen that the selected compounds

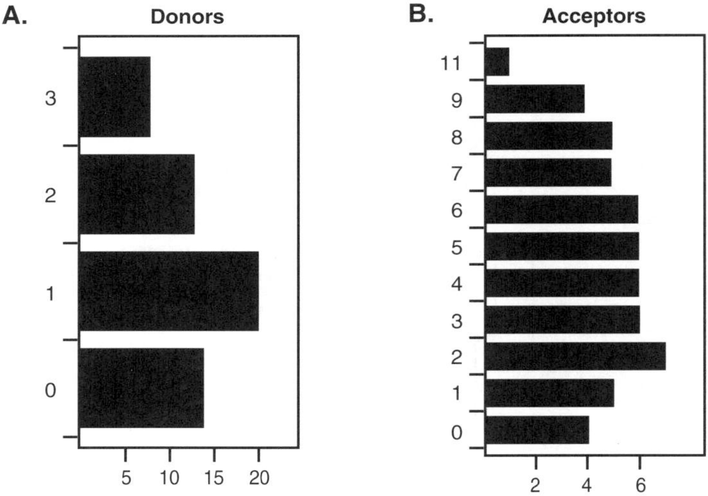

Frequencies

Level	Count	Probability	Cum. Prob.
0	14	0.25	0.25
1	20	0.36	0.62
2	13	0.24	0.85
3	8	0.15	1.00
Total	**55**		

Frequencies

Level	Count	Probability	Cum. Prob.
0	4	0.07	0.07
1	5	0.09	0.16
2	7	0.13	0.29
3	6	0.11	0.40
4	6	0.11	0.51
5	6	0.11	0.62
6	6	0.11	0.73
7	5	0.09	0.82
8	5	0.09	0.91
9	4	0.07	1.98
11	1	0.02	1.00

Fig. 8.28. Distributions and descriptive statistics (counts, quantiles and moments) of chemical descriptors for a set of 55 compounds selected using a D-optimal design from the candidate set shown in Fig. 8.27. (A) Number of hydrogen bond donors. (B) Number of hydrogen bond acceptors. (C) Molecular weight. (D) Calculated log *P*

make a reasonable approximation of the space. Areas that are potentially undersampled include the peaks of the univariate distributions in the original set of compounds. An advantage to the D-optimal designs is that they choose points that are spread out and are nearly orthogonal in the space that is defined by the descriptors. For the example, an orthogonal D-optimal design was used. Often, experimental designs of this type are augmented by specifying compounds or monomers that must be included in the final selection lists: monomers known to be associated with activity, or compounds that define the range of structure activity–relationships.

Choice of descriptors is an important consideration for these applications. Principal component analysis is useful if there are high correlations between

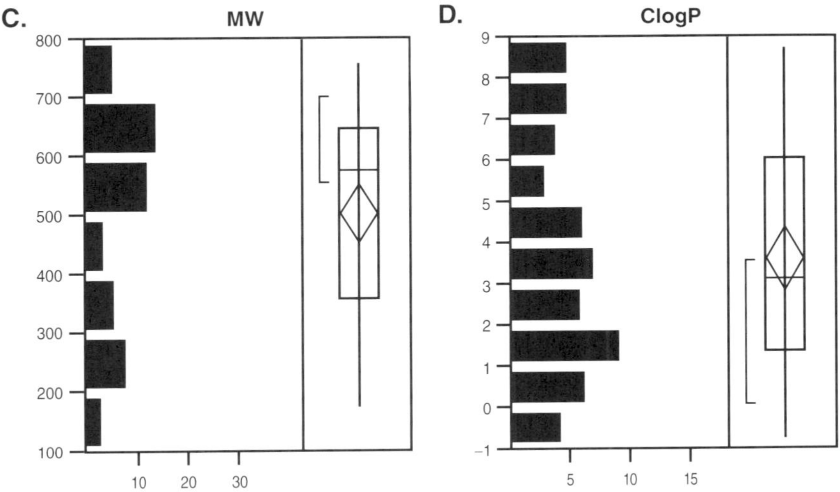

C. MW — Quantiles

maximum	100.0%	766.84
	99.5%	766.84
	97.5%	754.01
	90.0%	705.11
quartile	75.0%	653.71
median	50.0%	582.73
quartile	25.0%	361.44
	10.0%	210.08
	2.5%	178.42
	0.5%	175.23
minimum	0.0%	175.23

C. MW — Moments

Mean	507.69
Std. Dev.	182.69
Std. Err. Mean	24.63
Upper 95% Mean	557.08
Lower 95% Mean	458.31
N	55.00
Sum Weights	55.00

D. ClogP — Quantiles

maximum	100.0%	8.99
	99.5%	8.99
	97.5%	8.98
	90.0%	7.85
quartile	75.0%	6.21
median	50.0%	3.18
quartile	25.0%	1.37
	10.0%	0.17
	2.5%	-8.80
	0.5%	-0.88
minimum	0.0%	-0.88

D. ClogP — Moments

Mean	3.68
Std. Dev.	2.82
Std. Err. Mean	0.38
Upper 95% Mean	4.44
Lower 95% Mean	2.92
N	55.00
Sum Weights	55.00

Fig. 8.28. (*continued*)

the variables. For the example, the only high correlation was between the number of hydrogen bond acceptors and molecular weight ($R^2 = 0.67$); all other correlations are low ($R^2 < 0.20$). Martin *et al* (1995) describe the use of vectors of 16–18 descriptors or 12–14 principal components for monomer selection using D-optimal designs. The descriptors characterize lipophilicity, shape and branching, chemical functionality and receptor recognition features. Chemical functionality is based on a 2048-bit string that indicates the presence or absence of small fragments of up to seven bonds. The bit strings were converted into continuous descriptors by measuring dissimilarity between compounds using

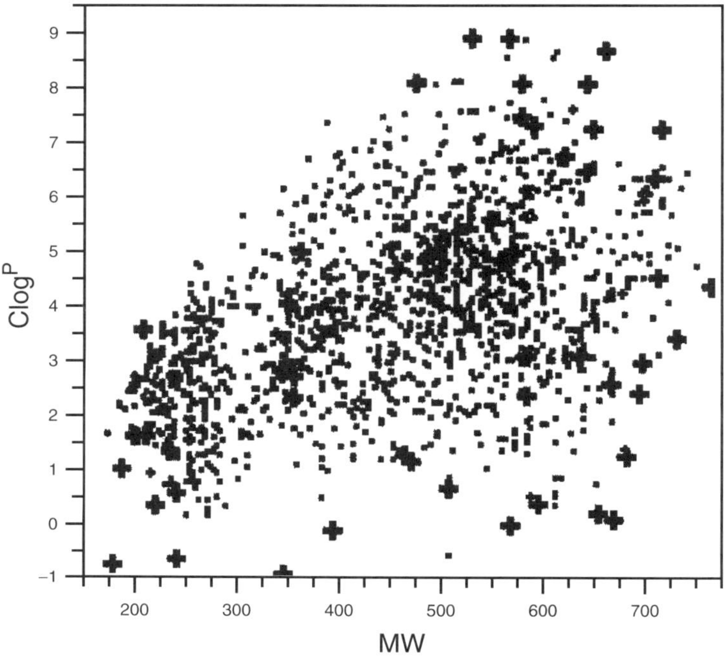

Fig. 8.29. Coverage of space for a D-optimal experimental design to select representative compounds. This plot shows two dimensions (molecular weight and calculated log *P*) of the original space of 1783 compounds. These compounds are shown with the cloud of points. The compounds selected by the D-optimal design are shown with the larger '+' symbol. The edges of the point cloud are sampled by the representative compounds, with the central regions of the cloud less densely sampled. The upper left region of the point cloud appears to be undersampled. Abscissa: molecular weight. Ordinate: calculated log *P*

multidimensional scaling based on matrices of Tanimoto distances. Receptor recognition descriptors are based on atomic properties that describe the chemical features that contribute to intermolecular interactions in receptor binding: for heteroatoms, atomic radius and characterization as acidic, basic, hydrogen bond donor, hydrogen bond acceptor and aromaticity. Multidimensional scaling was used to generate continuous descriptors. Other descriptors used in experimental designs and diversity/similarity analysis include Burden numbers (Burden 1989) and BCUTs (Stanton 1999). The choice of using multidimensional scaling or principal components to form linear functions of the descriptors is made based on analysis of the correlations between the descriptors. If they are highly correlated, then this step is used to create orthogonal descriptors.

D-optimal designs are one approach to the general problem of selecting a subset of compounds or substituents. The D-optimal design is considered an

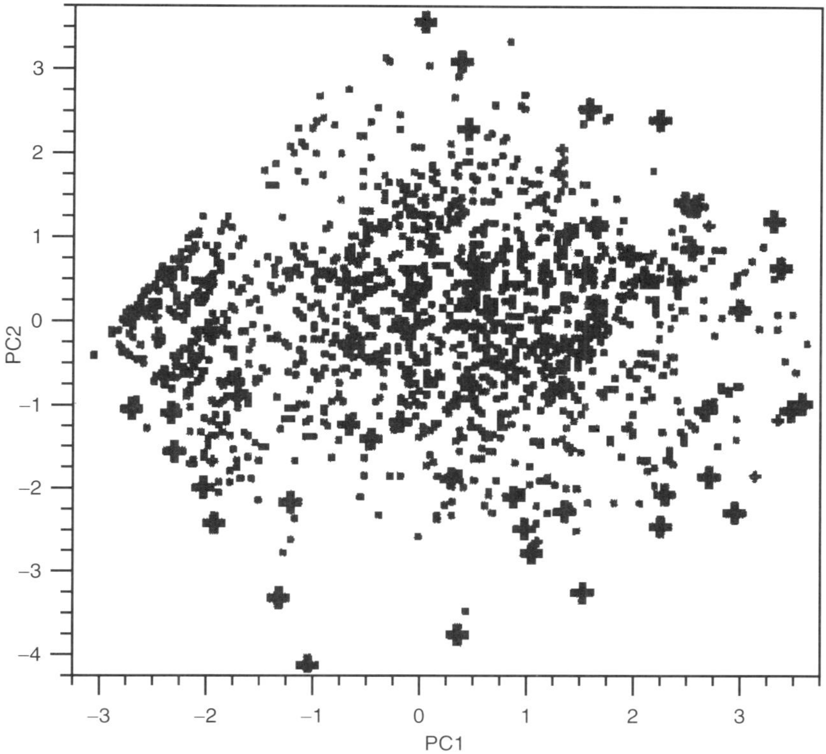

Fig. 8.30. Coverage of space for a D-optimal experimental design to select representative compounds. This plot shows two principal components from a principal component analysis of all of the descriptors used in the design. The first two principal component values for the original space of 1783 compounds are shown with the cloud of points. The compounds selected by the D-optimal design are shown with the larger '+' symbol. The edges of the point cloud are sampled by the representative compounds. The upper left region of the point cloud and central regions appear to be undersampled. Abscissa: first principal component value. Ordinate: second principal component value

edge design because the selections are made at the edges of the descriptor space. This design selects molecules from the corners first and then starts to fill the edges of the space. Regions of the space that contain molecules already selected are not avoided using this method, e.g. several molecules can be selected from a region of the space. Alternatives include spread designs and coverage designs (Higgs *et al* 1997) to fill better the design space. Figure 8.31 shows a simple two-dimensional example of these approaches for a set of 1400 different molecules. For spread designs, the subset is chosen by maximizing the dissimilarity between the molecules. These designs include molecules from the edges and from the

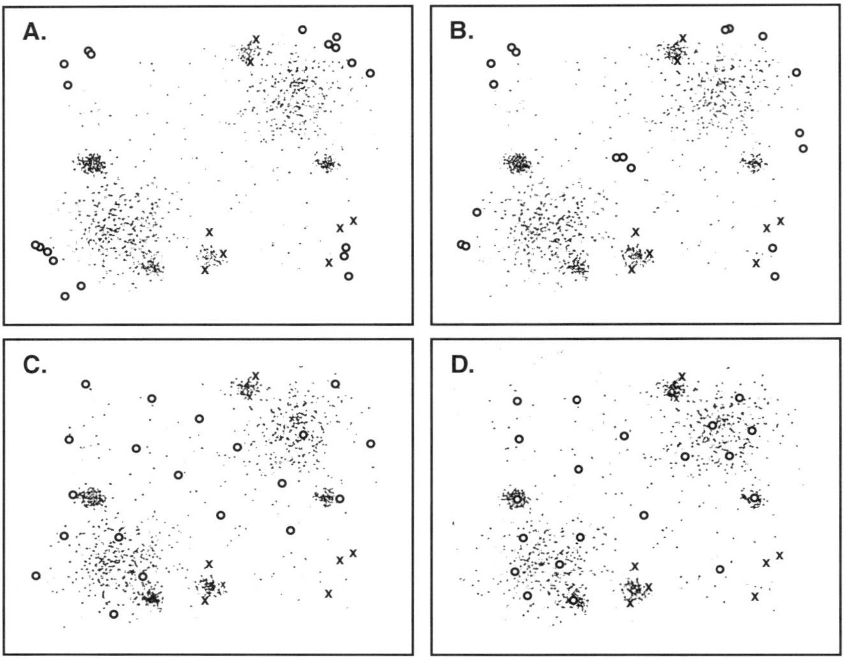

Fig. 8.31. Space coverage of different types of experimental designs. These plots show 1400 simulated molecules designed to have the properties of: five tightly packed clusters, two loosely packed clusters and other molecules uniformly distributed over the two-dimensional space. The purpose of the plot is to illustrate space coverage for the different designs. Eight molecules were chosen randomly to have been selected from a previous design and the objective is to complement these compounds with more compounds. (A) A D-optimal design is used to augment the original set. This design selects molecules from the corners first and then around the edges. Regions containing previously selected molecules are not avoided. (B) Quadratic and interaction terms are added to the D-optimal design. This selects molecules away from the corners but does not avoid previously populated regions or cover the space in the center. (C) A spread design selects a diverse subset of compounds near the edges and throughout the space. (D) The coverage design selects compounds near the center of clusters. The edges of the design are not usually sampled because it is unlikely that the center of a cluster will be near an edge. (Reprinted with permission from Higgs RE, *et al*, Experimental designs for selecting molecules from large chemical databases. *J Chem Inf Comput Sci* **37**: 861–870. Copyright 1997 American Chemical Society.)

entire design space. Coverage designs identify molecules that are maximally similar to the set of candidate molecules. If clusters of compounds defined by the design space are formed, then these designs choose molecules from the centers of the clusters. These designs avoid molecules from the edges of the design space.

Many implementations for spread and coverage designs are computationally intensive and are not practical for the selection of subsets from millions of compounds. Higgs *et al* (1997) describe implementations of experimental design algorithms that are efficient for use in subset selection from a large collection of molecules.

8.5 SUMMARY

High-throughput screening and synthesis of combinatorial libraries are major trends in drug discovery. Statistical experimental design is a powerful tool, both for development and optimization of the biological assay and for design of the combinatorial library. This chapter has described both classical methods and space-filling designs in chemistry and biochemistry applications. The case studies in this chapter are typical problems in the development of biological assays and in synthesis or compound selection. The availability of robotic systems for synthesis and biological screening is well suited for experimental design: completion of runs under different conditions, and randomization is often possible under software control.

Drug discovery involves decisions about which compounds to synthesize and test against an increasingly large array of potential biological targets. Experimental design is an efficient method to optimize chemical synthesis and to design robust and sensitive biochemical assays.

8.6 REFERENCES

Austel V (1983) Design of test series by combined application of 2^m factorial schemes and pattern recognition techniques. In *Quantitative Approaches to Drug Discovery*, ed. by JC Dearden. Elsevier, New York.

Box GEP and Behnken DW (1960) Some new three-level designs for the study of quantitative variables. *Technometrics* **2**: 455–475.

Box GE, Hunter WG and Hunter S (1978) *Statistics for Experimenters: an Introduction to Design, Data Analysis and Model Building*. Wiley, New York.

Burden FR (1989) Molecular identification number for substructure searches. *J Chem Inf Comput Sci* **29**: 225–227.

Carlson, R (1992) *Design and Optimization in Organic Synthesis*. Elsevier, New York.

Carroll RJ and Ruppert D (1982) A comparison between maximum likelihood and generalized least squares in a heteroscedastic linear model. *J Am Statist Assoc* **77**: 878–882.

Cochran WG and Cox GM (1957) *Experimental Designs*. Wiley, New York.

Daniel C (1976) *Applications of Statistics to Industrial Experimentation*. Wiley, New York.

Deming SN and Morgan SL (1987) *Experimental Design: a Chemometric Approach*. Elsevier, New York.

Efron B and Tibshirani RJ (1991) Statistical data analysis in the computer age. *Science* **253**: 390–395.

Efron B and Tibshirani RJ (1993) *An Introduction to the Bootstrap*. Chapman and Hall, New York.

Galil Z and Kiefer J (1980) Time and space-saving computer methods related to Mitchell's DETMAX for finding D-optimal designs. *Technometrics* **21**: 301–313.

Gallop MA, Barrett RW, Dower WJ, Fodor SPA and Gordon EM (1994) Applications of combinatorial technologies to drug discovery. 1. Background and peptide combinatorial libraries. *J Med Chem* **37**: 1233–1251.

Giltinan DM and Ruppert D (1989) Fitting heteroscedastic regression models to individual pharmacokinetic data using standard statistical software. *J Pharmacokin Biopharm* **17**: 601–614.

Goetz AS, Lutz MW, Queen K, Rimele TJ and Saussy DL (1994) Characterization of α_1-adrenoceptor subtypes in human and canine prostate membranes and rat aorta. *J Pharmacol Exp Ther* **271**: 1228–1233.

Gordon EM, Barrett RW, Dower WJ, Fodor SPA and Gallop MA (1994) Applications of combinatorial technologies to drug discovery. 2. Combinatorial organic synthesis, library screening strategies, and future directions. *J Med Chem* **37**: 1385–1401.

Haaland PD (1989) *Experimental Design in Biotechnology*. Marcel Dekker, New York.

Haaland P, McMillan N, Nychka D and Welch W (1994) Analysis of space-filling designs. *Comput Sci Stat* **26**: 111–120.

Hicks CR (1974) *Fundamental Concepts of Design of Experiments*. Holt, Rhinehart and Winston, New York.

Higgs RE, Bemis KG, Watson IA and Wikel JH (1997) Experimental designs for selecting molecules from large chemical databases. *J Chem Inf Comput Sci* **37**: 861–870.

Kennard RW and Stone LA (1969) Computer-aided design of experiments. *Technometrics* **11**: 137–148.

Kiefer J (1959) Optimal experimental designs. *J R Stat Soc Ser B* **21**: 272–319.

Lutz MW, Menius JA, Choi TD, *et al* (1995) Use of resampling techniques to estimate the variance of parameters in pharmacological assays when experimental protocols preclude independent replication: an example using Schild regressions. *J Pharmacol Toxicol Methods* **34**: 37–46.

Lutz MW, Menius JA, Choi TD, *et al* (1996) Experimental design for high-throughput screening. *Drug Discovery Today* **1**: 277–285.

Martin EJ, Blaney JM, Siani MA, Spellmeyer DC, Wong AK and Moo WH (1995) Measuring diversity: experimental design of combinatorial libraries for drug discovery. *J Med Chem* **38**: 1431–1436.

Menius JA, Rocque W, Emptage MR and Young SS (1994) Space-filling experimental design for determining protein construct storage conditions. *Comput Sci Stat* **26**: 106–110.

Montgomery DC (1984) *Design and Analysis of Experiments*. Wiley, New York.

Munson PJ and Rodbard D (1980) LIGAND: a versatile computerized approach for characterization of ligand-binding systems. *Anal Biochem* **107**: 220–239.

Murphy TD (1977) Design and analysis of industrial experiments, *Chem Eng* **June 6**, 168–182.

Nielsen SF, Christensen SB, Cruciani G, Kharazmi A and Lijefors T (1998) Antileishmanial chalcones: statistical design, synthesis and three-dimensional quantitative structure–activity relationship analysis. *J Med Chem* **41**: 4819–4832.

Plackett RL and Burman JP (1946) The design of optimal multifactorial experiments. *Biometrika* **33**: 305–325.

Sacks J, Welch WJ, Mitchell TJ and Winn HP (1989) Design and analysis of computer experiments. *Stat Sci* **4**: 409–435.

SAS/QC Software Reference (1993) SAS Institute, Cary, NC.

Snedecor GW and Cochran WG (1989) *Statistical Methods*. Iowa State University Press, Iowa.

Snee RD (1985) Computer-aided design of experiments — some practical experiences. *J Quali Technol* **17**: 222–236.
Stanton DT (1999) Evaluation and use of BCUT descriptors in QSAR and QSPR. *J Chem Inf Comput Sci* **39**: 11–20.
Wang J, DiPasquale AJ, Bray AM, Maeji NJ and Geysen HM (1993) Study of stereo-requirements of substance P binding to NK1 receptors using analogs with systematic D-amino acid replacements. *Bioorg Med Chem Lett* **3**: 451–456.
Welch WJ, Buck RJ, Sacks J, *et al* (1992) Screening, predicting and computer experiments. *Technometrics* **34**: 15–25.
Young SS and Hawkins DM (1995) Analysis of a 2^9 full factorial chemical library. *J Med Chem* **38**: 2784–2788.
Zemrock PJ (1986) Cluster analysis as an experimental design generator, with applications to gasoline blending experiments. *Technometrics* **28**: 39–49.

8.7 FURTHER READING

Box GEP, Hunter WG and Hunter JS (1978) *Statistics for Experimenters*. Wiley, New York.
John PM (1971) *Statistical Design and Analysis of Experiments*. Macmillan, New York.
Marengo E and Todeschini R (1992) A new algorithm for optimal, distance-based experimental design. *Chemomet Intell Lab Syst* **16**: 37–44.
Morris MD and Mitchell TJ (1995) Exploratory designs for computational experiments. *J Stat Planning Infer* **43**: 381–402.
Wynn HP (1972) Results in the theory and construction of D-optimum experimental designs. *J R Stat Soc* **34**: 133–147.

9 Analysis and Interpretation of Data

9.1 INTRODUCTION

Combinatorial libraries and automated systems for biological assay have lead to a dramatic increase in the amount of biological activity data available to drug discovery project teams. In the last decade, drug discovery project teams have shifted from working with hundreds of compounds tested against a single biochemical target to testing combinatorial libraries of thousands of compounds against a matrix of targets. Although many of the fundamental principles and methods historically used in structure–activity relationship analysis (SAR) and its more mathematical counterpart, quantitative structure–activity analysis (QSAR), are still applicable, the vast increase in the amount of data presents special challenges for analysis. Issues involved in working with screening data include detection of active compounds in pools and understanding the variability of the data.

This chapter introduces methods that are well suited to the types and volume of data that are available from high-throughput screening studies. Multivariate visualization includes methods that are used to identify general trends in large datasets in addition to methods that can graph complex relationships within the dataset. In addition to visualization methods, a variety of statistical techniques are available for fast, automatic analysis of data. Recursive partitioning is a statistical 'split and recombine' strategy for data that quickly identifies important features of a dataset, groups compounds that have similar activity and detects interactions and outliers. Neural networks and genetic algorithms are powerful pattern recognition or amplification tools that can be used to predict the activity of compounds planned for synthesis. In each case, the emphasis in this chapter is on the theoretical and practical considerations necessary to use the method in the context of problems in high-throughput screening and SAR. Finally, some general issues and suggestions for statistical analysis of high-throughput screening data are listed. These could be considered the statistical guide posts or issues to consider at different times in a drug discovery project: library design, assay design, screening, data analysis and compound optimization. Taken together, the analytical methods and guide posts illustrate effective ways of handling the vast amount of data available from high-throughput screening.

9.2 MULTIVARIATE VISUALIZATION

In the book *Envisioning Information*, Edward Tufte states that escaping the flat land of paper and computer terminals is the essential task of envisioning information (Tufte 1990). Multivariate visualization methods sharpen information resolution by increasing the number of dimensions that can be represented on a planar surface and by increasing the data density. Numerous examples exist of complex visualization problems, from maps of the London tube system to interactive maps of global population density. In all cases, the problem is encoding complex, multivariate data to a very limited two-dimensional display.

For high-throughput screening data, there are three main challenges: displaying large amounts of complex multivariate data for interpretation; retaining sufficient speed to be able to interactively change views; and developing the relevant links between data. Often, the first objectives for visualization are to identify trends in the data using simple exploratory methods. For exploratory data analysis, linked scatterplots (Fig. 9.1), where each point on a scatterplot can be linked by a mouse click to data tables, other scatterplots, other graphics, chemical structures and three-dimensional (3D) rotating scatterplots, are useful to examine relationships between data. Some interactive statistical software programs such as JMP (SAS Institute 1994) include the capability to link scatterplots to tabular data and to each other.

It is often necessary to examine correlations in datasets, e.g. correlations between chemical descriptors and biological activity or correlations between different biological activities. Changing aspects of the scatterplot points increases the dimensionality that can be displayed on scatterplots. Different colors, shapes and labels are assigned to the different points, often in the data table itself. Color and symbol encoding effectively divides the data into different chemical classes before any correlation analysis is performed. This step allows the analyst to determine whether the correlations are specific for chemical classes or perhaps a set of monomers. A graphical correlation matrix (Fig. 9.2) is a useful tool to examine the strength of different correlations and to identify which compounds are most influential in the correlation.

9.2.1 GRAPHICAL LINKS BETWEEN DIFFERENT DATA TYPES

Links between different types of graphical displays are useful. For example, histograms could be generated to show the distribution of different monomers in a library. By clicking on a monomer 'bar' in the histogram, all of the points associated with a specific monomer would be highlighted on scatterplots and scatterplot correlation matrices (Fig. 9.3). A particularly important link exists between a data point on a scatterplot and its associated chemical structure. The simplest type of link is to display the structure when the point is selected (Fig. 9.1). The display of chemical information from pooled data creates additional complexity because

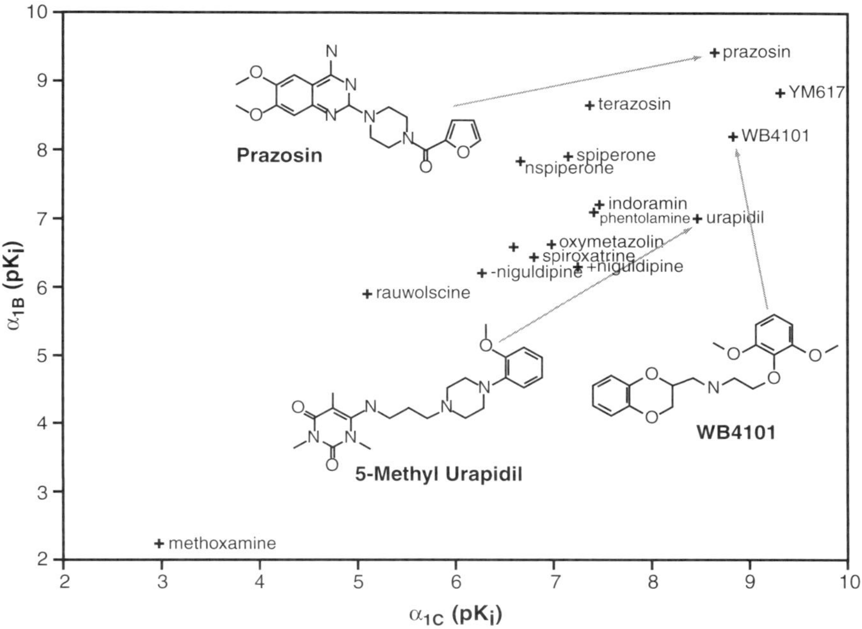

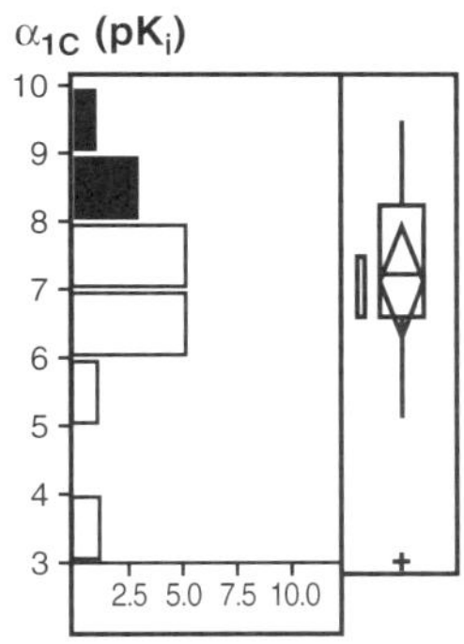

Compound	α_{1B} (pK$_i$)	α_{1C} (pK$_i$)	α_{1D} (pK$_i$)
prazosin	9.41	8.66	9.47
terazosin	8.62	7.39	8.45
YM617	8.81	9.48	9.36
WB4101	8.17	8.83	8.81
urapidil	6.98	8.48	7.34
+niguldipine	6.29	7.26	6.10
-niguldipine	6.19	6.39	5.97
indoramin	7.15	7.51	6.37
phentolamine	7.09	7.47	7.29
corynanthine	6.58	6.62	6.78
spiperone	7.9	7.17	7.54
nspiperone	7.83	6.68	7.25
spiroxatrine	6.43	6.81	7.31
oxymetazolin	6.6	6.99	5.74
methoxamine	2.24	3.02	3.82
rauwolscine	5.90	5.13	5.90

Fig. 9.1. Linkages between graphical information. The scatterplot, table, chemical structures and histogram are linked by the data. Clicking on the bars of the histogram where $-\log(K_i)$ for binding to the α_{1C}-receptor is >8 (10 nM) highlights the appropriate rows in the table (four compounds) and also the appropriate points in the scatterplot. In the scatterplot, the abscissa is $-\log(K_i)$ for the α_{1C}-receptor, and the ordinate is $-\log(K_i)$ for the α_{1B}-receptor. The points selected by clicking on the bar graph are automatically labeled

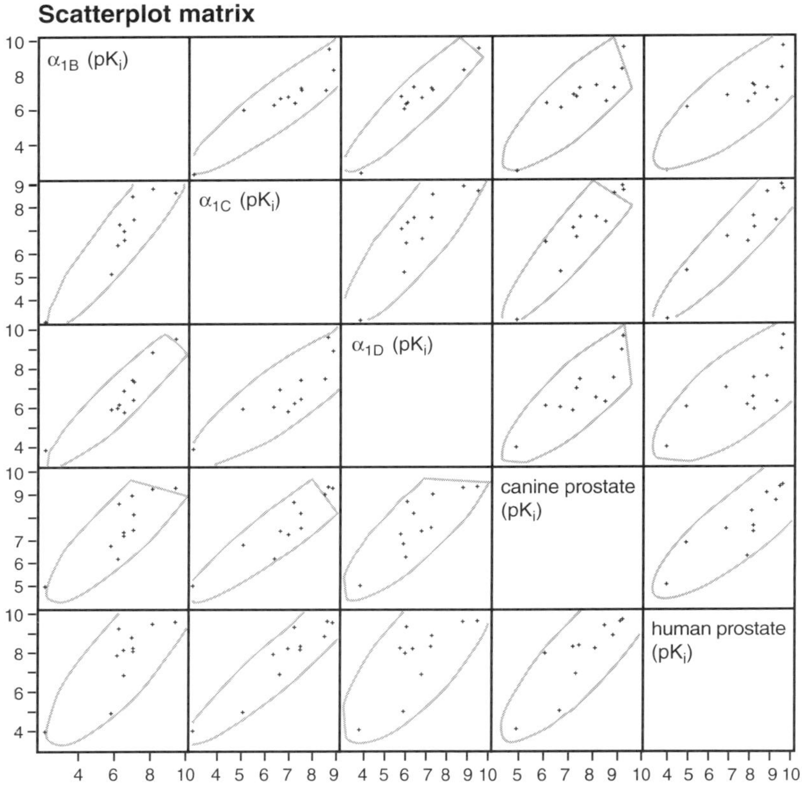

Variable	α_{1B} (pK_i)	α_{1C} (pK_i)	α_{1D} (pK_i)	canine prostate (pK_i)	human prostate (pK_i)
α1B (pk_i)	1.00	0.92	0.93	0.86	0.83
α1C (pk_i)	0.92	1.00	0.87	0.93	0.94
α1D (pk_i)	0.93	0.87	1.00	0.85	0.75
canine prostate (pK_i)	0.86	0.93	0.85	1.00	0.85
human prostate (pK_i)	0.83	0.94	0.75	0.85	1.00

Fig. 9.2. Graphical correlation matrix. Each cell is a graphical representation of the data that are used in the correlation coefficient calculation. The units of all axes are $-\log(K_i)$ for binding to various cloned human α_1-adrenoceptor cells and to canine and human prostate membranes. The width of the elipse indicates the strength of the correlation; a narrower elipse indicates a stronger correlation. (Data provided by David Saussy, Glaxo Wellcome.)

several structures are associated with the data point. The capabilities to scroll rapidly through large numbers of structures and to switch between modes where data are displayed both with and without structures is necessary for effective visual SAR.

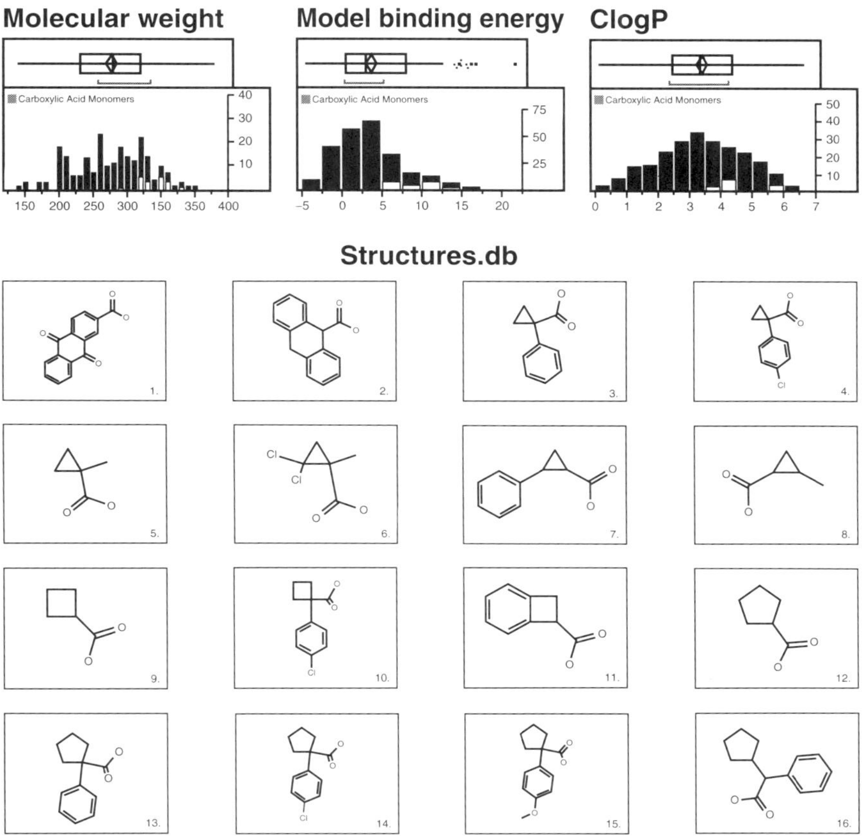

Fig. 9.3. Linking of monomers to histograms. A virtual library of 256 structures was created using two monomer sets: primary amines and carboxylic acids. Histograms are shown for three calculated properties of the library: molecular weight, a predicted binding energy (Andrews 1984) and calculated log *P*. The structures for 16 of the carboxylic acid monomers were selected and are displayed at the bottom of the figure. The values of the properties for these monomers appear shaded on the histogram, allowing the library designer to determine the distribution of properties for the subset within the library and to allow optimization of desired characteristics

9.2.2 COMPRESSION AND REDUCTION OF DIMENSIONALITY

Almost an infinite number of descriptors can be calculated for a molecule. These descriptors may be based on topology, computational chemistry or measured results. With more than five descriptors and one response, it becomes difficult to visualize the relationships between descriptors and the response. Statistical methods such as principal component regression or partial least-squares are used to compress a set of descriptors into a smaller set. These methods are most useful when there are high correlations between the descriptors, as is the case with topology-based parameters. Although it is possible to view the principal components, and indeed software like JMP allows an analyst to switch to principal component axes as an option, it is difficult to decode visually the contribution of each descriptor to a particular principal component loading. However, viewing data in principal component space is an effective means to identify outliers and to identify general trends in the data. Interpretation of a given axis is accomplished by examining the magnitude of the eigenvectors.

The output from a neural network with a special, two-parameter layer has been proposed as a method to reduce dimensionality (Livingstone *et al* 1991). The weights associated with these two parameters are used to calculate the x,y-coordinate values for a scatterplot. Although this method was shown to visualize effectively several QSAR datasets, it was later demonstrated not to be applicable in the general case (Reibnegger *et al* 1993). However, the method is effective for exploratory data analysis.

9.2.3 ENCODING, FILTERING AND SMOOTHING

Color has been mentioned as a possible means to encode an additional dimension in graphical displays. Color also can be used to filter or smooth data with fuzzy boundaries. In many cases, the goal of visualization is to identify or amplify a signal in noise. Figure 9.4 shows a plate of a combinatorial library tested against a biochemical target. Response is encoded as gray-scale blocks. Because the plate is organized by monomers with a regular structure, areas where there are white squares surrounded by lighter colored squares indicate a trend toward higher activity (greater percentage bound). Individual squares that are several shades darker or lighter than neighboring squares indicate that a specific combination of monomers reacts differently than the general trend, or that an interaction is present.

Numerous statistical methods exist for smoothing or filtering data. These methods are useful for cases where the objective is to discern trends in large amounts of noisy data. These methods are typically used in other disciplines such as signal processing and time-series analysis and include various forms of moving average filters and image processing techniques (e.g. to handle data from lawns of bead-encoded libraries).

Fuzzy logic is another possible approach to dealing with the problem of recognizing patterns in multivariate data from high-throughput screening. In fuzzy logic systems, variables are defined in terms of fuzzy sets, a mathematical formalism that allows the sets to be combined according to rules that allow for imprecision in the variables (Dubois and Prade, 1980). Fuzzy logic has its origin in control theory and has found some utility in QSAR problems. For example, it may not be possible to use a linear QSAR model for data from a high-throughput screen that provides only a qualitative readout. The data might be appropriate for a fuzzy

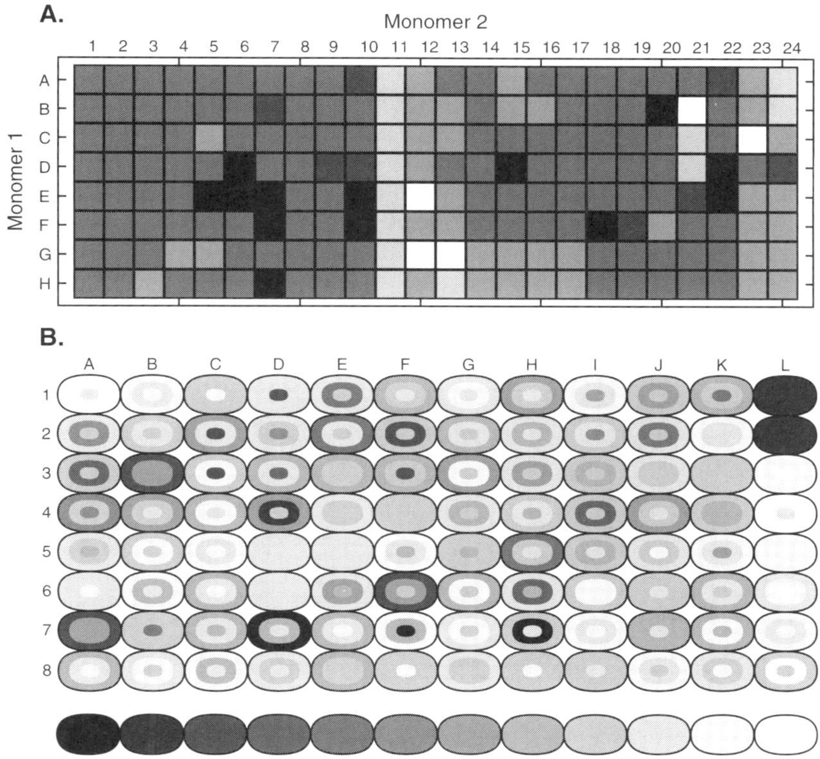

Fig. 9.4. (A) Gray-scale encoding of activity for a combinatorial library. The plate is organized by monomers with a regular structure. Lighter shadings indicate more activity. This view of data allows visual inspection of patterns of activity correlated to choice of monomer. (From Marc Navre, Affymax.) (B) Gray-scale encoding of activities for a plate of antimicrobial compounds. Each cell represents activities against three different microorganisms. Darker concentric circles indicate more activity. Concentric circles that are similar in intensity are non-selective. Concentric circles that show substantial contrast are selective for one or more microorganism. (From Marc Navre, Affymax.)

logic model that could identify common structural features that group compounds from a large set of data.

As the number of observations increases from thousands to millions, there are computational problems for data analysis and visualization. Calculation of descriptive statistics can take too much time if they are frequently updated. The approach taken in the software program Temple MVV (Temple University 1994) is to run a preprocessing step that computes all of the descriptive statistics needed for interactive visualization. Once this step is completed, changes to different displays and views of the data are fast because the descriptive statistics are available. TempleMVV uses a number of novel graphical displays that allow analysts to view the fine detail of library screening results on the 'flat land' of two-dimensional media. These types of displays are useful for examining the fine detail of a dataset after a statistical method such as multivariate partitioning or SAR modeling has identified the important features. Young and Hawkins (1995) have published a study of a 2^9 full-factorial library using recursive partitioning and multivariate visualization in TempleMVV. The original dataset was published by Wang *et al* (1993). Figure 9.5 shows biological activity as a function of changes at nine positions in an 11-amino acid polypeptide. Each position had

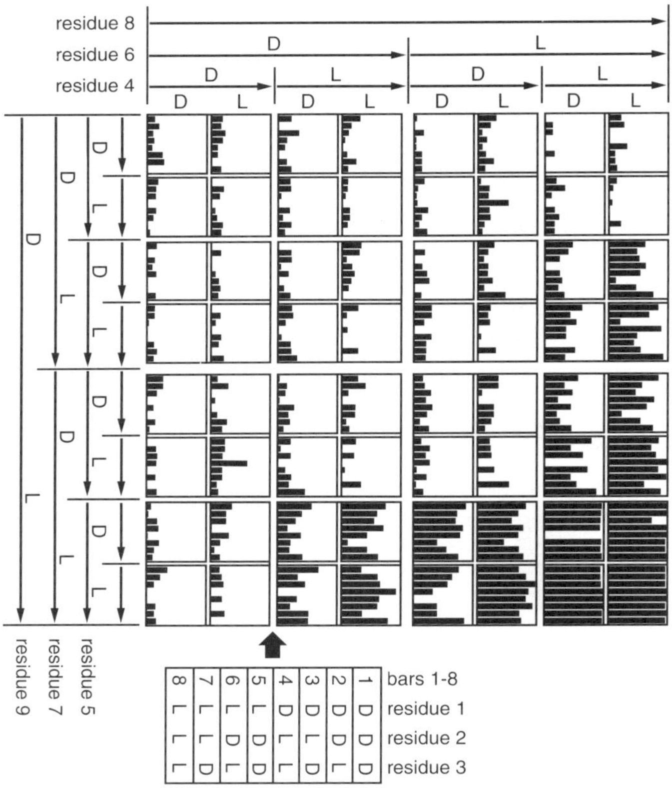

Fig. 9.5. Temple MVV plot of biological activity of a 2^9 polypeptide library tested against substance P. Each square indicates a particular combination of residues for the nine positions that were varied (D- or L-amino acid). The full peptide was length 11. Rows encode positions 4, 6 and 8. Columns encode positions 5, 7 and 9. Within a square, the eight bars encode positions 1, 2 and 3. The heights of the bars are proportional to activity. (Reprinted with permission from Young SS and Hawkins DM, Analysis of a 2^9 full factorial chemical library. *J Med Chem* **38**: 2784–2788. Copyright 1995 American Chemical Society.)

exactly two possible substituents, a D- or L-amino acid. Recursive partitioning indicated that positions 4–9 had the most significant effect on response, therefore they are used to organize the squares of bar graphs. Within a square, the eight bars represent a variable that encodes the three remaining positions. This encoding is done to reduce dimensionality because these positions were not considered to be important. The heights of the bars reflect potency. From inspection of the diagram, the most potent compounds are in the upper right corner. Generally, compounds in the last two columns and first two rows were more potent than the remainder. Clearly, a 1 (L-amino acid) in position 9 and a 1 in position 8 are most favorable to potency. Several outliers also can be identified in the plot: because the encoding of the individual bars uses factors that are not significant, the heights of bars within a square should be roughly uniform. Gaps or unusually large heights relative to neighboring bars indicate some systematic effect or outlier.

9.2.4 MODELING AND VISUALIZATION

Visualization tools can be used to recognize patterns in data after a mathematical or statistical model has been fitted to the data. Figure 9.6 shows several

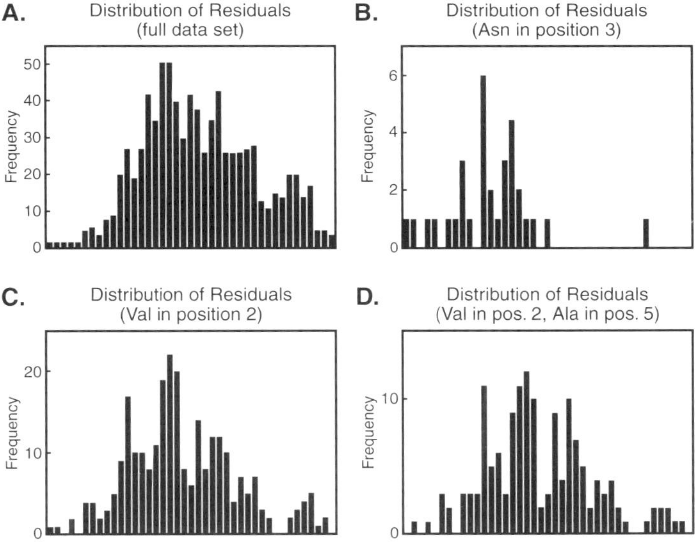

Fig. 9.6. Histograms of residuals obtained after fitting a linear model that includes main effects and first-order interactions for a pentapeptide library. The different histograms are for specific residues and interactions between residues. Abscissa: magnitude of residual. Ordinate: count of number of residuals in the bin. (A) Distribution of residuals for full dataset. (B) Distribution of residuals for asparagine in position 2. (C) Distribution of residuals for valine in position 2. (D) Distribution of residuals for valine in position 2 and alanine in position 5. (From Jeffrey S. Wiseman, Smith Kline Beecham.)

histograms of residuals obtained after a linear model (main effects and first-order interactions) has been fitted to biological activity data for a library of pentapeptides (J.S. Wiseman, personal communication, 1996). By looking at subsets of residuals and comparing their size to an estimate of pure error, an analyst can determine where the model is or is not fitting the data well and which interactions of specific residues at different positions are significant.

Figure 9.6A shows the distribution of residuals for the entire dataset. Clearly, there is some skew to the model fit. The distribution of residuals is not what would be expected for random error. For example, the right tail of the distribution is heavier than the left and the peak appears skewed to the left. Examination of Fig. 9.6B and Fig. 9.6C shows the differences in the model results for asparagine or valine in position 2 of the pentapeptide. Figure 9.6B shows that for the case of asparagine in position 2 there are six compounds that appear to group together in a common set. These compounds should be compared for structural similarity or dissimilarity. The bar at the far right side indicates that a compound has an unusually high residual value or may be an outlier. The distribution of residuals for valine in position 2 (Fig. 9.6C) appears to be closer to a normal distribution. Finally Fig. 9.6D shows the distribution of residuals for a residue–residue interaction. Although the linear model can provide a significance value for the interaction between sites on the peptide, it cannot identify which specific residues at two sites interact. The interaction plots help to identify which combinations of residues give results that are not predicted by the linear model.

9.2.5 LEVERAGE PLOTS

Leverage plots are a powerful tool for linear regression diagnostics (Sall 1990). The plots quickly identify colinear predictors, influential data points, outliers and the significance of a predictor or interaction after all the other terms in the model have had an effect on the response (Goetz *et al* 1994). Leverage plots are drawn in the scale of the original data. The points collapse toward the mean of the response (ordinate) if the predictor or interaction is not significant, and collapse to the mean of the predictor (abscissa) if the predictor is a linear combination of other predictors (Fig. 9.7). For large datasets with many predictors, leverage plots can quickly identify problems with the regression model being used and identify which compounds have the most significant effect on the regression equation parameters.

9.2.6 FLOWER PLOTS

Discrimination of patterns that are similar or dissimilar is something that humans can do very easily. Flower plots, as presented by Martin *et al* (1995) capitalize on this concept by placing all descriptors on a circular scaffold. Figure 9.8 shows the flower plots and structures for an oligo(*N*-substituted) glycine library. The 18 structures shown represent 18 specific side chains based on tyramine. The question that the flower plots attempt to address is whether the library exhibits molecular diversity in its structures. Each flower has one petal for each variable in the multidimensional space. For Fig. 9.8B, the petal lengths are based on the value of 16 chemical descriptors, including calculated log P, topological indices

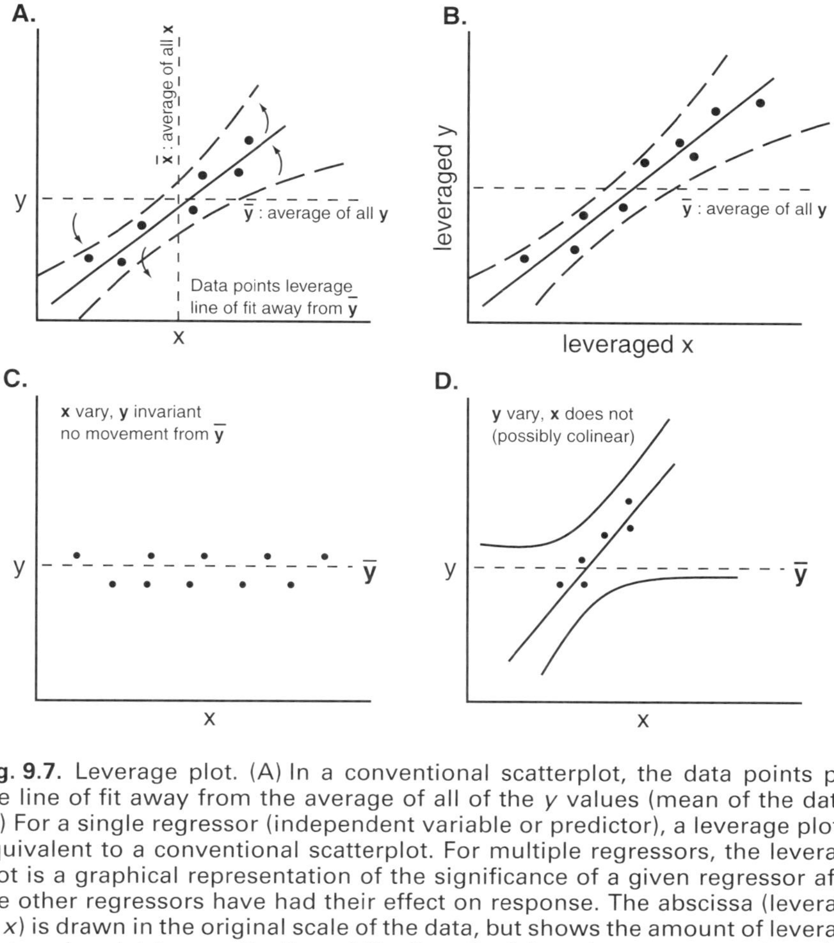

Fig. 9.7. Leverage plot. (A) In a conventional scatterplot, the data points pull the line of fit away from the average of all of the y values (mean of the data). (B) For a single regressor (independent variable or predictor), a leverage plot is equivalent to a conventional scatterplot. For multiple regressors, the leverage plot is a graphical representation of the significance of a given regressor after the other regressors have had their effect on response. The abscissa (leverage of x) is drawn in the original scale of the data, but shows the amount of leverage that each point has on the line of fit after all of the other regressors have had their effect on response. (C) In this leverage plot, the x values vary but there is no movement in the y direction. (D) In this case, there is variation in the y variable (response) but not in the x variable. Possibly, x is colinear with some other variable. (E) For this leverage plot, neither x nor y varies from the mean value. The confidence curves diverge from the center of the plot. (F) In this plot, one point is an outlier that has high leverage (upper left corner). This point is responsible for the significance of the fit (confidence curves cross the mean of y values — the center line)

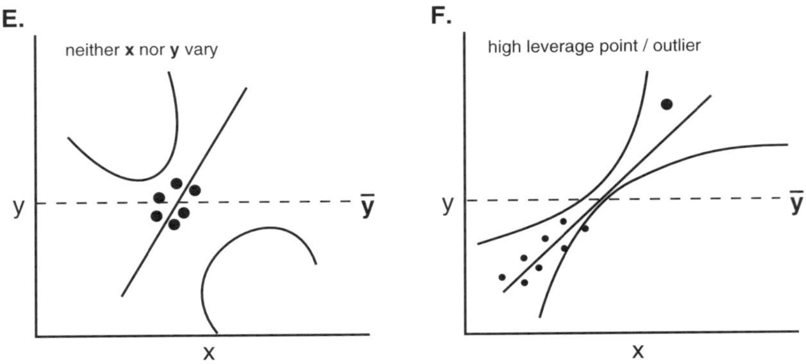

Fig. 9.7. *(continued)*

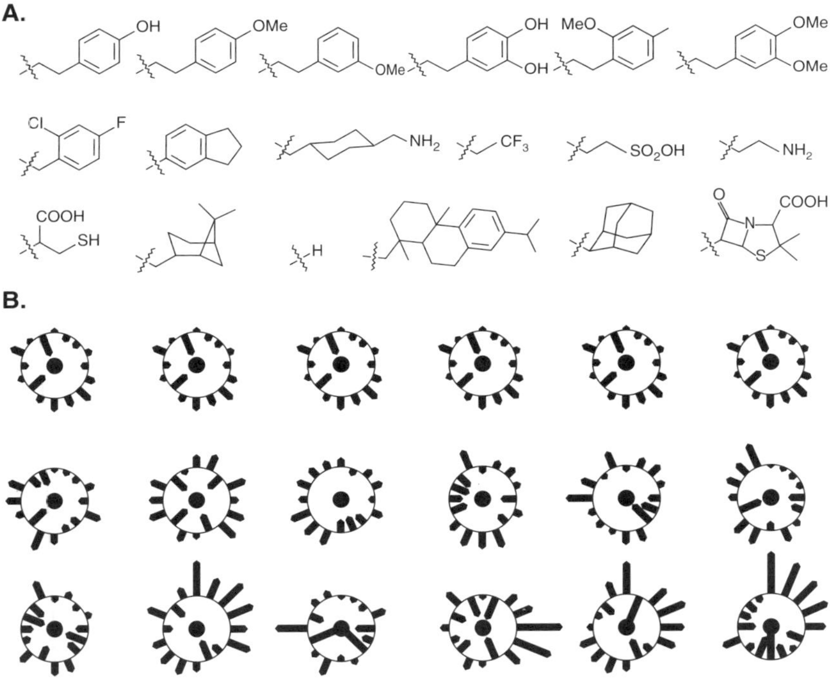

Fig. 9.8. Structures and flower plots for a combinatorial library based on the tyramine monomer. (A) Chemical structures for the tyramine analogs. (B) Corresponding flower plots. (Reprinted with permission from Martin EJ, *et al*, Measuring diversity: experimental design of combinatorial libraries for drug discovery. *J Med Chem* **38**: 2784–2788. Copyright 1995 American Chemical Society.)

and variables based on structural or atomic similarity measures. The striking feature about the flower plot is that the abscissa wraps around in a circle. Positive values for a variable point the petals outward and negative values point the petals inward. The radius of the circle was arbitrarily chosen as three standard deviation units for the first dimension. Figure 9.8 shows that the top rows of flower plots are very similar and these correspond to the top row of structures. The lower two rows of flower plots exhibit considerable dissimilarity.

The advantage of the flower plot is that it is possible to compare quickly different groups of monomers. Comparing multiple flower plots has the same disadvantage as comparing chemical structures: as the number of plots becomes larger, it becomes more difficult to make the pair-wise comparisons. Sorting by values of one or more of the descriptors can make scanning large numbers of plots easier.

A similar type of plot, commonly called a parallel plot, is constructed by drawing a vertical scale for each variable, perhaps after normalizing each dimension to a mean of zero and a standard deviation of unity. Lines are drawn for each compound, connecting adjacent points on the scales. Compounds that are similar will either overlap or be parallel in some sense. For large numbers of compounds, the density of the line segments connecting adjacent vertical scales is indicative of similar or dissimilar features.

9.2.7 PROBABILITY DISTRIBUTIONS

High-throughput screening data are effectively visualized as a probability distribution or histogram. The histogram bins, or abscissa, are units of activity, percentage inhibition or K_i. The height of the histogram bar shows the number of compounds with activity in the corresponding bin. Using histograms to view data allows you to: set cut-off criteria for re-test; compare responses for different assay types, such as scintillation proximity assay (SPA) or filter binding assay; investigate pooling strategy; and compare against theoretical results. Figure 9.9 shows the histograms for a random set of 90 000 compounds tested in four different assay systems. Note how, even qualitatively, the shapes of the distributions are different. Although the general shapes of the two SPA assays are similar, the scale shows very significant differences in assay performance for this set of compounds. The histogram for the filter binding assay is congruent with the distribution for a simulated high-throughput screening assay (Chapter 6, Fig. 6.9). Compounds in the right tail of the distribution are typically chosen for re-test.

9.2.8 SPREADPLOTS

So far, the linkages between graphs have been based on the data. A spreadplot extends this definition of a link by allowing the links to be based on the equations of an underlying data model (Young *et al* 1988, 1990). Spreadplots are similar

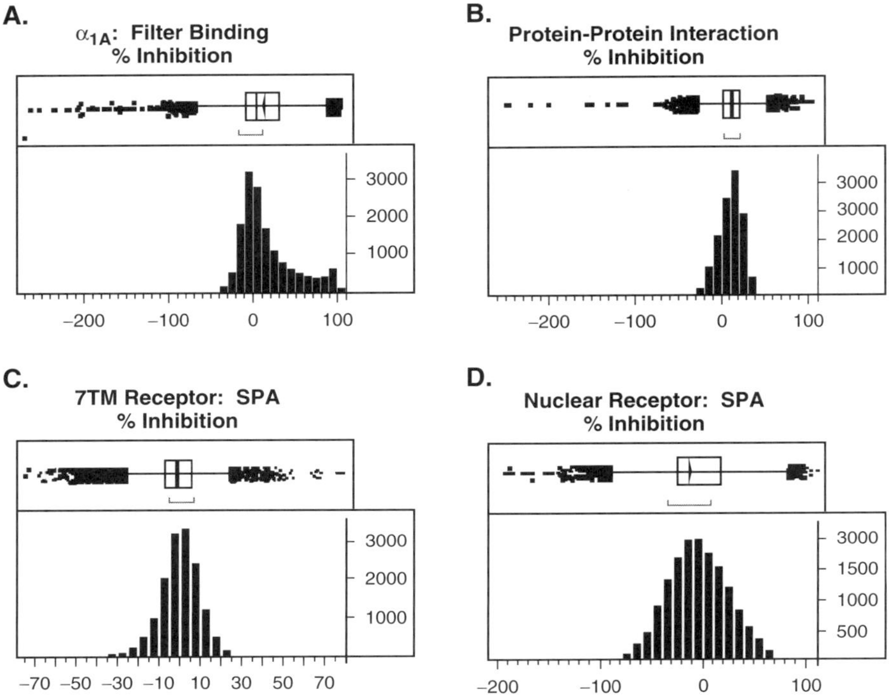

Fig. 9.9. Histograms of activities for four different biochemical assays. Abscissa: percentage inhibition in the assay. Ordinate: count of number of compounds having activity in the associated histogram bin. (A) A filter-binding assay for α_{1A} activity. (B) A protein–protein interaction assay. (C) A scintillation proximity assay (SPA) for a 7-transmembrane receptor. (D) An SPA assay for a nuclear receptor. (Data provided by Steve Blanchard, Glaxo Wellcome.)

to spreadsheets in that they support dynamic (values in spreadsheet cells or plots in a spreadplot) and static (equations in both spreadsheet and spreadplot) views. Spreadplots are similar to leverage plots in that they are a means to examine graphically the components of a regression model and to examine the structure of a dataset (Fig. 9.10). In the case of a spreadplot, the graphs are linked by the equations of a principal component model. Two of the seven graphs that form a principal component spreadplot are editable and allow the analyst either to modify the model or to estimate missing values. Changes in these editable plots are reflected in all plots. The remaining windows depict the different components of a principal components model: raw data (*X* by *X* matrices), scores, coefficients and the variance of each principal component. The spreadplot is an excellent tool for multivariate data analysis because it captures the easily interpretable aspects of principal component analysis.

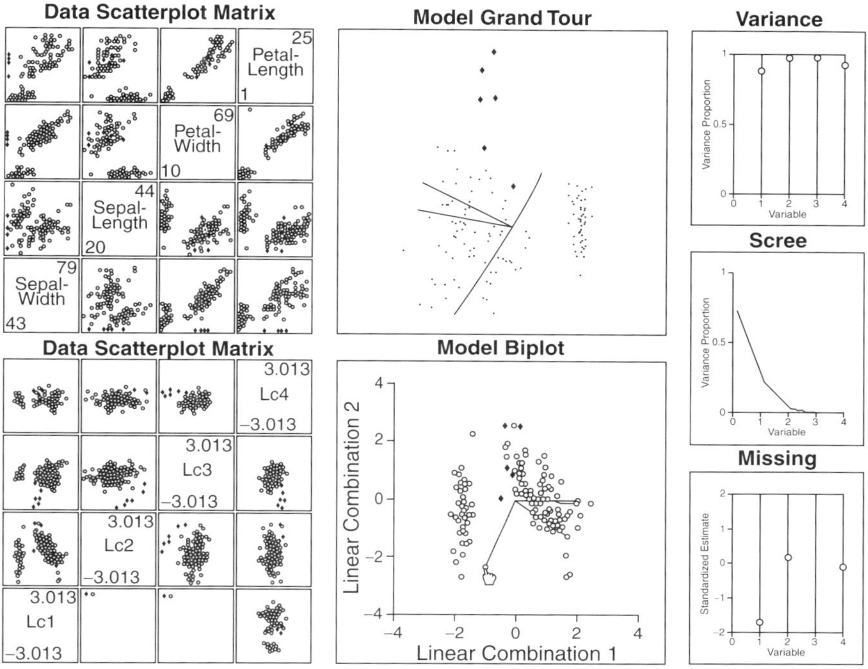

Fig. 9.10. Architecture of a spreadplot. There are seven panels to the spreadplot. The panels are linked by the algebraic equations that form a principal component regression. The original data are in the scatterplot matrix in the upper left column. (Reprinted with permission from Young FW, Faldowski RA and Harris DF, *The Spreadplot: Algebraically Linked Dynamic Plots for Visualizing High-Dimensional Structure in Multivariate Data; Psychometric Laboratory*. University of North Carolina, Chapel Hill, NC. Copyright 1990 by Forrest W. Young.)

9.3 STATISTICAL ANALYSIS OF HIGH-THROUGHPUT SCREENING DATA

9.3.1 RECURSIVE PARTITIONING

By definition, a combinatorial library is produced by reacting a series of chemical structures or monomers in sequence or attaching them at specific positions on a parent compound. The questions usually asked once the resulting compound library is tested in a biological assay are: which are the important monomer attachment positions and which monomers or sets of monomers in combination provide the best biological activity?

Because the size of these datasets is typically greater than 2000 compounds and the monomers at the different attachment positions may interact, these questions are not easily answered by repetitively sorting the data. Often molecules from libraries that have the highest activity may not share common monomers at specific positions.

Many tree regressions methods exist that are capable of analyzing combinatorial library results (Michie *et al* 1994). Although the results of these methods may seem similar, they utilize different algorithms for combining observations into predictor variable groups and building analysis trees. Software packages that perform these types of analysis include: CART, C4.5, FIRM, KnowledgeSEEKER and CHAID (Hawkins and Kass 1982; Breiman *et al* 1984; Kass 1990; Quinlan 1993; ANGOSS Software 1994). Recursive partitioning is an intuitive and robust method of tree regression. The particular versions of recursive partitioning found in FIRM and KnowledgeSEEKER use Bonferroni-adjusted hypothesis testing for combining observations and thus offer some protection from oversplitting the data.

9.3.1.1 Performing an Analysis

In order to answer the questions of which monomers and attachment sites are important for biological activity, the data describing a combinatorial library should be organized into vectors where each compound is defined by its biological activity and the monomers present at each attachment position. The analysis begins by first grouping together those monomers that do not differ significantly in biological activity for each attachment position. Once these groupings are made, the position that demonstrates the greatest statistically significant difference between groups is used to split the combinatorial library into smaller sets of compounds. This process is repeated for each of the smaller sets so that other statistically significant positions and interactions can be detected. The process ends when either the number of compounds remaining in a subset is too small for additional splits to occur or none of the remaining positions have enough statistical power to split the data further. Figure 9.11 depicts how a three-position combinatorial library would be analyzed using recursive partitioning. Similar monomers are grouped for each of the three positions and the library is split based upon these groupings. This reiterative process allows for the systematic analysis of the structure of the library with respect to biological activity.

9.3.1.2 Reading the Tree

The most common output from recursive partitioning is a dendrogram (Fig. 9.12). Each box or node of the dendrogram lists the mean, standard deviation and number of compounds contained in that node. The horizontal lines represent the position on which the previous node was split. The P value from the hypothesis test for this position is listed. The vertical lines that connect the horizontal line to the subsequent nodes contain the monomers common to a node. Finding the structural features that make up the most active molecules becomes a process

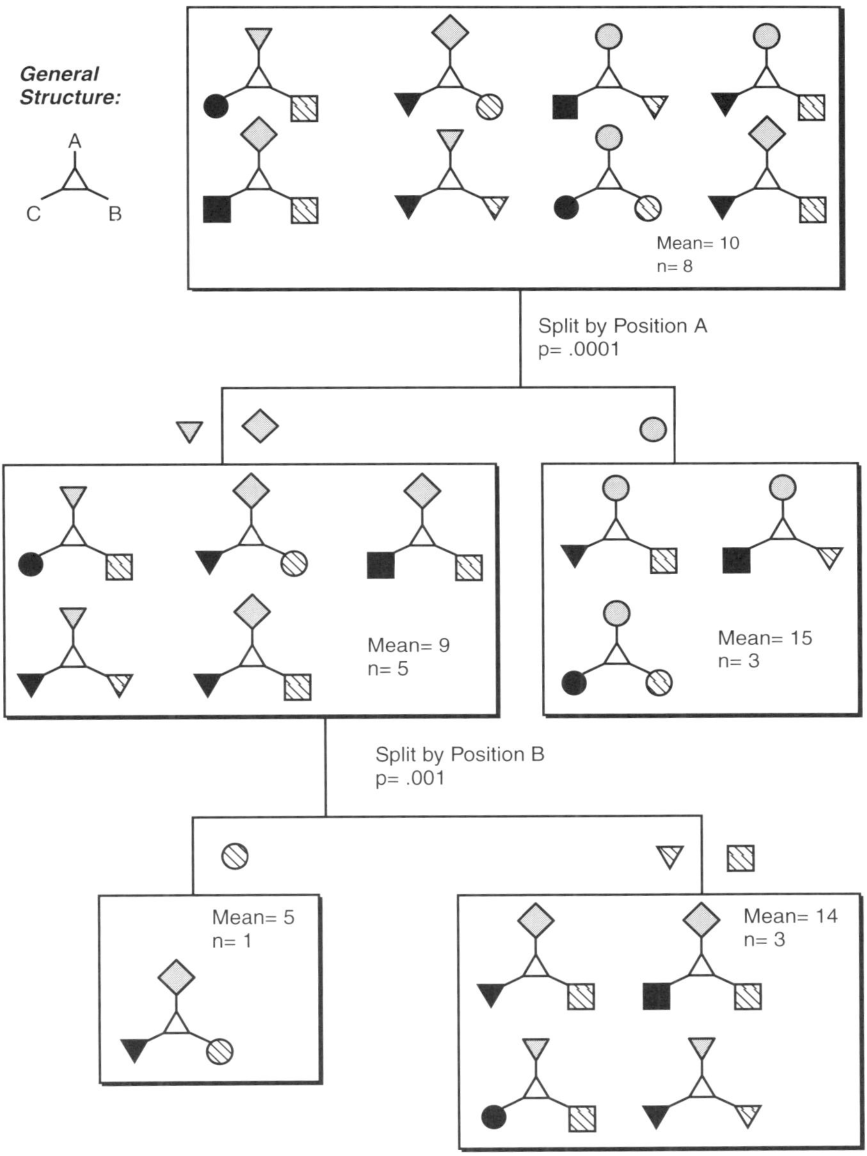

Fig. 9.11. Combinatorial library analysis using recursive partitioning. Monomers with similar potency are pooled by position. The most significant position is used to split the compounds into subsets. This process is repeated for subsequent positions. By 'walking down the tree' one can identify the monomers used to generate molecules with a specific potency. (From Alan Menius, Glaxo Wellcome.)

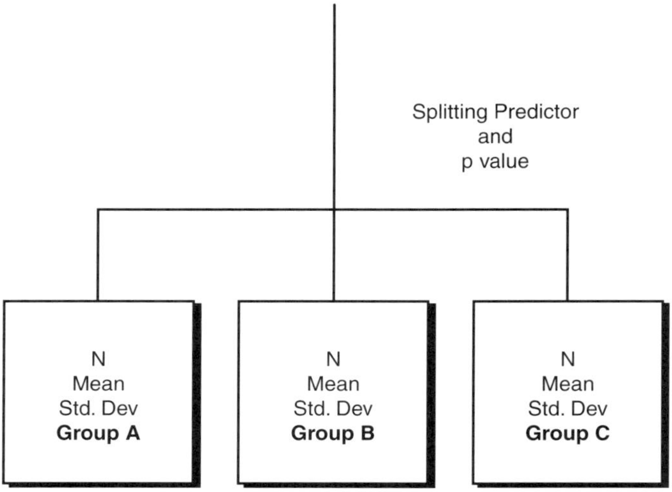

Fig. 9.12. Components of the dendrogram from recursive partitioning analysis. Each node or collection of observations (compounds, entries, results) is listed with the number of observations, the mean and the standard deviation. Members of the group in a node cannot be split further on the basis of the set of predictors in the analysis and the corresponding value of the dependent variable (y). At each node, the predictor used to form the split is identified, along with the P value associated with the split. Smaller values of the P value indicate more statistical significance for the split

of identifying the node or nodes that has the highest mean activity and then identifying the structural features of the molecules that make up that node by working backwards up the tree.

9.3.1.3 Analysis of an Example

Figure 9.13 shows a more complicated dendrogram resulting from the analysis of a small three-position library. Working from the top of the dendrogram, the mean activity, standard deviation and number of compounds in the library are listed. This group or node first splits into smaller groups based on the most significant attachment site or position on the molecule. For this set of compounds we find that position 'C' is most significant and that structures with monomer C10 at that site have the highest activity. This smaller group of 64 compounds split next on position A into two groups. The group of structures that now demonstrate superior activity have monomers A4 and A2 at position A. This group also splits on position B into four distinct monomer groups where the group of 15 compounds with a mean activity of 5.61 are identified. By 'walking down the tree' to this final node, the attributes that make up this class of compounds can be identified.

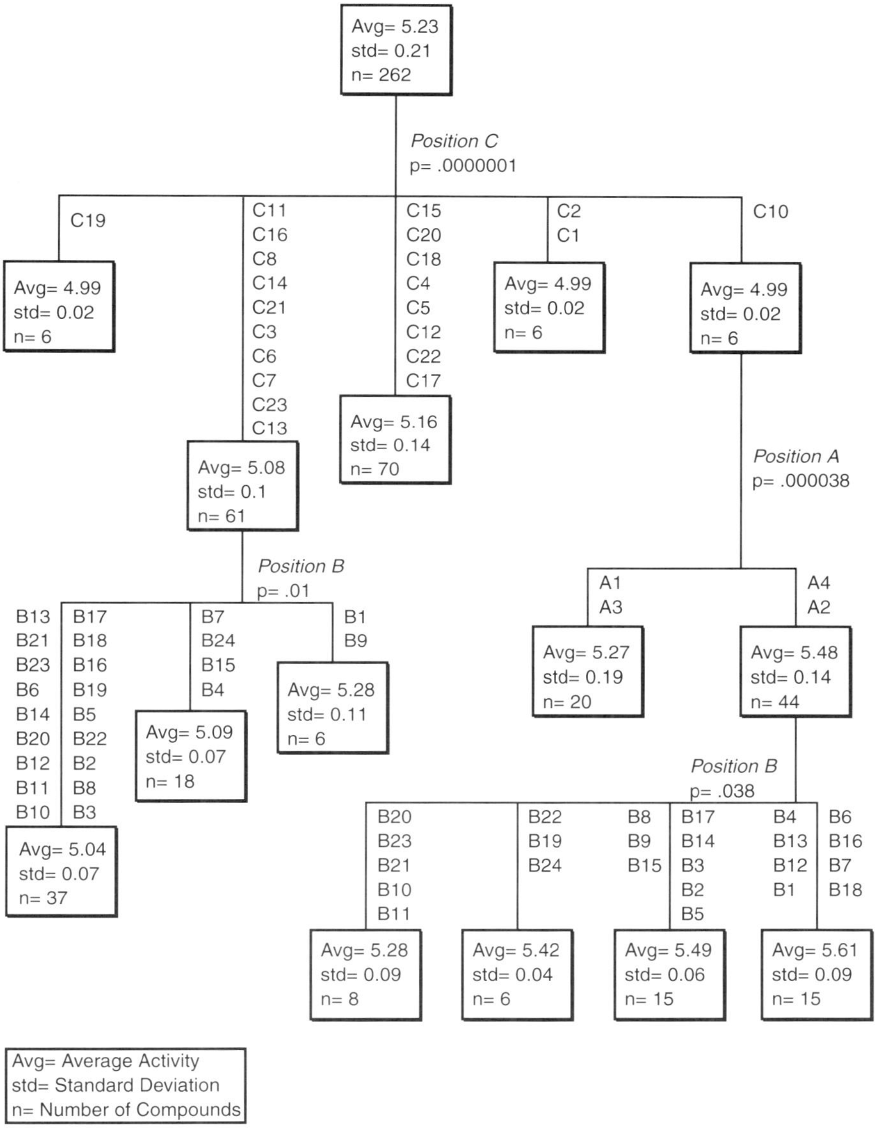

Fig. 9.13. Analysis of a three-position combinatorial library using recursive partitioning. A total of 262 compounds are first split by the monomer groups found at position C. Two subsequent subsets of compounds are split by position B and then positions A and B to produce classes of compounds whose structure correlates to a range of biological activity. (From Alan Menius, Glaxo Wellcome.)

9.3.1.4 Interaction Detection

In many instances, the biological activity for compounds containing a specific monomer at a certain position may depend on which monomer is present at another position. This is an example of an interaction between two positions. Recursive partitioning is capable of detecting these interactions because of the iterative splitting of the large group of compounds into smaller, more homogeneous groups. An example of an interaction between positions would be: a combinatorial library first splits into subgroups based on the monomer groupings at a position; an interaction between positions is detected if, during the next iteration, two of the subsets split on the same position but the resulting nodes have different sets of monomers in that position.

9.3.1.5 Non-linear Activity

In many instances, the association between chemical predictors and activity cannot be modeled successfully using linear regression techniques. This is because the combination of multiple interactions between positions and monomers creates pockets of molecules or localized maxima where molecules have high activity. Figure 9.13 demonstrates how recursive partitioning identifies these non-linear relationships. The combinatorial library splits first on position C. However, the first node that is able to split, splits further on position B and the next node splits on position A followed by B. This non-linear relationship creates distinct structural classes where one class depends only on two of the three positions and the other is affected by all three.

9.3.1.6 Applications and Extensions

Tree regression, specifically recursive partitioning, is a robust analysis method that is ideally suited for the analysis of large multivariate data. Recursive modeling can quickly determine any significant attachment positions and/or monomers in a large chemical library dataset. The analysis is easy to understand and yet can analyze complex datasets that would normally require the use of neural networks or other non-linear regression techniques. In some cases, recursive modeling is able to provide a meaningful analysis of a combinatorial library even though the range of the response may not be optimal. The reason for this is that a single monomer at a position is tested many times in a combinatorial library. Thus, the power to detect statistically significant differences between the tested monomers is greatly increased.

An extension of this type of analysis is to produce a number of chemical descriptors for a combinatorial library and then analyze these data using recursive partitioning. From these results the chemical descriptors important for biological activity can be identified and used to find similar molecules with those attributes. Recursive partitioning can be used in this way to reduce the number of predictor variables to be included in a more traditional linear or non-linear regression model.

Recursive partitioning can be performed on a multivariate vector of responses to correlate structural features with receptor or protein subtype selectivity. Figure 9.14 shows a simple recursive partitioning tree constructed using chemical descriptors based on topology (Hawkins *et al* 1997) to discriminate between selectivity for three different mutants of an HIV protease. Each node on the tree contains a small picture of the selectivity profile of the compounds represented by the node (C. Keefer, Glaxo Wellcome, personal communication, 1998). Differences between the bar height in terminal nodes of the tree indicate that the structural feature associated with splitting the pair of nodes has an effect on producing a different selectivity profile.

9.3.2 NEURAL NETWORKS

Articles on SAR and QSAR analysis done with the aid of neural networks are now common in the chemistry literature (Andrea and Kalayeh 1991; Gasteiger and Zupan 1993; Tetko *et al* 1993; Wikel and Dow 1993; Wikel *et al* 1996). However, it is often the experience of drug discovery project teams that, although the early stages of work with a neural network are promising, the initial optimism fades as what appeared to be a predictive model no longer appears valid as new data are tested. The theoretical basis for neural networks is different from classical statistical and rule-based methods. Many of the diagnostics and assumptions used in regression analysis do not apply to neural networks. Successful use of neural networks requires a good understanding of the input data, network architecture and appropriate diagnostics. Unfortunately, many commercial software packages available for neural network applications offer very limited control over network architecture and do not provide adequate diagnostics. New research has been published that identifies alternative architectures that could not be implemented using most of the commercial programs. Effective use of neural networks is achieved by understanding how neural networks work and by learning what diagnostics and tuning are required.

Neural networks are designed to model some of the learning and classification mechanisms of the brain. The brain quickly and effectively solves problems in pattern recognition, classification and qualitative comparisons. Compare the ease in visually sorting either the chemical structures or flower plots in Fig. 9.8 with writing a computer program to do the same. It is the parallel processing capabilities and memory inherent in the human brain and nervous system that allow it to recognize quickly those patterns based on past experience. Neural networks model brain architecture at the most rudimentary level by using parallel processing elements (neurons) with non-linear transfer functions.

9.3.2.1 Architecture and Implementation

The fundamental architecture for a network is layers of processing units connected together (Fig. 9.15). The input layer has a processing unit for each independent variable. The output layer usually has one processing unit: activity for QSAR studies. Intermediate layers connect the inputs to the output or to each other. Often, the network is fully interconnected. Neural networks must be trained on data where both input and output are provided. The network 'learns' by adjusting the weights on each processing unit to minimize the difference between predicted

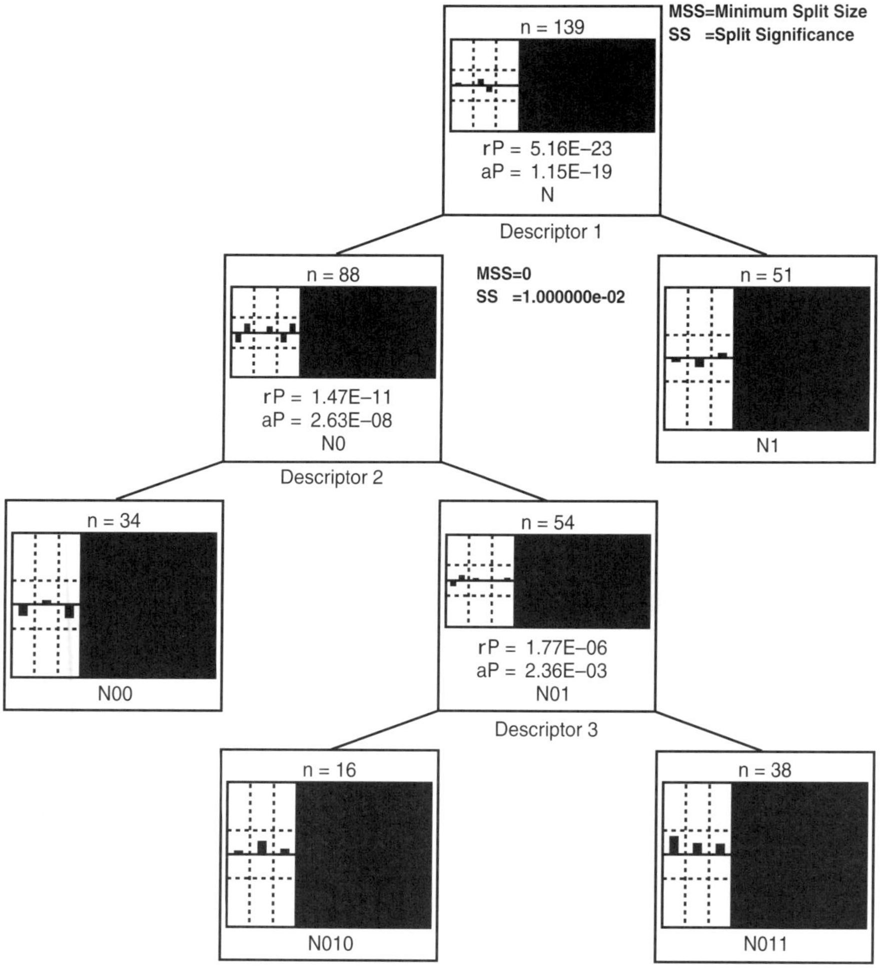

Subtype	Avg. pki.	SD (pki)	Min.	Max.
WT	8.43	0.92	6.27	10.77
M1	6.62	0.86	4.50	9.40
M2	6.41	0.73	4.70	7.82

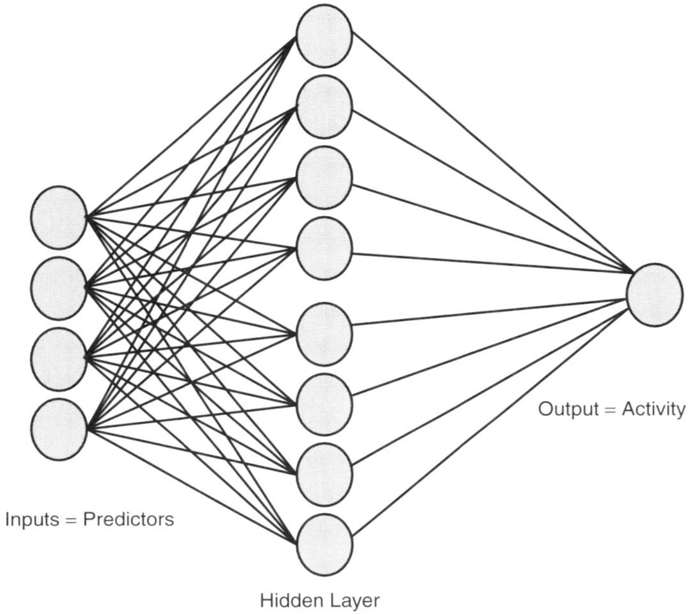

Fig. 9.15. Architecture of a typical neural network showing connectivity between the layers

and observed output. Aspects of training, the specific compounds used and their diversity, size of the training set and length of training time will have a profound impact on the results obtained when the network is used for prediction.

When networks are used for prediction, a compound is evaluated by placing the values of the predictor variables on the input layer and propagating the signal forward (feed-forward). Typically, networks are trained using a method known as back-propagation. In back-propagation, random weights assigned to all processing units are updated systematically by feeding back the error term from the output processing unit through the middle layers back to the inputs.

Fig. 9.14. Analysis of a multivariate response using recursive partitioning. This example shows a recursive partitioning tree for the affinity of a set of 139 different compounds for three mutants of an HIV protease. One structural feature is responsible for the first split of the tree. The 51 compounds in this node (N1) have slightly less affinity for the second mutant, as evidenced by the bar in the second column. The left side of the tree is split further by other chemical features. The node at the far left (N00) has relatively more selectivity for the second mutant than the first and third mutants. The node at the far right (N011) contains compounds that do not show selectivity; the bars are of roughly equal height. (Data from Eric Furfine, Glaxo Wellcome Graphic copyright (1999) Chris Keefer.)

It should be noted that mathematically, the utility of neural networks is derived from their ability to approximate highly non-linear surfaces. Traditional SAR and QSAR often demonstrate the need to model surfaces that are irregular or spiked. If there are sufficient processing units and there is adequate coverage from compounds in the training set, a neural network can approximate these surfaces and predict activity for other compounds on the surface.

9.3.2.2 Practical Considerations and Examples

Many examples of neural network application to SAR and QSAR are published, comparing predictive results obtained using neural networks with results obtained using multiple linear regression. The Selwood dataset of 31 antifilarial antimycin analogs and their associated 53 structure-based descriptors (Selwood *et al* 1990) is an example of a classic dataset, often tested against alternative neural network architectures. Examples of classes of compounds used in neural network SAR studies include benzodiazepines (Aoyama *et al* 1990), carbinoquinones (Aoyama *et al* 1990; Livingstone and Salt 1992), dihydropyridines (Viswanadhan *et al* 1995), HIV-1 reverse transcriptase inhibitors (Tetko *et al* 1994) and dihydrofolate reductase inhibitors, including triazines (So and Richards 1992).

There are some generally accepted practices associated with analysis using neural networks. Simulation studies are often done on network architecture to determine how the network responds to different types of data: linear, quadratic or non-linear. Cross-validation results, where the predictability of the network is assessed by systematically leaving out compounds and computing the correlation coefficient, are often reported.

9.3.2.3 Overfitting

Multiple hidden layers and large numbers of processing units can form complex networks that may faithfully reproduce patterns in the training set but are not robust when compounds that are not part of the training set are presented to the network. This phenomenon is known as overfitting, sometimes referred to as "memorizing the training data set" (Manallack and Livingstone 1992; Livingstone and Manallack 1993). From a statistical point of view, overfitting results from parameters that are unidentifiable. Owen has presented a diagnostic that describes the extent to which an arbitrary network can overfit (Owen 1994). Andrea and Kalayeh have published methods for the detection of overfitting and have derived some simple diagnostics that are worth generating for proposed network architecture (Andrea and Kalayeh 1991).

The diagnostics suggested by Andrea and Kalayeh (1991) are based on the parameter ρ, which is the ratio of the number of compounds to the number of network connections. In their work with dihydrofolate reductase inhibitors, they found that the optimal value for ρ was 2.0. Models with $\rho < 1.8$ were not predictive, and tended to overfit the data. In other words, there were too many connections for the number of compounds in the training set. Models with $\rho > 2.2$ also were not predictive, but for a different reason. These networks were not sufficiently complex to model the data. Manallack *et al* used a series of artificially generated structured datasets and actual QSAR datasets to study the relationship between ρ and prediction for different types of data (Manallack *et al* 1994). A series of graphs showing r^2 versus ρ for linear, quadratic and QSAR models that include indicator variables is helpful to determine where overfitting occurs (Fig. 9.16). The r^2 value obtained by leave-N-out cross-validation is also shown on the plot, as well as the r^2 value obtained by multiple linear regression.

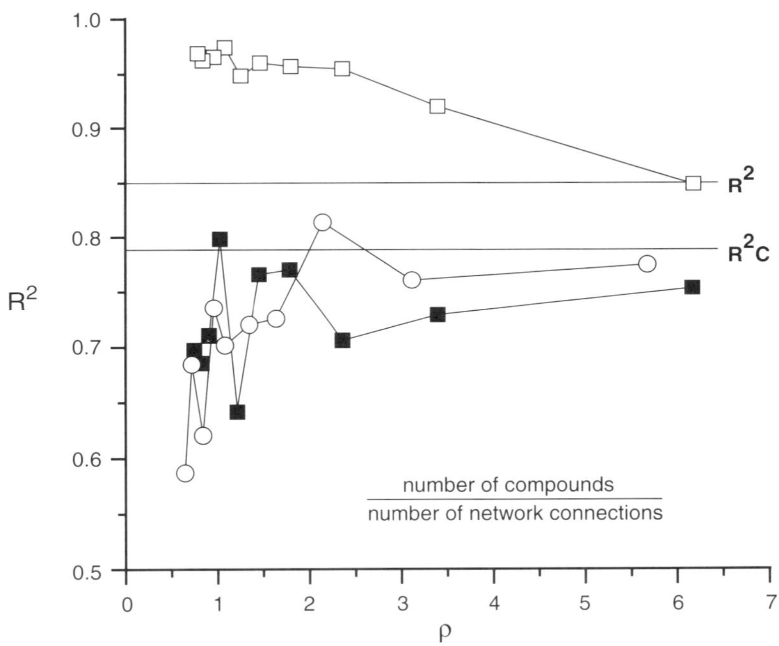

Fig. 9.16. Plot showing the results of multiple linear regression and neural network analysis for linear QSAR data. Abscissa: ratio of the number of compounds to the number of network connections ρ. Ordinate: correlation coefficient r^2. Open squares represent neural network results. Closed squares represent results for cross-validation using the leave-one-out method. Open circles show results for cross-validation using the leave-N-out method, where N is approximately 10%. The horizontal lines represent results for classical multiple linear regression. For multiple linear regression, both a standard r^2 and cross-validated r^2 (R^2C) based on a leave-one-out method are provided. (Reprinted with permission from Manallack DT, Ellis DD and Livingstone DJ, Analysis of linear and non-linear QSAR data using neural networks. *J Med Chem* **37**: 3758–3767. Copyright 1994 American Chemical Society.)

9.3.2.4 Overtraining

Overtraining is another problem that must be considered by users of neural networks. Overtraining is a function of the length of time that random sequences from the training set are applied to the network. Although the use of longer training times will always reduce the error of prediction, after a certain time the predictive ability of the network decreases as the network starts adjusting weights to model the noise in the data. Using a similar framework of synthetic and real QSAR datasets to that of Manallack (Manallack *et al* 1994) (linear, quadratic, indicator), Tetko *et al* examined the impact of overtraining on neural network predictability (Tetko *et al* 1995). They demonstrated a parabolic dependence of prediction on training time and suggested that a series of cross-validation steps be done during training to identify when overtraining occurs. An ensemble of training sets was used to partition the data into the sets needed for the multiple cross-validation steps. They also showed that the neural networks had a wide tolerance to overfitting.

9.3.2.5 Alternative Architectures

Although the vast majority of published reports on neural network applications in chemistry and software packages are based on feed-forward networks that are trained using the back-propagation algorithm, there are some alternative architectures that appear promising as well. For the feed-forward networks, architecture depends on the complexity: number of hidden layers and processing units. Several alternative transfer functions have also been reported in the literature.

Counter-propagation networks are an alternative architecture. These are four-layer networks: input, normalization, competitive and output. The first layer simply normalizes all of the predictors. The competitive layer utilizes Kohonen learning rules. For a given input vector, only one processing unit in the competitive unit will 'win' and provide input to the final, output layer. Peterson reported on the results of using counter-propagation networks for QSAR studies with compound sets of 39 carboquinones and 60 benzodiazepines (Peterson 1995). In comparison to back-propagation networks, the counter-propagation networks appeared to be more robust for training set results and had slightly larger errors of prediction for test sets. Peterson showed that the relative predictive abilities of traditional neural networks, counter-propagation neural networks and multiple linear regression were similar for the two datasets. For the carboquinones, which demonstrated a linear relationship between predictors and response, multiple linear regression was found to be more predictive than either neural network. In the case of the benzodiazepines, which was a more non-linear data set, the neural networks were more predictive.

A neural network framework that explicitly provides linear and non-linear components to the output has been proposed (Ajay 1993). The motivation for this architecture is to start with a linear model and add in non-linear elements if needed. One advantage to this approach is that it gives the modeler considerably more control over the complexity of the network, leading to more robust network architectures. In presenting this architecture as a framework for building QSAR models, the author also points out the advantages of using weight decay to systematically decrease the magnitude of all weights during training. This method resets processing units away from saturation levels of the transfer function and biases the network towards a less complex structure.

The neural network architectures and applications described in this chapter are classified as supervised learning. A training set is used to adjust the parameters of the network and determine the hidden layer. An alternative strategy is unsupervised learning, one example of which is the Kohonen neural network (Kohonen 1990). In these networks, training sets are not used. Instead, the network self-organizes to the patterns in a dataset; in other words, only an input vector is used instead of both input and output vectors. In a similar manner to recursive partitioning and classification and regression trees, self-organizing neural networks are used to identify molecular features that are important for biological activity from high-throughput screening data (Anzali *et al* 1998).

9.3.2.5 Summary

Neural networks are effective analysis tools for SAR and QSAR. Several considerations make them well suited as a method for analysis of data from high-throughput screening. By definition, large numbers of compounds are available for training. The benefit of this large sample size is greatly enhanced if the compounds are diverse. Indeed, space-filling designs could be utilized to select a diverse set of training compounds. Libraries often fill a highly non-linear parameter space. Activity may be highly segmented in a few areas. The predictors often are highly correlated, requiring biased regression methods, or the predictors may define an irregular surface. These characteristics suggest that the neural network approach may be better suited to problems in QSAR than multiple linear regression. Comparisons between different network architectures and multiple linear regression are common in the literature. A study by Reibnegger *et al* (1991) compared neural network methods with linear discriminant analysis and with classification and regression trees for a classification problem in clinical chemistry (classification of patients into three disease groups based on the results of four laboratory tests) and found that the neural network could detect features in the input data that were not explicitly formulated as input.

If a neural network is used, some of the diagnostics discussed in this section should be done to ascertain whether the network is overfitting the data and to what extent the network is predictive. Sufficient exploratory work should be done with the predictors to understand whether the training set of compounds adequately covers the surface to be modeled by the network. Neural network modelers should develop networks with consideration about the robustness of the model and understand that different architectures are available in addition to different methods for setting weights and characterizing the performance of the network. Object-oriented programming can be a powerful implementation method to develop the flexibility to construct and test alternative neural network architectures and methods (Welstead 1994).

9.3.3 GENETIC ALGORITHMS

Although neural networks have found utility as a computational tool for quantitative structure–activity relationship analysis, another method, genetic algorithms is useful for the selection of compounds for testing or to suggest monomers to be used in the construction of a combinatorial library. Both methods are discussed in this chapter because each is a method that predicts the activity of a compound. The ability to predict activity is important both for modeling and library design, in order to identify pharmacophoric features from screening data.

Combinatorial synthesis, by nature, is a process where fragments are combined to make a compound for testing. Sets of fragments or monomers are assembled around a linker or scaffold for the compounds. The sets are usually chosen from a collection of similar building blocks. Common building blocks include primary amines, secondary amines, carboxylic acids, amino acids and imines. A benzodiazepine library, for example, can be constructed from three building blocks: hydroxy or carboxy derivatives of 2-aminobenzophenone, amines and alkylating agents (Fig. 9.17).

Often, the number of potential fragments far exceeds the number of fragments that are used in synthesis. An approach for rational selection of fragments is to construct a virtual or computer library of all possible compounds and mathematically select fragments that construct compounds that are biased toward biological activity or have a superior pharmacokinetic or pharmacodynamic profile. This is an ideal problem for genetic algorithms because they rapidly search a space for combinations of fragments that are most likely to be active.

9.3.3.1 Theoretical Basis

Genetic algorithms are a computational optimization method based on the principles of biological evolution. Three genetically based operations characterize genetic algorithms: simple copying, mutation and crossover. As in the natural world, genetic algorithms maintain a population of parent genes that combine according to the three principles and evolve over a series of generations or iterations of the algorithm. At each generation, a fitness function is used to evaluate possible combinations and select those that are optimal. Over time, the average fitness of the population increases until an optimal set of a desired size is chosen.

For applications in combinatorial chemistry, the monomers must be encoded as genes. Each gene represents one position, building block or residue on the compound. Different monomers are equivalent to the allelles for the gene. An initial population is formed with randomly selected alleles for each gene. The

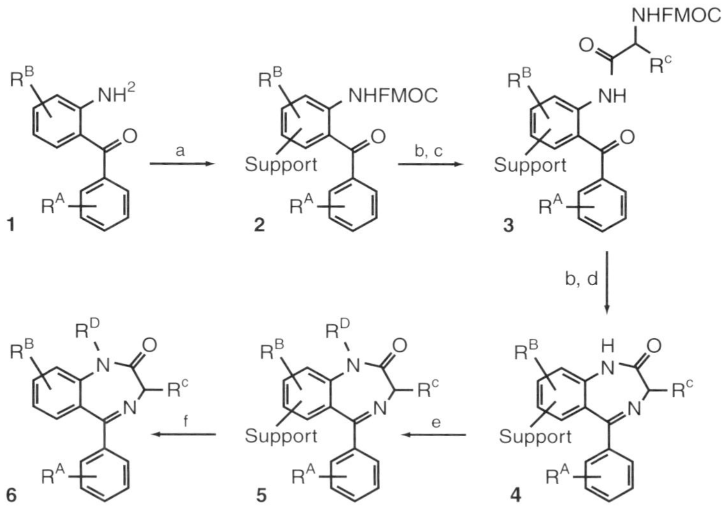

^{a}(a) See supplementary material;
(b) 20% piperidine in DMF;
(c) *N*-FMOC-amino acid fluoride, 4-methyl-2,6-di-*tert*-butylpyridine;
(d) 5% acetic acid in DMF, 60°C;
(e) lithiated 5-(phenylmethyl)-2-oxazolidinone in THF, −78°C, followed by alkylating agent and DMF;
(f) TFA/H_2O/Me_2S (95:5:10).

1,4-Benzodiazepine Derivatives 6 (Scheme I)

	Derivative				
entry	R^A	R^B	R^C	R^D	yield (5)^a
6a	4'-OH	5-Cl	CH_3	H	95
6b	4'-OH	5-Cl	CH_3	CH_3	100
6c	4'-OH	5-Cl	CH_3	CH_2CH_3	97
6d	4'-OH	5-Cl	CH_3	CH_2CHCH2	90
6e	4'-OH	5-Cl	$CH(CH_3)_2$	CH_2CH_3	85
6f	4'-OH	5-Cl	CH_2CO_2H	CH_2CH_3	95
6g	4'-OH	5-Cl	$(CH_2)_4NH_2$	CH_2CH_3	95
6h	4'-OH	5-Cl	CH_2Ph(4-OH)	CH_2CH_3	98
6i		4-CO_2H,5-Cl	CH_2Ph	CH_3	100
6j		4-CO_2H,5-Cl	CH_3	CH_2Ph	93

Fig. 9.17. Illustration of the synthesis of a combinatorial library of benzodiazepines on a solid phase. This figure shows how a combinatorial library of 1,4-benzodiazepines can be synthesized from 2-aminobenzophenones, amino acids and alkylating agents. The figure is included to demonstrate combinatorial synthesis and to illustrate how a fragment-based approach to synthesis corresponds to a genetic algorithm approach to virtual libraries. (Reprinted with permission from Bunin BA and Ellman JA, A general and expedient method for the solid-phase synthesis of 1,4-benzodiazepine derivatives. *J Am Chem Soc* **114**: 10997–10998. Copyright 1992 American Chemical Society.)

fitness for each individual or compound is computed and a new population is formed from the old. Some of the compounds are taken intact from the old population. In others, a mutation or change is made in one or more genes. Finally, in some compounds, crossover occurs where the compound is formed from genes of a pair of parent compounds.

Note that random selection is an important part of the algorithm. The operation (copy, mutate, crossover), the compounds (individual or parents for crossover)

Fragment Libraries

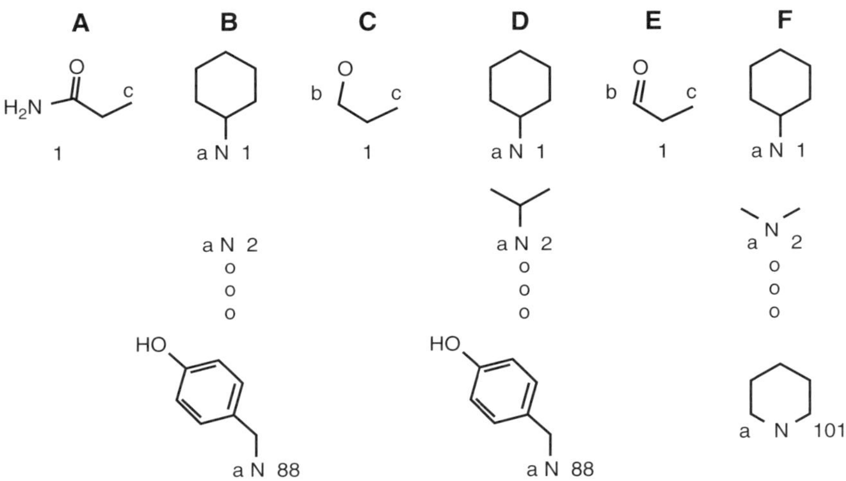

Fusion instructions to produce a tripeptoid:

First fragment	A	1		
Fusion 1	B	2	c-a	blank-keep
Fusion 2	C	1	a-b	blank-blank
Fusion 3	D	88	c-a	blank-keep
Fusion 4	E	1	a-b	blank-blank
Fusion 5	F	101	c-a	blank-blank

Example:

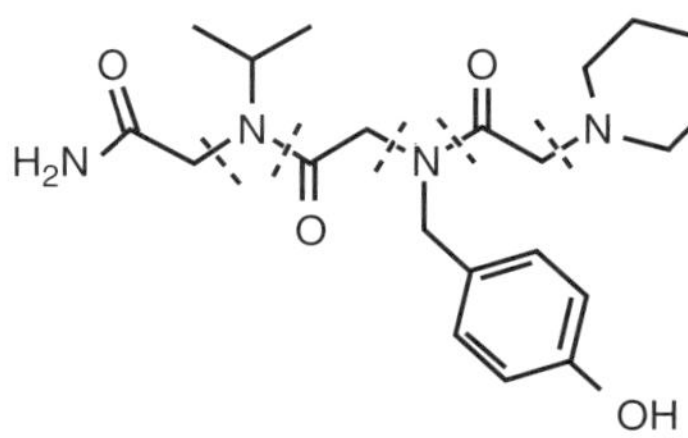

Genome: 1 2 1 88 1 101

and the monomers (alleles) are chosen at random. There is no guarantee that the best compound will be found. Genetic algorithms are computer intensive for large problems but are usually faster than a systematic search. In the case of combinatorial libraries, often a systematic search of the entire space is impractical.

9.3.3.2 Practical Considerations and Examples

For chemistry applications of genetic algorithms, the rules for combining the fragments or monomers must be defined. Sheridan and Kearsley address this problem by marking the atoms that may form a covalent bond for each fragment, compiling the fragments into sets called fragment libraries and finally by defining fusion instructions that indicate how to generate a connection table for the entire molecule from the fragments (Sheridan and Kearsley 1995). Figure 9.18 shows an example of the use of a genetic algorithm to build a tripeptoid library. Six genes (or residues) comprise each tripeptoid. Three of the genes (A, C and E) contain only a single allele. The remaining three residues have sets of 88–101 alleles. Assembling a tripeptoid involves stepwise fusion of the six residues.

For this example, several different scoring functions are used: similarity to a specific tripeptoid target; similarity to two tetrapeptide CCK antagonists; and a trend vector derived from activity on ACE inhibitors. A flow diagram of the method is shown in Fig. 9.19. Scoring itself is based on topological descriptors, specifically atom pairs (Carhart *et al* 1985) and topological torsions (Nilakantan *et al* 1987).

In this example, three test cases for the tripeptide library are constructed using genetic algorithms. The first case tested the various library construction

---

Fig. 9.18. Using a genetic algorithm to build a tripeptoid library. The six residues, labeled A–F, each have an associated set of fragments called a fragment library, e.g. 1–88 for residue B. The fragments are combined to produce the product by identifying fusion atoms on each fragment (*a, b, c*). Fusion instructions describe how the fragments are joined. For example, the first fusion instruction is to take fragment 1 from library A. The second instruction is to take fragment 2 from library B and bond an atom in the A fragment with fusion type *c* to an atom in the B fragment with fusion type *a* (noted *c-a* in the instruction: methyl *c* to amine nitrogen *a*). The word 'blank' indicates that the fusion type of the bonded atom in the A fragment is to be erased so that it is not available for further bonding. The word 'keep' indicates that the atom in the B fragment (an amine nitrogen) is to keep its type. By carefully determining the fusion types and assigning the types to the atoms in the fragments, the chemistry for constructing all potential products can be supported. The structure at the bottom of the figure is the result of the five fusion instructions and is represented in genetic algorithm terms by the 'genome' or list of fragments (1, 2, 1, 88, 1, 101). Bonds created by the fusion instructions are indicated by the dotted lines. (Reprinted with permission from Sheridan RP and Kearsley SK, Using a genetic algorithm to suggest combinatorial libraries. *J Chem Inf Comput Sci* **35**: 310–320. Copyright 1995 American Chemical Society.)

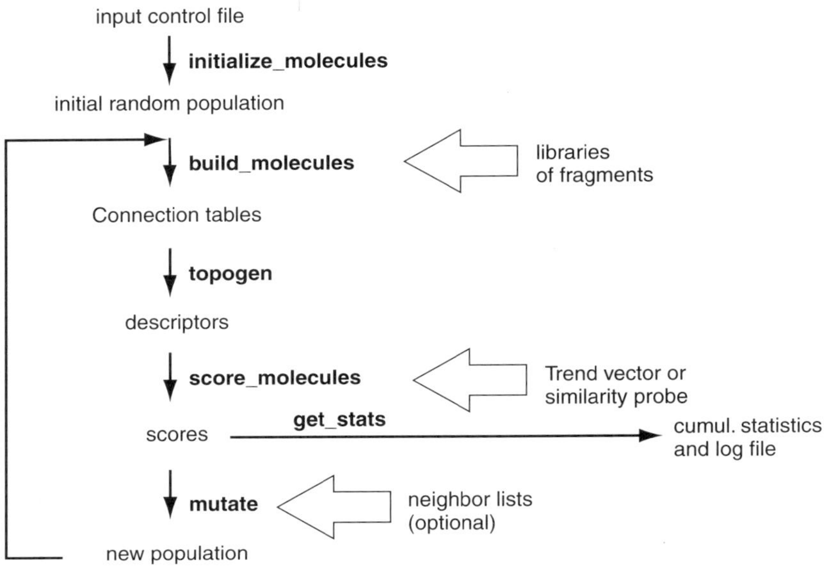

Fig. 9.19. Flow diagram of the use of a genetic algorithm to build trend vectors for ACE inhibitors. The diagram shows the relationship between the different steps in the process: building the fragment libraries, constructing products, calculating descriptors, scoring the product molecules and using the genetic operators to make a new population. (Reprinted with permission from Sheridan RP and Kearsley SK, Using a genetic algorithm to suggest combinatorial libraries. *J Chem Inf Comput Sci* **35**: 310–320. Copyright 1995 American Chemical Society.)

and scoring protocols by comparing the library against a peptoid target. In this case, an exact match to the target is possible. Using various selection methods, the target was identified in 11–23 generations of the algorithm. The second case involved a more realistic problem for drug discovery. In this case, two tetrapeptide CCK antagonists were chosen from a commercial database of drug-like structures. Commonly occurring fragments are shown in Fig. 9.20. It is clear that the chosen fragments resemble the targets. Sheridan and Kearsley point out that residue F corresponds to the N-terminus of the probes, with a preference for an indole-containing fragment. Residue D selects for a set of fragments containing carboxylate and aliphatic groups. Residue B selects for phenyl-containing fragments, similar to the C-terminus. From these rules, it is possible to select monomers for a library that will be topologically biased toward the target peptides.

The third and most complex example for this design was to take compounds from the drug database that are classified as angiotensinase (ACE) inhibitors. In their example, Sheridan and Kearsley found 223 compounds that have this classification in the MDDR database. These compounds were identified as target compounds. The set was augmented with 2000 randomly selected compounds that do not have angiotensinase activity. The genetic algorithm was run and typical active compounds are shown in Fig. 9.21. The trend vector summarizes features of the molecules that appear to be shared by ACE inhibitors. The most common fragments are shown in Fig. 9.22. Residue F selects for the Ala–Pro dipeptide, whereas residue B selects for aromatic rings at the end of short alkyl chains. Residue D selects for carboxylates and phosphates. Many ACE inhibitors contain carboxylates, esters and phosphoesters.

9.3.3.3 Summary

Experience with genetic algorithms for compound selection and library design is in the very early stages. As library sizes become larger, the promise of virtual screening or selecting libraries to synthesize from a larger set of potential molecules becomes promising. Genetic algorithms are one approach to making this selection. The examples chosen in this chapter address topological similarity. Various biochemical measurements can likewise be used as descriptors. In all cases, the goal is to design a library with a higher probability of containing active compounds. The genetic algorithm is used as the selection mechanism. For the approach to be successful, the predictors must adequately define the SAR for the compounds. As more examples are published, choices about which targets are most amenable to this approach, descriptor choices and scoring methods become clearer.

9.3.4 INTERPRETATION OF DATA FROM POOLS

Interpretation of data from experiments where compounds are pooled is complicated by a variety of biochemical and experimental factors. Technically, if the compounds are all competitive inhibitors, the activity of the pool is the following sum of the individual activities:

$$\frac{[Bound]}{B_{\max}} = \frac{L^*}{L^* + K_{\mathrm{d}}\left(1 + \dfrac{I_1}{K_{\mathrm{I}1}} + \dfrac{I_1}{K_{\mathrm{I}2}} + \ldots + \dfrac{I_n}{K_{\mathrm{I}n}}\right)}$$

where $[Bound]$ is concentration bound, $B_{\max}$ is the concentration bound in the absence of any inhibitor, L^* is the concentration of the radioligand, K_{d} is the dissociation constant for the radioligand, I_1–I_n are inhibitor concentrations and $K_{\mathrm{I}1}$–$K_{\mathrm{I}n}$ are the dissociation constants for the inhibitors.

In screening, the assumption is often made that only one or two compounds in the pool will have significant activity. Therefore, in stepwise screening, pools that test positive at a given concentration are 'decoded' or tested as discrete compounds. Orthogonal pooling is an alternative to this approach (Fig. 9.23). In orthogonal pooling, each discrete compound appears once and only once in each

Probes:

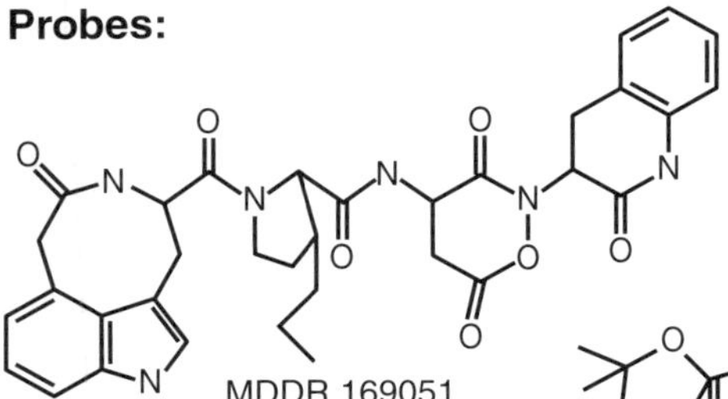

MDDR 169051

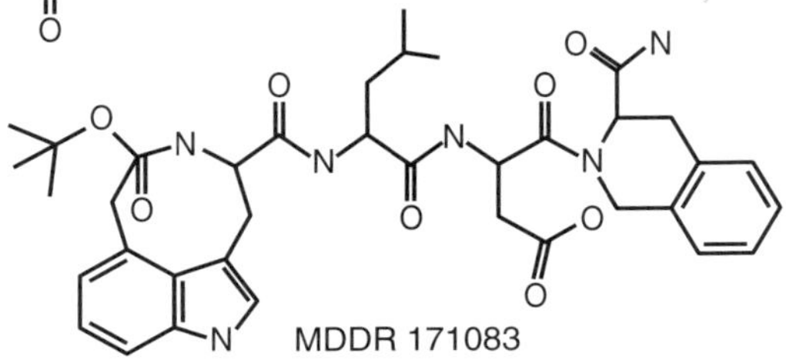

MDDR 171083

Most frequent fragments:

B

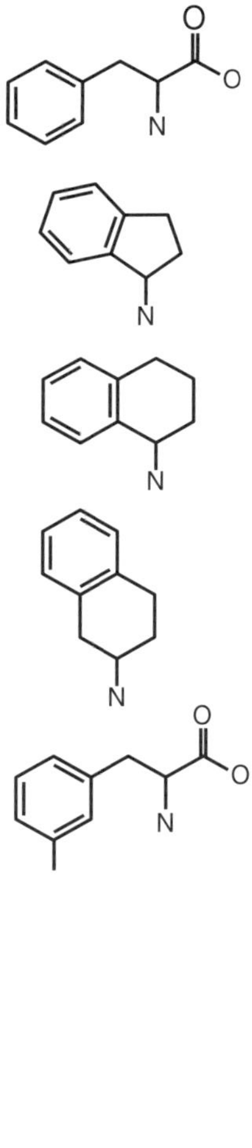

D

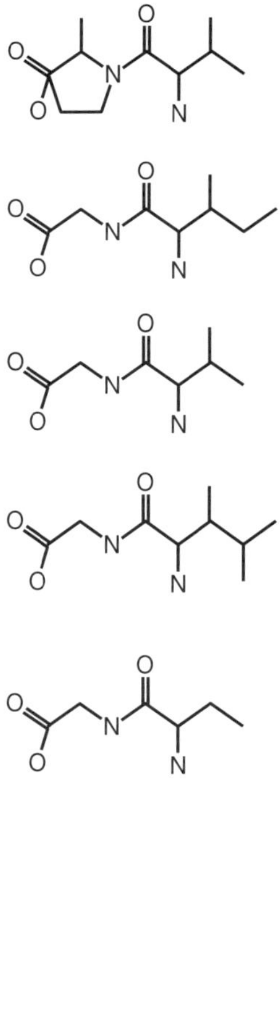

F

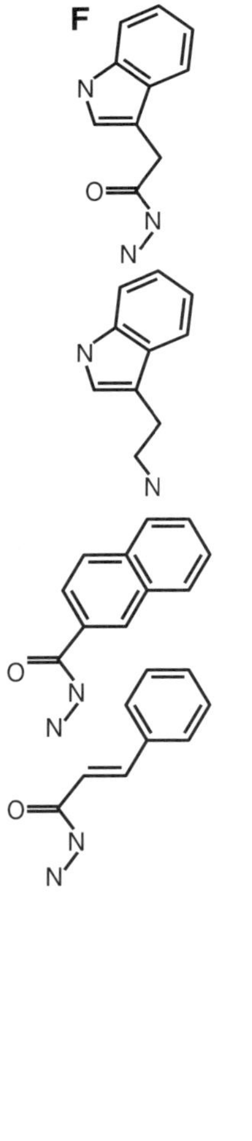

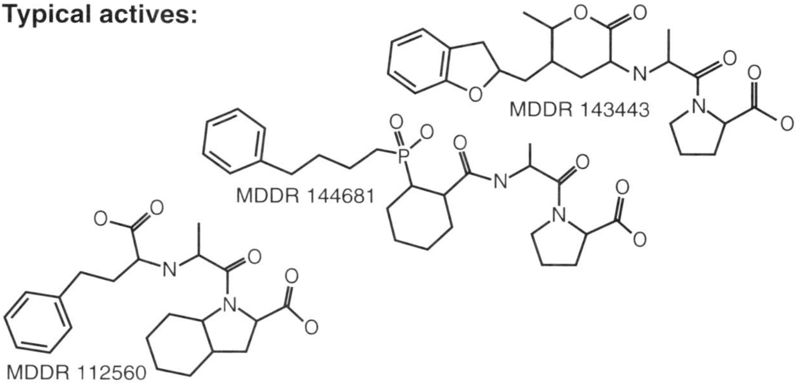

Fig. 9.21. Active compounds from genetic algorithm selection of ACE inhibitors. These compounds are well-known ACE inhibitors used to train the genetic algorithm. (Reprinted with permission from Sheridan RP and Kearsley SK, Using a genetic algorithm to suggest combinatorial libraries. *J Chem Inf Comput Sci* **35**: 310–320. Copyright 1995 American Chemical Society.)

of two pools, usually on different plates. Decoding uses a computer database to identify compounds that show activity in both pools. In practice, some ambiguity is present in the data. A compound will test positive in one pool but not the other, or a set of pools will test positive, making it necessary to re-test several compounds. The success of a pooling strategy depends on the compounds in the test set, the biological assay and the stringency of the percentage inhibition threshold.

Figure 9.24 shows pooling ambiguity for two different enzyme assays. The pooling strategy was orthogonal. Each compound appeared once in two pools of ten compounds each. A successful decode is a case where you can identify a single compound because it appears as a unique positive in both wells. There are also ambiguous decodes, where 2–10 compounds would have to be tested in order to confirm a positive discrete compound. In practice, ambiguity higher than two or three compounds is uncommon. Finally, there are cases where there is no

---

Fig. 9.20. Commonly occurring structural fragments from a CCK antagonist library. The two structures identified as probes are well-known CCK antagonists. The average of descriptor scores for these probes is calculated and used to score the compounds generated by the genetic algorithm (virtual compounds). In this case, the product compounds are tripeptoids. Tripeptoids with descriptor scores similar to the probes will appear most often as generations of the genetic algorithm progress. Fragments that appear most often are shown at the bottom of the figure. (Reprinted with permission from Sheridan RP and Kearsley SK, Using a genetic algorithm to suggest combinatorial libraries. *J Chem Inf Comput Sci* **35**: 310–320. Copyright 1995 American Chemical Society.)

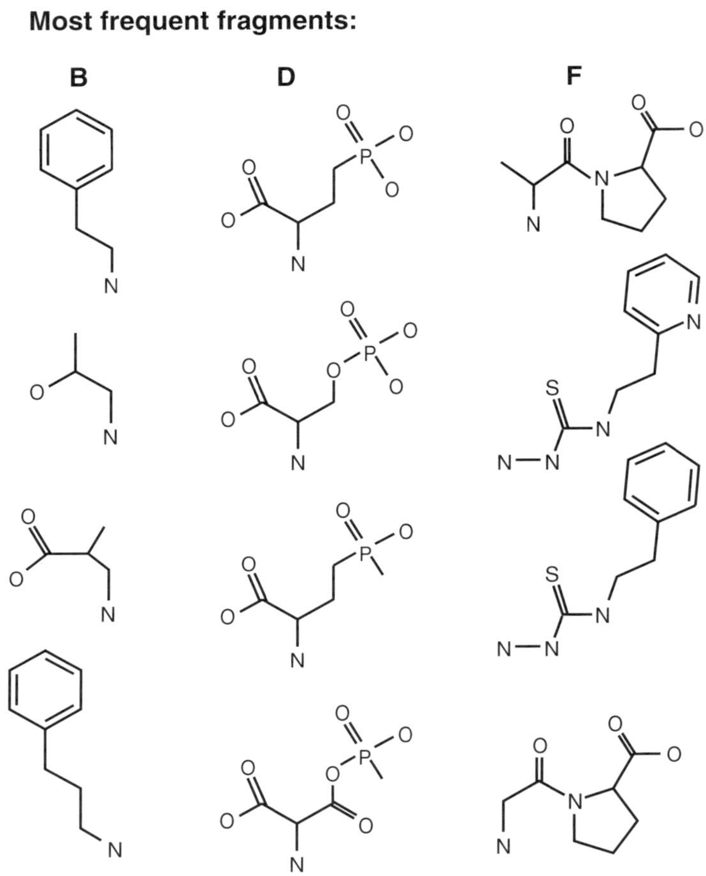

Fig. 9.22. Most commonly occurring fragments from an ACE inhibitor genetic algorithm study. (Reprinted with permission from Sheridan RP and Kearsley SK, Using a genetic algorithm to suggest combinatorial libraries. *J Chem Inf Comput Sci* **35**: 310–320. Copyright 1995 American Chemical Society.)

correspondence between the two pools, e.g. a positive in one well has no positive result in the corresponding well. The assay on the right in Fig. 9.24 is well suited for pooling. The number of decodes falls off sharply as a function of threshold (abscissa: percentage inhibition). The ambiguous and no-correspondence cases are much lower and fall off as a function of threshold, indicating that we could pick a suitable threshold for decoding. For the assay on the right, there is not as marked a decrease with threshold and it is particularly disconcerting that the ambiguous and no-correspondence cases remain high, even with an increasingly stringent threshold.

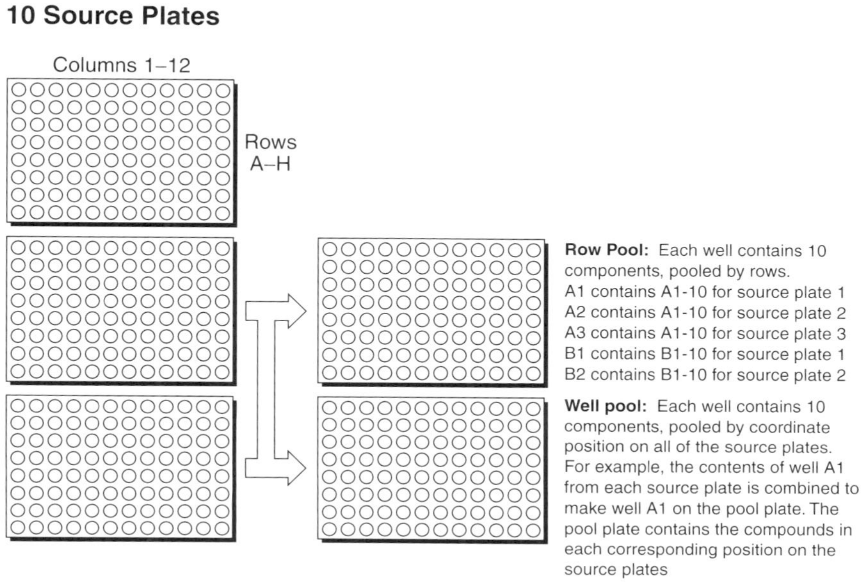

Fig. 9.23. Schematic for orthogonal pooling. In the orthogonal pooling strategy, each compound appears in two different pools (a row and a well pool) combined with nine different compounds. Decoding is theoretically possible because an active compound should be associated with an active row and well pool. In this example, 10 source plates are pooled. Columns 1–10 contain compounds and columns 11 and 12 are used for controls. Each row of each source plate is combined into a single well on the row-pool plate. The column coordinate of the row-pool plate is therefore the source plate number, and the row coordinate of the row-pool plate is the corresponding row coordinate on the source plate (e.g. row coordinates remain the same). For the well-pool plates, the compounds in each coordinate position from the collection of source plates are pooled. Therefore, the row and column coordinates on the well-pool plate are identical to the row and column coordinates on the source plates. In other words, if the source plates were stacked on top of each other, the well-pool plate would be the result of pooling at each coordinate position in the set of stacked plates

The histograms of pooled data are excellent diagnostics for experimental factors that influence the success of a pooling strategy. Figure 9.25 shows histograms for pool sizes of 10 and 20 compounds. An orthogonal pooling strategy was utilized to construct the pools. The pooled compounds were tested against three different biological targets. The data show that the three targets responded differently to the pooled data. In the case of the first assay, there is little change in response for the different pool sizes: both distributions are comparable with respect to their mean and variance. The second and forth assays show a pattern that can be explained by dilution effects. The larger pool size leads to distributions that are more compressed. There are fewer data points outside the central distribution and

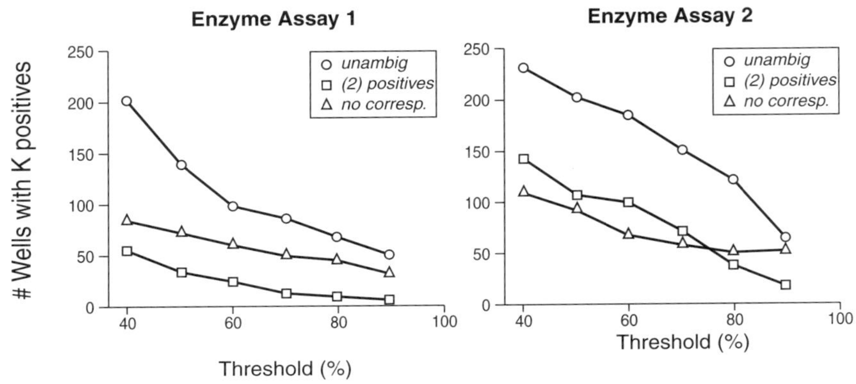

Fig. 9.24. Pooling ambiguity for two biochemical assays. Abscissa: percentage inhibition threshold used to select positive (active) compounds. Ordinate: number of wells with at least one active compound as a result of decoding the pooled results. The number of active compounds is referred to as 'K'. Open circles: one and only one active compound, $K = 1$. Open squares: two and only two active compounds, $K = 2$. Open triangles: either row or well pool, but not both, show activity at the threshold level, e.g. it is not possible to decode the active compound

the distribution itself is much narrower. Finally, the third assay demonstrates the effect of additivity. The larger pool size leads to a distribution that is wider and contains a larger count of active compounds than from the smaller pool size.

9.4 STATISTICAL GUIDE POSTS

The previous sections of this chapter have discussed methods that show promise for the analysis of data from high-throughput screening. In this final section, we consider some of the more general issues involved in data analysis.

First, it is important to start with high-quality data. Experimental design should be done to optimize biological assays so that they provide robust and reproducible results. The integrity and variability of the assay data must be calculated and used at different stages in the analysis, for example, to compare with the variability of groups of compounds identified by recursive partitioning. Libraries and compound sets must be chosen to provide diversity in terms of predictor variables if the compound set is to provide a reasonable surface for modeling.

It is important to recognize that experimental designs and protocols are often chosen for different objectives: screening for hits, SAR or lead optimization. Unfortunately, it is not possible to extract more information out of a poorly designed experiment. For example, it is difficult to detect outliers without replication, but often concentration–response curves are run with more replication at each concentration than is needed. Choosing an optimal sampling protocol

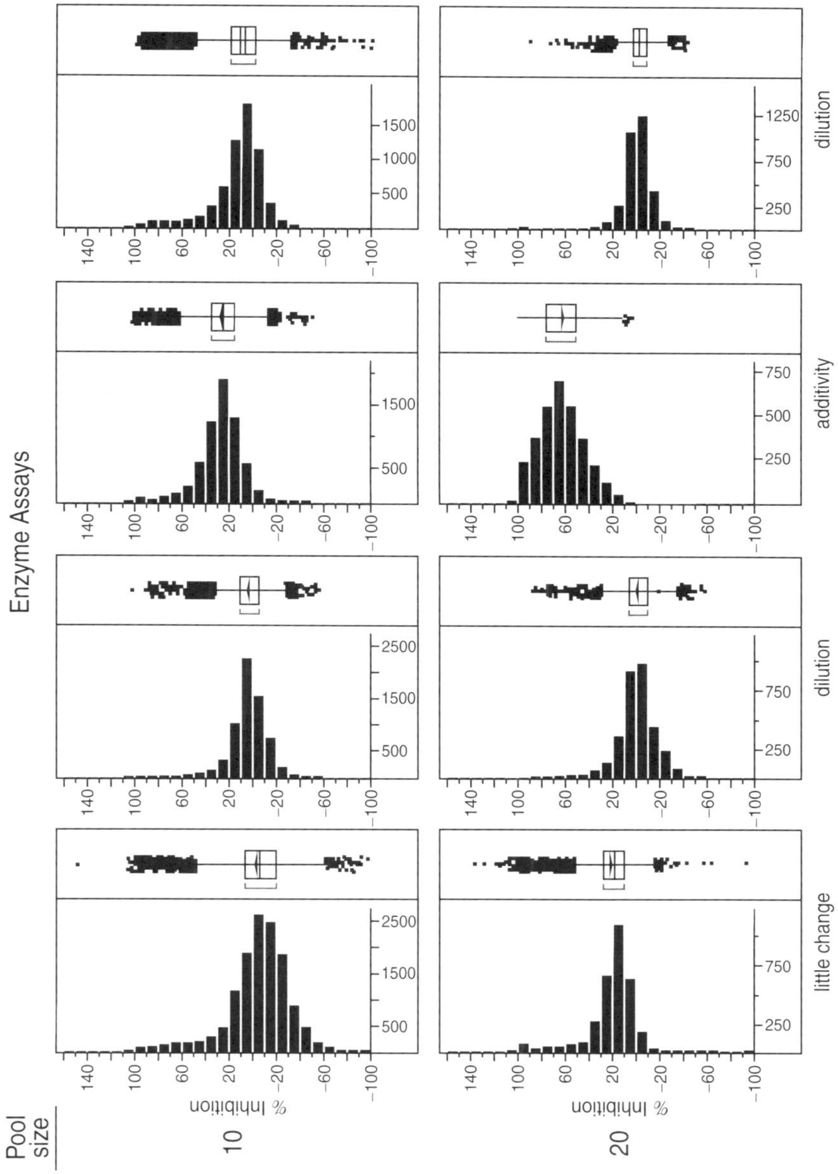

Fig. 9.25. Histograms comparing pools of 10 and 20 compounds for four different assays. Abscissa: percentage inhibition in the assay. Ordinate: count of number of compounds having activity in the associated histogram bin. For the top row, pools of 10 compounds were constructed, for the bottom row, pools of 20 compounds were used. (From Steve Blanchard, Glaxo Wellcome.)

can be done by partitioning assay variability. Systematic error is a possibility, particularly with robotic systems. Data analysts must take care to test for these effects.

Computer simulation is a powerful tool to gain an understanding of data and data analysis tools. Many of the neural network examples cited earlier included details on computer simulations of different types of data: linear, non-linear and different combinations with indicator variables. Computer simulation can also be used to solve some of the sampling problems of large combinatorial libraries.

It is important to cast a wide net to find potentially useful methodologies in other disciplines. Visualization and analysis of large datasets are common problems in other areas of science as well as finance. In many cases, the data from these disciplines are similar to data from high-throughput screening assays in terms of the semi-qualitative nature. Some of the methods and software may be appropriate for analogous problems in analysis of high-throughput screening data.

Perhaps the most important guide post is the knowledge that these methods are complementary and not mutually exclusive. Multivariate visualization is useful during the initial phases of a study to explore the distributions of predictor and response data and to assess the diversity of the compound set. Later, it is used to examine specific features of the data that were identified by recursive partitioning or a neural network. Recursive partitioning can be used for several different objectives: to categorize quickly data by key predictive features, to identify interactions and to detect outliers. The speed and versatility of this method and performance with large numbers of observations make it well suited for analysis of combinatorial library data. Neural networks are useful to amplify patterns in complex, non-linear data. However, they require tuning and flexible implementation for practical use. A stepwise approach to building neural network models is conducive to building predictive, robust QSAR models (Ajay 1993).

The volume of data generated by high-throughput screening of combinatorial libraries presents great challenges and opportunities for data analysis, SAR and QSAR. Taken together, these methods plus additional methods not discussed (e.g. fuzzy logic systems) begin to address the problem of translating data into useful information that guides the development of future drugs.

9.5 REFERENCES

Ajay (1993) A unified framework for using neural networks to build QSARs. *J Med Chem* **36**: 3565–3571.

Andrea TA and Kalayeh HJ (1991) Applications of neural networks in quantitative structure–activity relationships of dihydrofolate reductase inhibitors. *J Med Chem* **34**: 2583–2590.

Andrews PR, Craik DT and Martin JL (1984) Functional group contributions to drug–receptor interactions. *J Med Chem* **27**: 1648–1657.

ANGOSS Software (1994) *ANGOSS KnowledgeSEEKER for Windows Users Guide*, Version 3.0. ANGOSS Software International Limited, Toronto, Canada.

Anzali S, Gasteiger J, Holzgrabe U, *et al* (1998) The use of self-organizing neural networks in drug design. Perspectives in *Drug Discovery Design* **9/10/11**: 273–299.

Aoyama T, Suzuki Y and Ichikawa HJ (1990) Neural networks applied to quantitative structure-activity relationship analysis. *J Med Chem* **33**: 2583–2590.

Breiman L, Friedman JH, Olshen RA and Stone CJ (1984) *Classification and Regression Trees*. Wadsworth, Belmont, CA.

Bunin BA and Ellman JA (1992) A general and expedient method for the solid-phase synthesis of 1,4-benzodiazepine derivatives. *J Am Chem Soc* **114**: 10997–10998.

Carhart RE, Smith DH and Venkataraghavan R (1985) Atom pairs as molecular features in structure-activity studies: definition and applications. *J Chem Inf Comput Sci* **25**: 64–73.

Dubois D and Prade H (1980) *Fuzzy Sets and Systems: Theory and Applications*. Academic Press, New York.

Gasteiger J and Zupan J (1993) Neural networks in chemistry. *Angew Chem Int Ed* **32**: 503–527.

Goetz AS, Lutz MW, Queen KL, Rimele TJ and Saussy DL (1994) Characterization of alpha-1 adrenoceptor subtypes in human and canine prostate membranes. *J Pharmacol Exp Ther* **271**: 1228–1233.

Hawkins DM and Kass GV (1982) Automatic interaction detection. In *Topics in Applied Multivariate Analysis*, ed. by DM Hawkins, pp. 269–302. Cambridge University Press, Cambridge.

Hawkins DM, Young SS and Rusinko A (1997) Analysis of a large structure–activity data set using recursive partitioning. *Quant Struct-Act Relat* **16**: 296–302.

Kass GV (1990) An exploratory technique for investigating large quantities of categorical data. *Appl Stat* **29**: 199–127.

Kohonen T (1990) Self-organized formation of topologically correct feature maps. *Proc IEEE* **78**: 1460–1480.

Livingstone DJ and Salt DW (1992) Regression analysis for QSAR using neural networks. *Bioorg Med Chem Lett* **2**: 213–218.

Livingstone DJ and Manallack DT (1993) Statistics using neural networks: chance effects. *J Med Chem* **36**: 1295–1297.

Livingstone DJ, Hesketh G and Clayworth D (1991) Novel method for the display of multivariate data using neural networks. *J Mol Graphics* **9**: 115–118.

Manallack DT and Livingstone DJ (1992) Artificial neural networks: applications and chance effects for QSAR analysis. *Med Chem Res* **2**: 181–190.

Manallack DT, Ellis DD and Livingstone DJ (1994) Analysis of linear and non-linear QSAR data using neural networks. *J Med Chem* **37**: 3758–3767.

Martin EJ, Blaney JM, Siani MA, Spellmeyer DC, Wong AK and Moos WH (1995) Measuring diversity: experimental design of combinatorial libraries for drug discovery. *J Med Chem* **38**: 1431–1436.

Michie D, Spiegelhalter DJ and Taylor CC (1994) *Machine Learning, Neural and Statistical Classification*. Ellis Horwood, New York.

Nilakantan R, Bauman N and Dixon JS (1987) Topological torsion: a new molecular descriptor for SAR applications. Comparison with other descriptors. *J Chem Inf Comput Sci* **27**: 82–85.

Owen AB (1994) Overfitting in neural networks. *Comput Sci Stat* **26**: 57–62.

Peterson KL (1995) Quantitative structure-activity relationships in carboquinones and benzodiazepines using counter-propagation neural networks. *J Chem Inf Comput Sci* **35**: 896–904.

Quinlan JR (1993) *C4.5: Programs for Machine Learning*. Morgan Kaufman, San Mateo, CA.

Reibnegger G, Weiss G, Werner-Felmayer G, Judmaier G and Wachter H (1991) Neural networks as a tool for utilizing laboratory information: comparison with linear discriminant analysis and with classification and regression trees. *Proc Natl Acad Sci USA* **88**: 11426–11430.

Reibnegger G, Werner-Felmayer G and Wachter HJ (1993) A note on the low-dimensional display of multivariate data using neural networks. *J Mol Graphics* **11**: 129–133.

Sall J (1990) Leverage plots for general linear hypotheses. *Am Stat* **44**: 308–315.

SAS Institute (1994) *JMP User's Guide*, Version 3. SAS Institute, Cany, NC.

Selwood DL, Livingstone DJ, Comley JC, *et al* (1990) Structure–activity relationships of antifilarial antimycin analogues: a multivariate pattern recognition study. *J Med Chem* **33**: 136–142.

Sheridan RP and Kearsley SK (1995) Using a genetic algorithm to suggest combinatorial libraries. *J Chem Inf Comput Sci* **35**: 310–320.

So S and Richards WG (1992) Application of neural networks: quantitative structure–activity relationships of the derivatives of 2,4–diamino-5-(substituted-benzyl)pyrimidines as DHFR inhibitors. *J Med Chem* **35**: 3201–3207.

Temple University (1994) *TempleMVV*, Version 1.3. Temple University and Mihalisin Associates Inc., Amber, Pa.

Tetko IV, Luik AI and Poda GI (1993) Application of neural networks in structure–activity relationships of a small number of molecules. *J Med Chem* **36**: 811–814.

Tetko IV, Tanchuk VY, Chentsova NP, Antonenko SV, Poda GI, Kukhar VP and Luik AI (1994) HIV-1 reverse transcriptase inhibitor design using artificial neural networks. *J Med Chem* **37**: 2520–2526.

Tetko IV, Livingstone DJ and Luik AI (1995) Neural network studies: comparison of overfitting and overtraining. *J Chem Inf Comput Sci* **35**: 826–833.

Tufte ER (1990) *Envisioning Information*. Graphic Press, Cheshire, Ct.

Viswanadhan VN, Mueller GA, Basak SC and Weinstein JN (1995) A new QSAR algorithm combining principal component analysis with a neural network: application to calcium channel antagonists. *Net Sci www.netsci.org*.

Wang J, DiPasquale AJ, Bray AM, Maeji NJ and Geysen HM (1993) Study of stereo-requirements of substance P binding to NK1 receptors using analogs with systematic D-amino acid replacements. *Bioorg Med Chem Lett* **3**: 451–456.

Welstead ST (1994) *Neural Network and Fuzzy Logic Applications in C/C++*. Wiley, New York.

Wikel JH and Dow ER (1993) The use of neural networks for variable selection in QSAR. *BioMed Chem Lett* **3**: 645–651.

Wikel JH, Dow ER and Heathman M (1996) Interpretative neural networks for QSAR. *Net Sci www.netsci.org*.

Young FW, Kent DP and Kuhfeld WF (1988) *Dynamic Graphics For Statistics*. Wadsworth, Belmont, CA.

Young FW, Faldowski RA and Harris DF (1990) *The Spreadplot: Algebraically Linked Dynamic Plots for Visualizing High-Dimensional Structure in Multivariate Data Psychometric Laboratory*. University of North Carolina, Chapel Hill, NC.

Young SS and Hawkins DM (1995) Analysis of a 2^9 full factorial chemical library. *J Med Chem* **38**: 2784–2788.

9.6 FURTHER READING

Multivariate Visualization

Cleveland WS (1993) *Visualizing Data*. Hobart Press, Summit, NJ.

Recursive Partitioning

Brieman L, Friedman JH, Olshen RA and Stone CJ (1984) *Classification and Regression Trees*. Wadsworth, Belmont, CA.

Brodley CE and Utgoff PE (1995) Multivariate decision trees. *Machine Learning*. **19**: 45–76.

Buntine W (1992) Learning classification trees. *Stat Comput* **2**: 63–73.

Neural Networks

Ajay, Walter WP and Murcko MA (1998) Can we learn to distinguish between drug-like and nondrug-like molecules? *J Med Chem* **41**: 3314–3324.

Genetic Algorithms

Davis L (1991) *Handbook of Genetic Algonthms*. Van Nostrand Reinhold, New York.

Forrest S (1993) Genetic algorithms — principles of natural-selection applied to computation. *Science* **261**: 872–878.

Gillet VJ, Willett P and Bradshaw G (1997) The effectiveness of reactant pools for generating structurally-diverse combinatorial libraries. *J Chem Inf Comput Sci* **37**: 731–740.

Gillet VJ, Willett P and Bradshaw J (1998) Identification of biological activity profiles using substructural analysis and genetic algorithms. *J Chem Inf Comput Sci* **38**: 165–179.

Goldberg DE (1989) *Genetic algorithms in Search, Optimization and Machine Learning*. Addison-Wesley, Reading, MA.

Interpretation of Data from Pools

Federer WT (1994) Pooling and other designs for analyzing laboratory samples more efficiently. *Statistician* **43**: 413–422.

Other Topics

Ajay, Walter WP and Murcko MA (1998) Can we learn to distinguish between drug-like and nondrug-like molecules? *J Med Chem* **41**: 3314–3324.

Cummins DJ, Andrews CW, Bentley JA and Cory M (1996) Molecular diversity in commercial databases: comparison of medicinal chemistry knowledge bases and databases of commercially available compounds. *J Chem Inf Comput Sci* **36**: 750–763.

Finn P (1996) Computer-based screening of compound databases for the identification of novel leads. *Drug Discovery Today* **1**: 363–370.

Finn P, Muggleton S, Page D and Srinivasan A (1998) Pharmacophore discovery using the inductive logic programming system Progol. *Machine Learning* **22**: 1–33.

Freier SM, Konings DAM, Wyatt JR and Ecker DJ (1995) Deconvolution of combinatorial libraries for drug discovery: a model system. *J Med Chem* **38**: 344–352.

Gordon EM, Barrett RW, Dower W, Fodor SPA and Gallop MA (1994) Applications of combinatorial technologies to drug discovery. 2. Combinatorial organic synthesis, library screening strategies and future directions. *J Med Chem* **37**: 1385–1401.

Gray NS, Wodicka L, Thunnissen AMWH, *et al* (1998) Exploiting chemical libraries, structure and genomics in the search for kinase inhibitors. *Science* **281**: 533–538.

King R, Muggleton S, Lewis R and Sternberg M (1992) Drug design by machine learning: the use of inductive logic programming to model the structure–activity relationships of trimethoprim analogues to dihydrofolate reductase. *Proc Natl Acad Sci USA* **89**: 11322–11326.

King R, Muggleton S, Srinivasan A and Sternberg M (1996) Structure–activity relationships derived by machine learning: the use of atoms and their bond connectives to predict mutagenicity by inductive logic programming. *Proc Natl Acad Sci USA* **93**: 438–442.

Klopman G (1984) Artificial intelligence approach to structure–activity studies. Computer automated structure evaluation of biological activity of organic molecules. *J Am Chem Soc* **106**: 7315–7321.

Lewis RA, Mason JS and McLay IM (1997) Similarity measures for rational set selection and analysis of combinatorial libraries: the diverse property-derived approach. *J Chem Inf Comput Sci* **37**: 599–614.

Weinstein JN, Myers TG, O'Connor PM, *et al* (1997) An information-intensive approach to the molecular pharmacology of cancer. *Science* **275**: 343–349.

Young SS, Farmen M and Rusinko A (1996) Random versus rational — which is better for general compound screening? *Net Sci www.netsci.org/science/screening/feature09.html.*

INDEX

Index compiled by Geoffrey C. Jones